The Tropical Turn

The Tropical Turn

Agricultural Innovation in the Ancient Middle East and the Mediterranean

---

Sureshkumar Muthukumaran

UNIVERSITY OF CALIFORNIA PRESS

University of California Press
Oakland, California

Library of Congress Cataloging-in-Publication Data

Names: Muthukumaran, Sureshkumar, author.
Title: The tropical turn : agricultural innovation in the ancient Middle East and the Mediterranean / Sureshkumar Muthukumaran.
Description: [Oakland, California] : University of California Press, [2023] | Includes bibliographical references and index.
Identifiers: LCCN 2022034721 (print) | LCCN 2022034722 (ebook) | ISBN 9780520390836 (cloth) | ISBN 9780520390843 (paperback) | ISBN 9780520390850 (ebook)
Subjects: LCSH: Agricultural innovations—Mediterranean Region—History—To 1500. | Tropical crops—Mediterranean Region—History—To 1500.
Classification: LCC S494.5.I5 M88 2023 (print) | LCC S494.5.I5 (ebook) | DDC 338.1/6091822—dc23/eng/20220829

LC record available at https://lccn.loc.gov/2022034721
LC ebook record available at https://lccn.loc.gov/2022034722

28 27 26 25 24 23
10 9 8 7 6 5 4 3 2 1

CONTENTS

ILLUSTRATIONS

FIGURES

MAPS

PREFACE AND ACKNOWLEDGMENTS

This book responds to the growing demand to understand Afro-Eurasia as a cohesive and meaningful historical unit underlain by recurring patterns of connectivity by land and sea. It uses the anthropogenic mobility of cultivated plants and other biological materials as a roadmap for understanding cross-cultural exchanges and the movements of people through space and time. It is my hope that this model of writing history, which uses diverse data sets and organic materials as proxies for human endeavors, will be applied to other temporal and spatial units. It is also my intention that this publication sustain and develop ongoing dialogues between practitioners and readers of the historical, social, and biological sciences.

As a historian by training, it is also my desire that this book dissolves barriers within the field of ancient history. Too often the practice of ancient history is compartmentalized and parochialized along regional and linguistic lines. This not infrequently results in the failure to observe organic interconnections between societies, whether close together or far apart. I especially owe this book's expansive vision of history to my undergraduate and postgraduate education at University College London.

A book attempting a synthesis of this order naturally stands on the shoulders of giants, be they my own teachers and mentors or the many gurus through space and time with whom my only contact was through the printed word. I owe a particular debt of gratitude to Karen Radner (Munich) for her unstinting support as my primary PhD supervisor and her enthusiasm for all things Assyrian. I would also like to thank a number of brilliant educators and mentors I had the pleasure of interacting with at UCL, Oxford, and Yale-NUS College, including Amelie Kuhrt, Riet van Bremen, Eleanor Robson, Dorian Fuller, John Ma, Nicholas Purcell, David d'Avray, Rajeev Patke, and Pattaratorn Chirapravati. For the many fond memories of conversational Sanskrit in both Heidelberg and Varanasi, a special thanks to the Heidelberg Sanskritists, especially Ute Hüsken and Sadananda Das.

This book has also immensely benefited from the comments and assistance of its astute and kind reviewers, Robert Spengler (Jena) and Daniel Fuks (Cambridge). Many thanks to both of them. Any remaining errors of interpretation remain entirely my own. I should also like to thank Peter Palm (Berlin), who drew all the maps in this book on short notice. For trusting in the premise of the book and painlessly guiding the publication process, I am indebted to editor Eric Schmidt and editorial assistant LeKeisha Hughes. My gratitude is also due to the meticulous copyeditor Roy Sablosky, production editor Cindy Fulton, and others on the publishing team. I am especially grateful to the University of California Press for supporting the publishing process through its FirstGen program. Finally, a massive thank you to my parents, Indira and Muthukumaran; siblings, Rathika and Geetha; and friends for their unfailing encouragement.

ABBREVIATIONS

ASSYRIOLOGY

ARM 21	J-M. Durand, *Archives royales de Mari XXI: Textes administratifs des salles 134 et 160 du Palais de Mari*. Paris: Librairie Orientaliste Paul Geuthner, 1983.
BAM 2	F. Köcher, *Die babylonisch-assyrische Medizin in Texten und Untersuchungen, Band 2: Keilschrifttexte aus Assur 2*. Berlin: De Gruyter, 1963.
BBS	L. W. King, *Babylonian Boundary-Stones and Memorial Tablets in the British Museum*. London: British Museum, 1912.
BdI Adab	F. Pomponio, G. Visicato, and A. Westenholz, *Le Tavolette Cuneiformi di Adab delle Collezioni della Banca d'Italia*. Rome: Banca d'Italia, 2006.
BE 9	A. T. Clay and H. V. Hilprecht, *Business Documents of Murashû Sons of Nippur Dated in the Reign of Artaxerxes I (The Babylonian Expedition of the University of Pennsylvania, vol. 9)*. Philadelphia: University of Pennsylvania, 1898.
BM	British Museum Tablets.
Cam	J. N. Strassmaier, *Inschriften von Cambyses, König von Babylon*. Leipzig: Pfeiffer, 1890.
CBS	Museum siglum of the University of Pennsylvania Museum of Archaeology and Anthropology (Catalogue of the Babylonian Section).
CT 2	T. G. Pinches, *Cuneiform Texts from Babylonian Tablets in the British Museum 2*. London: British Museum, 1896.

CT 14	R. C. Thompson, *Cuneiform Texts from Babylonian Tablets in the British Museum 14*. London: British Museum, 1902.
CT 37	S. Smith, *Cuneiform Texts from Babylonian Tablets in the British Museum 37*. London: British Museum, 1923.
CT 49	D. A. Kennedy, *Cuneiform Texts from Babylonian Tablets in the British Museum 49: Late Babylonian Economic Texts*. London: British Museum, 1968.
CT 55	T. G. Pinches, *Cuneiform Texts from Babylonian Tablets in the British Museum 55*. London: British Museum, 1982.
CT 56	T. G. Pinches, *Cuneiform Texts from Babylonian Tablets in the British Museum 56*. London: British Museum, 1982.
Dar.	J. N. Strassmaier, *Inschriften von Darius, König von Babylon*. Leipzig: Pfeiffer, 1897.
EA	A. Rainey, *The El-Amarna Correspondence*, 2 vols. Leiden: Brill, 2015.
ETCSL	J. A. Black, Cunningham, J. Ebeling, E. Flückiger-Hawker, E. Robson, J. Taylor, and G. Zólyomi, *The Electronic Text Corpus of Sumerian Literature*. Faculty of Oriental Studies, University of Oxford (http://etcsl.orinst.ox.ac.uk/), 1998–2006.
GCCI 2	R. P. Dougherty, *Archives from Erech, Neo-Babylonian and Persian Periods (Goucher College Cuneiform Inscriptions 2)*. New Haven, CT: Yale University Press, 1933.
IBK	K. Oberhuber, *Sumerische und akkadische Keilschriftdenkmäler des Archäologischen Museums zu* Florenz (Innsbrucker Beiträge zur Kulturwissenschaft). Institut für Sprachwissenschaft der Universität Innsbruck, 1960.
K	Museum siglum of the British Museum (Kuyunjik).
KADP	F. Köcher, *Keilschrifttexte zur assyrisch-babylonischen Drogen- und Pflanzenkunde*. Berlin: Akademie, 1955.
KAR	E. Ebeling, *Keilschrifttexte aus Assur: religiösen Inhalts*. Leipzig: J. C. Hinrichs, 1919–23.
Kent	R. G. Kent, *Old Persian: Grammar, Texts, Lexicon*. New Haven, CT: American Oriental Society, 1953.
MBLET	O. R. Gurney, *The Middle Babylonian Legal and Economic Texts from Ur*. London: British School of Archaeology in Iraq, 1983.
MSL	B. Landsberger et al., *Materials for the Sumerian Lexicon*. Rome: Pontificium Institutum Biblicum, 1957–74.
Nbn	J. N. Strassmaier, *Inschriften von Nabonidus, König von Babylon*. Leipzig: Pfeiffer, 1887.
NCBT	Newell Collection of Babylonian Tablets, Yale University, New Haven, CT.

ND	Field numbers of tablets excavated at Nimrud.
PBS 2/1	A. T. Clay, *Business Documents of Murashû sons of Nippur Dated in the Reign of Darius II.* Babylonian Section, University of Pennsylvania Museum of Archaeology and Anthropology, 1912.
PF	R. T. Hallock, *Persepolis Fortification Tablets.* University of Chicago Press, 1969.
PFNN	Persepolis Fortification Tablets. Unpublished Elamite texts, numbered by Hallock.
RAcc	F. Thureau-Dangin, *Rituels accadiens.* Paris: Éditions Ernest Leroux, 1921.
RIMA I	A. K. Grayson, *Assyrian Rulers of the Third and Second Millennia BC (Royal Inscriptions of Mesopotamia. Assyrian Periods. Volume I).* University of Toronto Press, 1987.
RIMA II	A. K. Grayson, *Assyrian Rulers of the Early First Millennium BC I (1114–859 BC) (Royal Inscriptions of Mesopotamia. Assyrian Periods. Volume II).* University of Toronto Press, 1991.
RIMA III	A. K. Grayson, *Assyrian Rulers of the Early First Millennium BC II (858–745 BC) (Royal Inscriptions of Mesopotamia. Assyrian Periods. Volume III).* University of Toronto Press, 1996.
RIMB II	G. Frame, *Rulers of Babylonia: From the Second Dynasty of Isin to the End of Assyrian Domination (1157–612 BC) (Royal Inscriptions of Mesopotamia. Babylonian Periods. Volume II).* University of Toronto Press, 1995.
RIME I	D. R. Frayne, *Presargonic Period (2700–2350 BC) (Royal Inscriptions of Mesopotamia. Early Periods. Volume I).* University of Toronto Press, 2008.
RIME II	D. R. Frayne, *Sargonic and Gutian Periods (2334–2113 BC) (Royal Inscriptions of Mesopotamia. Early Periods. Volume II).* University of Toronto Press, 1993.
RIME III/1	D. O. Edzard, *Gudea and His Dynasty (Royal Inscriptions of Mesopotamia. Early Periods. Volume III/1).* University of Toronto Press, 1997.
RIME III/2	D. R. Frayne, *Ur III Period (2112–2004 BC) (Royal Inscriptions of Mesopotamia. Early Periods. Volume III/2).* University of Toronto Press, 1997.
RINAP 1	H. Tadmor and S. Yamada, *The Royal Inscriptions of Tiglath-pileser III and Shalmaneser V, Kings of Assyria (Royal Inscriptions of the Neo-Assyrian Period 1).* Winona Lake, IN: Eisenbrauns, 2011.
RINAP 3/1	A. K. Grayson and J. Novotny, *The Royal Inscriptions of Sennacherib, King of Assyria (704–681 BC) Part 1 (Royal Inscriptions of the Neo-Assyrian Period 3/1).* Winona Lake, IN: Eisenbrauns, 2012.

RINAP 3/2 A. K. Grayson and J. Novotny, *The Royal Inscriptions of Sennacherib, King of Assyria (704–681 BC) Part 2 (Royal Inscriptions of the Neo-Assyrian Period 3/2).* Winona Lake, IN: Eisenbrauns, 2014.

RINAP 4 E. Leichty, *The Royal Inscriptions of Esarhaddon, King of Assyria. 680–669 BC (Royal Inscriptions of the Neo-Assyrian Period 4).* Winona Lake, IN: Eisenbrauns, 2011.

RINAP 5/1 J. Novotny and J. Jeffers, *The Royal Inscriptions of Ashurbanipal (668-631 BC), Aššur-etel-ilāni (630-627 BC), and Sîn-šarra-iškun (626-612 BC), Kings of Assyria, Part 1* (Royal Inscriptions of the Neo-Assyrian Period 5/1). University Park, PA: Eisenbrauns, 2018.

Rm. Museum siglum of the British Museum (Rassam).

SAA 2 S. Parpola and K. Watanabe, *Neo-Assyrian Treaties and Loyalty Oaths (State Archives of Assyria 2).* Helsinki University Press, 1988.

SAA 5 G. B. Lanfranchi and S. Parpola, *The Correspondence of Sargon II, Part II: Letters from the Northern and Northeastern Provinces (State Archives of Assyria 5).* Helsinki University Press, 1990.

SAA 8 H. Hunger, *Astrological Reports to Assyrian Kings (State Archives of Assyria 8).* Helsinki University Press, 1992.

SAA 12 L. Kataja and R. Whiting, *Grants, Decrees and Gifts of the Neo-Assyrian Period (State Archives of Assyria 12).* Helsinki University Press, 1995.

SAA 15 A. Fuchs and S. Parpola, *The Correspondence of Sargon II, Part III: Letters from Babylonia and the Eastern Provinces (State Archives of Assyria 15).* Helsinki University Press, 2001.

SAA 17 M. Dietrich, *The Babylonian Correspondence of Sargon and Sennacherib (State Archives of Assyria 17).* Helsinki University Press, 2003.

SAA 19 M. Luukko, *The Correspondence of Tiglath-Pileser III and Sargon II from Calah/Nimrud (State Archives of Assyria 19)*. Helsinki University Press, 2012.

SAACT 10 J. Novotny, *Selected Royal Inscriptions of Assurbanipal: L³, L⁴, LET, Prism I, Prism T, and Related Texts (State Archives of Assyria Cuneiform Texts 10).* Winona Lake, IN: Eisenbrauns, 2014.

SpTU H. Hunger and E. von Weiher, *Spätbabylonische Texte aus Uruk.* Berlin: Gebr. Mann and von Zabern, 1976–98.

SU Sultantepe Tablets at the Archaeological Museum, Ankara.

TFS S. Dalley and J. N. Postgate, *The Tablets from Fort Shalmaneser.* London: British School of Archaeology in Iraq, 1984.

VAT Vorderasiatische Abteilung, Tontafeln, tablet signature, Vorderasiatisches Museum, Berlin.

CLASSICAL AND HEBREW TEXTS

Aelian *NA*	*De natura animalium*
Aelian *VH*	*Varia historia*
Aeschylus *Supp.*	*Suppliants*
Aet. Amid. *Med.*	Aetius of Amida, *Libri medicinales*
Amm. Marc.	Ammianus Marcellinus
Aretaeus *De curat. acut. morb.*	*On Therapy of Acute and Chronic Diseases*
Arist. *Hist. an.*	Aristotle, *Historia animalium*
Arr. *Anab.*	Arrian, *Anabasis Alexandri*
Arr. *Ind.*	Arrian, *Indica*
Ath.	Athenaeus, *Deipnosophistai (The Learned Banqueters)*
BNJ	*Brill's New Jacoby*, 2007– (accessed at https://scholarlyeditions.brill.com/bnjo/)
Celsus *Med.*	Celsus, *De medicina*
CMG	*Corpus Medicorum Graecorum* (accessed at http://galen.bbaw.de/epubl/online/editionen.html), Berlin-Brandenburgischen Akademie der Wissenschaften
Columella *Rust.*	*On Agriculture*
Diod. Sic.	Diodorus Siculus, *Library of History*
Dioscorides *Mat. Med.*	*De materia medica*
Ed. Diocl.	*Diocletian's Edict on Prices*
Galen *Aliment. fac.*	*On the Properties of Foodstuffs*
Galen *Antid.*	*Antidotes*
Galen *Comp. Med. sec. Loc*	*Composition of Medicines according to Places*
Galen *Gloss.*	*Glossarium*
Galen *Simpl. med.*	*On the Powers of Simple Medicines*
HA	*Historia Augusta*
Hdt.	Herodotus, *The Histories*
Hippoc. *Acut. Sp.*	Hippocrates, *Regimen in Acute Diseases*
Hippoc. *Epid.*	Hippocrates, *Epidemics*
Hippoc. *Morb.*	Hippocrates, *Diseases*
Hippoc. *Morb. Mul.*	Hippocrates, *Diseases of Women*
Hippoc. *Nat. Mul.*	Hippocrates, *Nature of Women*
Hippoc. *Vict.*	Hippocrates, *Regimen*
Hom. *Od.*	Homer, *Odyssey*
Horace *Sat.*	*Satirae*
Josephus *Ant.*	*Jewish Antiquities*
ME	*Metz Epitome*

Palladius *Agr.*	*Opus agriculturae*
PEG I	A. Bernabé. *Poetae Epici Graeci* I. Stuttgart: Teubner, 1996.
Philostr. *Vita Apoll.*	Philostratus, *Life of Apollonius of Tyana*
Phrynichus *Praep. soph.*	*Praeparatio sophistica*
Pliny *HN*	*Natural History*
Plut. *Quaest. Conv.*	Plutarch, *Quaestiones convivales (Table Talk)*
P. Freib.	*Mitteilungen aus der Freiburger Papyrussammlung*, 1914–86
P. Hawara	*Demotische Urkunden aus Hawara,* ed. E. Lüddeckens et al., 1998
P. Mich.	*Michigan Papyri* (1931–2011)
P. Oxy.	*The Oxyrhynchus Papyri* (1898–)
P. Tebt.	*The Tebtunis Papyri* (1902–2005)
PME	*Periplus Maris Erythraei*
Pollux *Onom.*	*Onomasticon*
Ptol. *Geog.*	Ptolemy, *Geography*
SB	*Sammelbuch griechischer Urkunden aus Ägypten*, ed. Preisigke et al., 1915–
Schol. Ar. Eq.	Scholium to Aristophanes, *Equites*
Serv. Aen.	Servius, *Commentary on Vergil's Aeneid*
Suet. Aug.	Suetonius, *Life of Augustus*
Tab. Vindol. II	A. K. Bowman and J. D. Thomas, *The Vindolanda Writing-Tablets* (*Tabulae Vindolandenses II*). London: British Museum, 1994.
TB	*Babylonian Talmud*
TJ	*Jerusalem Talmud*
Theophr. *Hist. pl.*	Theophrastus, *Enquiry into Plants*
Theophr. *Caus. pl.*	Theophrastus, *De causis plantarum*
Xen. Cyr.	Xenophon, *Cyropaedia*
Vergil *G.*	Vergil, *Georgics*
Vergil *Ecl.*	Vergil, *Eclogues*
Vinidarius *Exc.*	Vinidarius, *Excerpta Apicii*

DICTIONARIES

CAD	*Assyrian Dictionary of the Oriental Institute of the University of Chicago, Vols. 1–21*, 1956–2010.
CDIAL	R. L. Turner, *A Comparative Dictionary of Indo-Aryan languages.* Oxford University Press, 1962–85.

DEDR	T. Burrow and M. B. Emeneau, *A Dravidian Etymological Dictionary*, 2nd ed. Oxford: Clarendon Press.
LSJ	H. G. Liddell, R. Scott, H. S. Jones, et al., *A Greek-English Lexicon, 9th ed. with Revised Supplement.* Oxford: Clarendon Press, 1940–96.
PSD	A. W. Sjöberg et al., *The Sumerian Dictionary.* University of Pennsylvania Museum of Archaeology and Anthropology, 1984– (http://oracc.museum.upenn.edu/epsd2/).

Introduction

TRACING A GENEALOGY FOR CROP EXCHANGES

The past is perhaps most foreign in the sensory experience of quotidian life. Half a millennium ago, the world not only looked different but also smelled and tasted different. Imagine Italy without tomatoes, Australia without cattle, Florida without oranges, India without chilies, France without tobacco, Colombia without coffee, and Switzerland without chocolate. The regime of diseases afflicting humans was also markedly different. The peoples of the New World had no experience of smallpox, the now-extinct dreadful pustular rash, and other viral and bacterial infections endemic to the Old World.[1] The Afro-European colonization of the Americas brought little short of a revolution in the biosphere, irrevocably fusing the ecosystems, agricultural regimes, and dietary habits of the Old and New Worlds. Voyaging across the Atlantic from the New World were pumpkins, squashes, maize, peanuts, pineapples, guavas, cacao, chili peppers, cashews, cassavas, tomatoes, papayas, sunflowers, and potatoes, among other crops; while wheat, barley, rice, oats, sugarcane, coffee, bananas, citruses, and other Old World mainstays traveled to the Americas to become part of a labor-intensive, often slavery-based, cash-cropping system.[2] The year 1492 inaugurated a world without biological borders—and, inadvertently, one of the worst ecological-demographic disasters. The influx of peoples, livestock, and food crops also opened a Pandora's box of free-ranging weeds, pests, commensals, and microbes (and their attendant diseases) whose impact on the native

1. Guzmán-Solís et al. 2021.
2. Heywood 2012, 72–73; McNeill 2014, 444–47. For a fuller list of crops, see Hawkes 1998.

populations of the Americas was calamitous.[3] Coining a term for this pivotal biological diffusion, the environmental historian Alfred Crosby called his 1972 book, now held as one of the foundational texts of environmental history, *The Columbian Exchange*.

But the Columbian Exchange is only the most recent and best known of a series of intercontinental biological dialogues involving floral and faunal exchanges. Both archeologists and historians have identified major thresholds in agricultural history, marked by the introduction of new crops and modifications in labor and capital inputs, and have described them as either a revolution or an exchange. The most fundamental "revolution" was of course the Neolithic Revolution, which witnessed the domestication of plants and animals and the related evolution of hunter-gatherers into sedentary farmers some 12,000 years ago in multiple regions of the Old World.[4] The other notable prehistoric "revolution" was the Secondary Products Revolution, which entailed modifications in the management of domesticated animals to obtain secondary products like milk, cheese, yoghurt, ghee, wool, and leather—not to mention traction.[5] More recent agrarian "revolutions" familiar to students of modern history include the eighteenth-century British Agricultural Revolution, a prelude to the Industrial Revolution, and the Green Revolution of the 1940s to 1960s; both led to higher crop yields and considerable demographic growth.[6] In most cases, however, *revolution* is a misnomer, since these watersheds in agricultural history were often culminating points of a process long in the making or, as the archaeobotanist Marijke van der Veen puts it, a "time

3. New World populations were not immune to a host of Old World diseases, including smallpox, measles, mumps, whooping cough, influenza, yellow fever, and malaria (McNeill 2014, 442). A genetic study by Llamas et al. 2016, based on mitochondrial DNA from the osteological remains of 92 pre-Columbian South American individuals, draws attention to mass mortality and extinction of lineages during the early phase of European colonization.

4. Barker 2006; Bellwood 2004; Childe 1936. On the issue of multiple centers and protracted processes of crop domestication, see Fuller, Willcox, and Allaby 2011; Meyer, Duval, and Jensen 2012. A DNA study of 44 Middle Eastern individuals dating between 12,000 and 1400 BCE demonstrates that the lineages of the earliest farming communities of the southern Levant and the Zagros were distinct, suggesting that the transition from hunter-gatherer to farmer developed independently in the two regions (Lazaridis et al. 2016).

5. Sherratt 1983, 1999.

6. On the British Agricultural Revolution, see Chambers and Mingay 1966; Overton 1996; Toynbee 1884. On the Green Revolution, see Gaud 1968. The Green Revolution (a term coined by officials at the US Agency for International Development) was marked by the introduction of high-yield, disease-resistant, genetically modified crops to populous developing countries. The most conspicuous feature of more recent agrarian advances is the mechanization of farming, which led to the shift away from peasant agriculture to the agrarian capitalism of the present day. Mechanization freed labor for non-agricultural pursuits, which then, for the first time, became more significant than subsistence production in the local and global economy.

period when changes reached critical mass."[7] Admittedly, there are scholars who espouse more gradualist approaches, deploying the notion of a "revolution" as a metonym for a seminal transitional process.[8]

The post-Neolithic agricultural regime of the Old World was remarkably fluid, absorbing new cultivars from different biogeographic zones and adapting them to local social and cultural complexes.[9] The histories of anthropogenic crop dispersals are, however, as much about resistance inspired by cultural preferences as they are about absorption. Exotica were fêted in some circles, but met in others with scorn and suspicion. In most cases, the process of crop nativization was long-drawn and dictated not only by climatic-environmental limitations but also, more crucially, by conservative agricultural practices and foodways. It took two centuries for early modern European farmers and consumers to be convinced that the New World tomato and potato were not toxic.[10] Yet the very presence of new crops, and new varieties of old crops, indicates that some traditional agriculturists did experiment and innovate.[11]

Crop dispersals in the post-Neolithic world encompassed calorific staples like grains, pulses, and tubers; cash crops like fiber and oil plants, aromatics, and spices; ornamental plants; and non-staple calorific supplements like fruits and vegetables.[12] Broomcorn millet (*Panicum miliaceum*), an East Asian grain, reached Europe by the late second millennium BCE through the mediation of Central Asian agropastoral communities.[13] While Europe has a species of wild apple (*Malus sylvestris*), a species native to Central Asia (*Malus sieversii*) is the main contributor to the gene pool of the domesticated apple (*Malus pumila*).[14] Other surprising Central and East Asian contributions to the European agricultural landscape of the Iron Age (c. 1200–550 BCE) include apricots (*Prunus armeniaca*), peaches (*Prunus persica*), hemp (*Cannabis sativa*), pistachios (*Pistacia vera*), and carrots (*Daucus carota* subsp. *sativus*).[15]

7. Van der Veen 2010, 8; see also Scott 2017, 10–12, 18–19; Squatriti 2014, 1208.

8. Bar-Yosef 1998; Squatriti 2014.

9. Boivin, Fuller, and Crowther 2015; Harris 1998; McNeill 2014; Sherratt 1999, 26; Zohary, Hopf, and Weiss 2012, 7–8.

10. Albala 2002, 236–37.

11. Johnson 1972.

12. Sherratt 1999, 27.

13. Boivin, Fuller, and Crowther 2012, 459; Cunliffe 2015, 67; Filipović et al. 2020; Herrscher et al. 2018; Hunt et al. 2008; Miller, Spengler, and Frachetti 2016; Motuzaite-Matuzeviciute et al. 2013; Spengler 2019, 59–88; Stevens et al. 2016; Zohary, Hopf, and Weiss 2012, 7, 69–72.

14. Cornille et al. 2012, 2019; Zohary, Hopf, and Weiss 2012, 136.

15. Boivin 2017, 366–68; Dalby 2003, 20; Daryaee 2006–07, 76; Fuller and Madella 2001, 341; Heywood 2012, 71; Iorizzo et al. 2013; Sadori et al. 2009; Stevens et al. 2016; Spengler 2019; Stolarczyk and Janick 2011; Weisskopf and Fuller 2013a, 2013b; Zohary 1998, 126–27; Zohary, Hopf, and Weiss 2012, 7, 106–07, 144–45, 151–52.

Sub-Saharan African crops like sorghum (*Sorghum bicolor*), finger millet (*Eleusine coracana*), pearl millet (*Pennisetum glaucum*), hyacinth bean (*Lablab purpureus*), cowpea (*Vigna unguiculata*), and castor (*Ricinus communis*) made their way into South India via maritime routes in a piecemeal process by the early second millennium BCE.[16] The archaeobotanist Dorian Fuller and colleagues dub this process the "Bronze Age inter-savannah translocations."[17] Somewhat later, in the first millennium BCE, bananas (*Musa × paradisiaca*), taro (*Colocasia esculenta*), and yams (*Dioscorea alata*) were transmitted in the reverse direction from tropical Asia to sub-Saharan Africa, where they remain important calorific sources.[18] These spatially and temporally distinct examples underscore the sheer scale, complexity, and multidirectionality of crop movements across the post-Neolithic Old World.

In this book, I trace the origins of and examine one pivotal trajectory within the dynamic and multidirectional Old World exchanges of crops and fauna. The Indian subcontinent encompasses a great variety of ecosystems and is extraordinarily diverse agriculturally, with native crops being supplemented by cultivars from Africa, the Middle East, and Central, East, and Southeast Asia. It is specifically the transmission of crops and fauna of tropical and subtropical Asian origin from the Indian subcontinent to the Middle East and the Mediterranean which the archaeologist Andrew Sherratt describes as the "most important movement of crops before the Columbian Exchange."[19] The region of origin for the crops under consideration in this book falls in the tropical and subtropical zone extending between latitude 28° N and 9° S from Papua New Guinea to northwestern India. Broadly speaking, these crops are frost and drought intolerant and hence are predominantly summer irrigated cultivars in the Middle East and the Mediterranean. Tropical and subtropical South and Southeast Asian crops collectively constitute the single "largest group of introduced aliens" in the Middle East and the Mediterranean.[20] This biological diffusion was responsible for the creation of a "unified ecological contact zone" of overlapping sets of cultivars and livestock across the southern Eurasian landmass.[21] The growing agro-ecological coherence of Afro-Eurasia in turn abetted processes of cultural coalescence and facilitated the spread of ideas and technologies.

The movement of crops from India to the Middle East and the Mediterranean has been variously styled in the scholarly literature depending on the agency,

16. Blench 2003; Boivin and Fuller 2009; Boivin et al. 2014, 551–52, 561; Fuller 2002, 288–92, 310–12, 2003a; Fuller and Boivin 2009; Fuller, Boivin, Hoogervorst, and Allaby 2011; Fuller and Madella 2001, 334, 342–44; Fuller et al. 2015; Possehl 1998.

17. Fuller, Boivin, Hoogervorst, and Allaby 2011.

18. Boivin et al. 2013, 215, 257–59, 2014, 554; Fuller and Boivin 2009; Rangan et al. 2015, 144–51.

19. Sherratt 1999, 28.

20. Zohary 1998, 127.

21. Mikhail 2011, 952.

routes, and chronology scholars have chosen to privilege. The historian Andrew Watson speaks of this exchange as an Arab Agricultural Revolution or the Medieval Green Revolution (c. 700–1100 CE).[22] The archaeologist Andrew Sherratt opts for a more neutral "trans-Eurasian exchange" but elsewhere describes the same process as part of the "orientalization" of the Mediterranean.[23] John McNeill, a global historian, privileges the maritime routes across the Indian Ocean and names it the Monsoon Exchange.[24] Lynda Shaffer, another global historian, situates the crop exchange within a broader exchange of ideas and technology and describes it as the "southernization" of the Northern Hemisphere.[25] In terms of chronology, scholarship has mostly failed to appreciate the lengthy process of crop transfers and has unhelpfully attributed agency to one or more political actors, including the Romans, the Sasanian Persians, and the medieval Arabs.[26] But this book will show that South Asian crop movements to the Middle East and the Mediterranean extend back to the Bronze Age horizon, and no single political actor can be credited with initiating and sustaining this process.

The "tropical turn" of the Middle East and Mediterranean under consideration here first entered scholarly purview with Andrew Watson's groundbreaking 1974 article, "The Arab Agricultural Revolution and Its Diffusion." This was followed by another article, "A Medieval Green Revolution" (1981), and finally a monograph, *Agricultural Innovation in the Early Islamic World: The Diffusion of Crops and Farming Techniques* (1983). In these works, Watson argued for an agricultural "revolution" in the early Islamic Middle East and the Mediterranean (c. 700–1100 CE) fostered by the diffusion of new Indian crops and improved farming techniques. The tropical South Asian summer crop package discussed by Watson included rice, sorghum, durum wheat, sugarcane, cotton, sour orange, lemon, lime, pomelo, banana, coconut, watermelon, spinach, artichoke, taro, eggplant, and mango. Watson identified the cosmopolitan, Islamicate populations of the Middle East and the Mediterranean as the principal agents in this biodiffusion. In Boserupian fashion, he held that the privatized labor-intensive farming of the Islamic lands, which involved equitable distribution of water, enriched irrigation methods (extensive use of water-lifting devices, underground canals, dams, and so on), crop rotation, and greater application of fertilizer, enabled greater crop yields, demographic expansion, and urban development.

But Watson's work is plagued by what one reviewer calls a "profound lack of interest in the pre-Islamic landscape and a host of flawed assumptions."[27] Most of

22. Watson 1974, 1981, 1983.
23. Sherratt 1999: 27; Sherratt 2006.
24. McNeill 2001.
25. Shaffer 1994.
26. King 2015.
27. Decker 2009, 191.

the crops cited by Watson were already introduced well before the Islamic period and certainly did not arrive in a package.[28] Nor was there anything astoundingly novel about the hydraulic engineering of the early Islamic world, which inherited and expanded pre-existing Roman and Sasanian hydraulic technologies.[29] The dispersal histories of individual crops are varied and complex. The notion that tropical South Asian crops were an "exogenous *deus ex machina* to kick-start change" in hitherto static Middle Eastern and Mediterranean economies is misleading.[30] The diffusion of South Asian flora and fauna to the Middle East and beyond was not a linear, uninterrupted event but an episodic process with its roots in the interconnected world of the Late Bronze Age. While Watson's study provides the stimulus for thinking about biological diffusions in the Old World, the so-called Arab Agricultural Revolution is far older than supposed and a much more gradual process.

SCOPE AND LIMITS

This book, in essence, offers an ecological reading of long-distance connectivity in the ancient world by investigating tropical and subtropical botanical transfers via maritime and overland routes, from South Asia to the Middle East and the Mediterranean, with the aim of assessing the motivations behind and impact of this phenomenon on ancient Middle Eastern and Mediterranean societies. The discussion of individual crops will be prefaced by a detailed and much-needed sketch of the Middle East and the Mediterranean's relations with South Asia from prehistory to the late centuries BCE. This *longue durée* narrative of interactions between the Middle East, the Mediterranean, and South Asia, whose full contours have yet to be appreciated by ancient historians and archaeologists alike, forms an essential backdrop to both early and later crop movements.

The textual and archaeological materials assembled in this book have yielded evidence for the introduction and naturalization of several South Asian cultivars in the ancient Middle East and the Mediterranean. The strategy employed here has been to maximize the available evidence while offering a selection of different

28. While Watson's case for the introduction of new tropical crops in the Middle East and the Mediterranean during the early centuries of Islam has been shown to be an invalid causal association (Decker 2009; Kelley 2019; King 2015; Ruas et al. 2015; Samuel 2001, 418–22), there are many scholars who continue to cite his conclusions, albeit with modifications (e.g., Amar and Lev 2017, 49–53; Fuks, Amichay, and Weiss 2020; McNeill 2014, 439; Mears 2011, 153; Zohary 1998). See Squatriti 2014 for a discussion of the lasting influence of the "Watson thesis" on historical studies and other academic disciplines. There have, however, been strong critics of the "Watson thesis" from its inception. See Johns 1984 and Aubaile-Sallenave 1984 for early critiques of Watson's work which argue that it is shoddy in the details.

29. Angelakis et al. 2020, 24–25; Avni 2018; Decker 2009, 190; Kamash 2012.

30. Squatriti 2014, 1212.

types of economic plants: cereal (rice), fiber (cotton), timber (sissoo), tuber (taro), legume (lotus), and fruits and vegetables (citruses and cucurbits). Of course a good number of these plants had non-comestible functions as well. And this list of new crops is far from exhaustive. A variety of Indian millet, sugarcane, eggplant, sebesten plum, and sambac jasmine are known from pre-Islamic Middle Eastern and Mediterranean contexts, but the textual and archaeological evidence for their dispersal and cultivation is meager.[31] Yet crop transfers that were seemingly insignificant in their earliest phases could hold great potential in the future, as is the case with sugarcane. As our focus is strictly on economic plants from tropical Asia which were cultivated in the ancient Middle East and the Mediterranean, the survey excludes tropical species which were available as long-distance commodities but not taken into cultivation. Dozens of spices, for instance, have been moving between South Asia, the Middle East, and the Mediterranean throughout the ages, among them economically important ones like black pepper, ginger, cardamom, cinnamon, and cassia. These will not be discussed in any detail as they deserve a separate treatment of their own.

Heading in the opposite direction, Mediterranean and Middle Eastern cultivars, most prominently barley and wheat, were also transmitted to South Asia in antiquity.[32] A number of commonly used spice and aromatic plants in South Asia, including coriander, cumin, black cumin, ajwain, fenugreek, saffron, marjoram, and licorice, originate in temperate zones west of the Indian subcontinent.[33] Although coriander (*Coriandrum sativum*), a native of the Middle East, was already identified as seeds in a late-third-millennium BCE context in Miri Qalat in Pakistani Baluchistan, its widespread use in South Asia is probably due to culinary exchanges with the Achaemenid Empire in the first millennium BCE.[34] This is suggested by the use of an Akkadian or Aramaic loanword for the plant in Sanskrit: *kustumburu* (Akkadian *kusibirru*, Aramaic *kusbara*).[35] Similarly, almonds, walnuts, pomegranates, and apples are not native to the Indian subcontinent but derive from Central Asia and the Middle East.[36] Asafoetida (*Ferula asafoetida*), whose

31. Indian millet: Pliny *HN* XVIII.55; Dalby 2003: 306. Sugarcane: Dioscorides *Mat. Med.* II.82.5; Pliny *HN* XII.32; Brust 2005, 563–64; Floor 2009; Watson 1983, 26, 160. Eggplant: Aet. Amid. *Med.* I.210; Aubaile-Sallenave 1988. Sebesten plums (*Cordia myxa*): van der Veen 2011, 151–53. Sambac jasmine (*Jasminum sambac*): Agatharchides of Knidos, ap. Diod. Sic. III.46.2 (plausible reference); TB *Shabbat* 50b; Amar and Lev 2017, 155; Germer 1985, 153; Newberry 1890, 47; Schweinfurth 1884, 314; Woelk 1965, 238.

32. Boivin 2017, 356–58; Fuller and Lucas 2017, 313, 316.

33. See Dalby 2003 for further literature on individual cultivars.

34. Tengberg 1999, 6, 10; Zohary, Hopf, and Weiss 2012, 163. On fenugreek at Harappan sites, see Bates 2019, 881.

35. References to coriander in the early Sanskrit medical corpus have been collected in Singh 1999, 113.

36. Archaeobotanical evidence indicates familiarity with almonds and walnuts in northwestern India by the late Harappan period, c. 1800 BCE (Bates 2019, 881; Fuller and Madella 2001, 340).

pungent resin is a popular spice across South Asia, was (until trial cultivation in Himachal Pradesh in 2020) imported to India from the arid regions of Iran and Afghanistan, where it grows wild.[37] The movement of cultivars from the Mediterranean and the Middle East to South Asia is, on the whole, poorly documented. While these crop movements in the reverse direction shared the same routes and mechanisms with the tropical Asian crops moving to temperate zones, the Middle Eastern and Mediterranean contribution to South Asian agriculture, diet, and culture is beyond the scope of the present book.

Like the Columbian Exchange, the tropical "Indian" exchange entailed the transfer of human populations, animals, commensals, pests, microbes, and diseases which left an irrevocable ecological imprint on host landscapes in the Middle East and the Mediterranean. The animals transmitted to the Middle East and the Mediterranean from South and Southeast Asia included both domesticates and exotica: chickens, peafowl, parakeets, Asian elephants, tigers, mongooses, langurs, water buffaloes, caprids, and even cat and dog breeds.[38] The movement of fauna had a significant impact on agricultural practice, communications, warfare, and leisurely pursuits. The westward dispersal of chickens (figures 1 and 2) ensured that egg consumption could become regular rather than seasonal as is generally the case with eggs produced by geese and ducks.[39] But calorific needs were not the only motivation for the chicken's westward march. In classical Greece, cockerels appear as important courtship gifts in pederastic relationships and were intimately connected to one of the most common modes of gambling across the Old World: cockfighting.[40]

Pests and commensals trailing humans, animals, and plants along Indian Ocean routes included the black rat (*Rattus rattus*), house mouse (*Mus musculus*), Asian house shrew (*Suncus murinus*), house crow (*Corvus splendens*), gecko (*Hemidactylus* spp.), and a great variety of insects.[41] Trade in grain and farinaceous products

37. Sood 2020.

38. Bodson 1999, 75–78; Boivin 2017, 359–62; Çakırlar and Ikram 2016; Laursen and Steinkeller 2017, 83–88; Ottoni et al. 2017, 5; Pareja et al. 2020a, 2020b; Secord 2016.

39. The wild red junglefowl (*Gallus gallus*) is the main contributor to the domesticated chicken's gene pool. The native habitat of this species stretches from northeastern India to southern China and Southeast Asia. Introgression with another species of wild fowl in South Asia, the gray junglefowl (*Gallus sonneratii*), was responsible for the yellow-legged feature of many varieties of modern chicken. Chickens were known in the Middle East from the mid-to-late second millennium BCE on but only became common in the first millennium BCE. For zooarchaeological, iconographic, and literary materials on the dispersal of chickens from tropical Asia to the Middle East and Europe, see Borowski 1998, 156–58; Carter 1923; Coltherd 1966; Corbino et al. 2022; Ehrenberg 2002; Fuller et al. 2011, 551; Laursen and Steinkeller 2017, 87–88; Lawal and Hanotte 2021; Perry-Gal et al. 2015; Peters 1913; Peters et al. 2022; Trentacoste 2020, 8–11. On the dispersal of chickens in Africa, see Boivin et al. 2013, 252–54, 2014, 553, 556.

40. Dalby 2003, 83. On the importance of social and cultural factors (with an emphasis on cockfighting) in the westward spread of the chicken, see Sykes 2012.

41. Boivin 2017, 359, 362, 374–75; Boivin et al. 2013, 246–49, 264; Fuller and Boivin 2009, 29–31.

FIGURE 1. Terra cotta *askos* (flask) in the form of a rooster, Etruria, Italy, fourth century BCE. Metropolitan Museum of Art, New York (public domain).

FIGURE 2. Terra cotta neck-amphora with two roosters, attributed to the Painter of London B 76, Attica, Greece, c. 570–560 BCE. Metropolitan Museum of Art, New York (public domain).

across overland and maritime Afro-Eurasian routes led to the spread of agricultural pests like the red flour beetle (*Tribolium castaneum*), lesser grain borer (*Rhyzopertha dominica*) and khapra beetle (*Trogoderma granarium*), natives of India which are unambiguously present in the palaeoentomological records of the Middle East and the Mediterranean from the second millennium BCE on.[42] A number of weed species, like horse purslane (*Trianthema* spp.), buttonweed (*Spermacoce* spp.), and creeping woodsorrel (*Oxalis corniculata*), accompanied the trans–Indian Ocean shipment of grain.[43] The other significant unintentional dispersal was that of microbial pathogenic entities like protozoa, bacteria, and viruses, which had a central role in the formation of familiar Old World diseases (like smallpox, measles, and influenza) and epidemics.[44] The sheer complexity of this biological exchange cannot be pursued in any great depth here, and we will restrict our scope to the movement of crops (and, where relevant, agricultural pests) from South Asia to the Middle East and the Mediterranean.

It may be wise to embrace another caveat at this juncture. While I do not deny the importance of the "tropical turn" for the Middle East and the Mediterranean, the exceptionality of this exchange should not be exaggerated. The "Indian" tropical bioexchange stands out from other early crop movements not only on account of its wide geographical scope and diverse repertoire of calorifically and culturally significant crops but also because the cultures which were directly involved have bequeathed us written records, lending greater visibility to agrarian changes here compared to similar processes elsewhere in the Afro-Eurasian landmass. The stories of the latter have to be chronicled by archaeologists rather than historians. The only other ancient transcontinental Eurasian bioexchange with a comparable impression on documentary sources is perhaps the exchange of crops along the maritime and terrestrial Silk Roads between China and Central, West, South, and Southeast Asia between the Han and Tang dynasties (second century BCE to ninth century CE).[45]

SOURCES AND METHODS

Owing to the wide spatial and temporal scope of the present study, the sources consulted here appear in a range of genres and in several ancient languages, including Sumerian, Akkadian, Elamite, Hebrew, Aramaic, Greek, Latin, Sanskrit, and the

42. Edde 2012, 2; Kislev 2015, 88–91; Panagiotakopulu 1998, 232–33, 2001, 1239–42, 2003, 356; Panagiotakopulu and van der Veen 1997, 201–02.

43. Boivin et al. 2013, 215–16; Fuller and Boivin 2009, 27–28; Groom, Van der Straeten, and Hoste 2019.

44. Achtman 2017; Green 2017.

45. Boivin, Fuller, and Crowther 2015, 353–55; Buell et al. 2020; Laufer 1919; Schafer 1963, 1967; Spengler 2019.

Prakrits. References to plants and their byproducts in ancient texts are often incidental to the work's purpose, whether literary, spiritual, pharmacological, geographical, administrative, fiscal, or otherwise. While the diversity of sources reflects the multiplicity of applications for botanical materials, the cultivation context and modes of transmission are often invisibilized by all-too-casual references. The piecemeal nature of the surviving data is, however, compensated for by a cumulative reading of the textual sources.

The documentary sources in Sumerian, Akkadian, and Elamite, preserved on clay tablets in the cuneiform script, have rarely been privileged in discussions of ancient crop translocations despite the wealth of references to wild and cultivated plants in these languages. This hesitation is a result of the imprecise identification of ancient plant names with modern botanical equivalents, especially in cases where there are no clear lexical cognates in other surviving languages.[46] Even where cognates are present in related and neighboring languages, the sheer volatility of ancient naming conventions means that there can be no certainty in the identification of flora and fauna on the basis of textual data alone. The same plant and its constituent parts may go by several names within the same language. There may also be diachronic semantic shifts and expansion prompted by loss of familiarity or the usage of other plant species with similar morphologies or ethnobotanical applications.

Furthermore, contextual data by way of extensive descriptions is often wanting in Mesopotamian contexts, in part due to the laconic nature of ancient scholarship on clay tablets. Some cuneiform texts, however, do afford brief notes on morphological features of plants and their medical applications.[47] From the late second millennium BCE on, there was also a rudimentary attempt to systematize botanical and pharmaceutical knowledge in scholarly compendia and annotated lexical series.[48] At present, most plant names in cuneiform records still merit generic descriptors like "a medicinal plant" or "a briar" in cautious philological studies, and only around sixty plants can be matched with modern botanical equivalents with any degree of certainty.[49]

While Greco-Roman records offer more substantial descriptions of plants, it must be borne in mind that not all authors saw the plant being described and some authors plagiarized and garbled older accounts, with the result that little useful information on the contemporary cultivation status of the plant can be extracted. Pliny (c. 24–79 CE), the author of the *Natural History* which holds a significant

46. On the difficulty of identifying botanical species in the cuneiform record, see Postgate 1984, 5–7; Watson 2004, 110–12.

47. Böck 2013, 129–63, 2015, 22–28; Tavernier 2008.

48. Frahm 2011, 253–254; Rumor 2018.

49. Bleibtreu 1980, 16; Böck 2011, 696.

influence over discussions of botany in antiquity, is notorious for plagiarism by modern standards.[50] He claims, for instance, that the citron refused to grow anywhere but in Persia and Media, a statement which is a misreading of the Greek scholar Theophrastus (whom he copies) and completely at odds with the relatively rich archaeological data on citrus fruits in Roman Italy (see chapter 4).[51] The agricultural historian Andrews's appraisal of Solinus (third century CE), who in turn plagiarized Pliny's description of citrons (46.6), is perhaps apt for Pliny as well: "Solinus . . . was such an unobservant individual that he was capable of being oblivious to a citron tree growing in his neighbor's backyard."[52] And as with cuneiform texts in Sumerian and Akkadian, lexical discrepancies and fluid nomenclatures are not unknown in Greek and Latin sources.[53] An empirical and systematic botanical taxonomy of the modern Linnaean type was singularly lacking across the ancient world, resulting in substantial semantic fluidity among some botanical terms. Take for example the word *malum* in Latin, a prodigiously well-documented ancient language. While *malum* is regularly translated as "apple," its semantic field encompasses a whole range of fleshy arboreal fruits like peaches, quinces, pomegranates, and citruses.

In light of the high mutability of botanical nomenclature in antiquity, the fields of archaeobotany and plant genetics are therefore of great importance in assessing interpretations of plant names in textual sources. As new crops fall outside the traditional taxation and tithing regimes, their frequency in administrative documents is also expected to be low or zero. In such instances, the archaeobotanical evidence, which typically appears as seeds, bolls, charcoal, plant impressions, pollen, starch granules, phytoliths (silica deposits in plant cells), or even whole desiccated fruits, is indispensable in understanding the spatial and chronological distribution of new cultivars.[54] But the archaeobotanical record is far from exhaustive, and some countries (like Egypt, Israel, Italy, and Greece) have benefited from more extensive sampling of archaeological plant remains compared to their neighbors in the Middle East and the Mediterranean.

Botanical remains are also subject to the fortuitous coincidence of deposition and preservation. The procurement, processing, consumption, and disposal practices of ancient peoples are crucial in determining what survives in the first place. Archaeobotanical assemblages are differentially affected by climate, soil types, deposition conditions, and other taphonomic processes like charring, desiccation,

50. For a discussion of plagiarism in Latin literature, including Pliny, see McGill 2012.

51. Pliny *HN* XII.16.

52. Andrews 1961, 41. See also Purcell 2012 on Pliny: "He not infrequently garbles his information through haste or insufficient thought."

53. Totelin 2016, 13–14.

54. For an overview of archaeobotanical data sets, see Farahani 2021, 11–20; Fuller 2002, 248–49; Lodwick and Rowan 2022; Zohary, Hopf, and Weiss 2012, 9–13.

and waterlogging, resulting in an uneven sampling of cultivars. Not all organic materials survive well in the archaeological record. The fibers of cotton, one of the cultivars under consideration in this book, are largely composed of cellulose, which is easily degraded by cellulolytic microorganisms. Fleshy fruits which are not subject to charring rarely endure the passage of time, while others, like leafy vegetables, leave almost no macro-remains. This is also the case for most starchy tuberous crops, although the scrutiny of microfossil remains like charred parenchymatous tissue (soft "filler" tissue) and starch grains has facilitated their identification.

Further avenues for research on ancient human diets and agricultural regimes are afforded by biomolecular studies on human, faunal, and plant remains, including ancient DNA, proteomic, lipid, and stable isotope analyses.[55] The surfaces of ancient cooking, storage, and serving wares offer ample opportunities for residue analysis of biochemical compounds of botanical origin. Palaeoentomological records, especially those concerning invasive agricultural pests, can serve as proxies for anthropic crop movements. But the application of scientific methods in determining ancient organic remains has not been without challenges. The study of ancient proteins, for example, has been stymied by chemical degradation and contamination of samples, leading to polarizing, if not outlandish, results in some cases.[56] Nonetheless, the convergence of diverse data sets remains the only viable option for establishing the stability and comparability of botanical nomenclature in a field where identifications of botanical materials solely on the basis of ancient textual materials can be highly contentious. Such an integrated approach also provides complementary data on cultivation contexts and modes of transmission which may not be explicit in textual sources.

55. Brown and Brown 2011; Schrader 2019, 127–48; Twiss 2019, 24–38.

56. For an overview of paleoproteomics and the existing challenges in this field, see Hendy et al. 2018; Warinner, Richter, and Collins 2022. See Chowdhury, Campbell, and Buckley 2021 and Scott et al. 2020 for proteomic studies which unconvincingly argue for the presence of soybean, an East Asian domesticate, in the Late Bronze Age Middle East.

1

The Historical Context

SOUTH ASIA, THE MIDDLE EAST, AND THE MEDITERRANEAN FROM THE MID-FOURTH MILLENNIUM BCE TO THE LATE CENTURIES BCE

The ease with which tropical crops were transmitted to the Middle East and the Mediterranean from the Bronze Age on underscores the long-standing connectivity between South Asia, the Middle East, and the Mediterranean. The diversity of terrestrial and maritime routes, middlemen, and centers of exchange connecting South Asia, the Middle East, and the Mediterranean matches the complex peregrinations of individual crops. The arrival of a new crop was, in most cases, preceded by trade in related botanical produce.[1] Increasing demand eventually made it profitable to replace imports with local cultivation. Other cases of crop introductions may be altogether inadvertent: piggybacking as part of a merchant's subsistence needs before being bartered for local foods at the end destination. Traders apart, the foodways of porters, guides, nomads, deportees, refugees, mercenaries, fisherfolk, and other mobile communities would have been equally crucial in the spread of cultivated crops. To understand how the societies of the Middle East and the Mediterranean were transformed by exchange networks, commercial or otherwise, it is necessary to sketch a *longue durée* history of "Middle Asian" mobility across land and sea, from its prehistoric beginnings to the late centuries BCE. I have deliberately minimized any discussion of botanical translocations in this chapter, since they occupy the rest of the book. Our primary purpose in this chapter is to establish

1. Sherratt 1999, 19.

beyond doubt that there were impactful linkages, from distant antiquity, between South Asia, the Middle East, and the Mediterranean.

ROUTES AND MODES OF TRANSPORTATION

The origins of east–west connectivity in the Eurasian landmass lie in the distant prehistoric horizon. The earliest transactional networks between South Asia and the Middle East, archaeologically discernible at least by the fourth millennium BCE, were little more than unorganized and random down-the-line transmissions of small portable goods and perishables like precious and semiprecious stones, shells, resins, reeds, hides, and cured foods, and finished crafts like baskets and stone and wood artifacts.[2] These local and regional circulations of goods involved both mobile pastoralists and settled communities.[3]

Local deficiencies in vital raw materials probably played a key role in stimulating trade in bulky commodities over longer distances. The scarcity of materials like metal, timber, and stone in southern Mesopotamia made connectivity across the Persian Gulf and the Iranian plateau a recurring structural feature of the region's historical development.[4] In the earliest phases, no single agency shaped the contours of long-distance connections since there were no professional merchants responding to distant demand with premeditated commercial strategies.

In the absence of paved roads and imperial highways, natural impediments to communication in the form of mountains and deserts funneled the prehistoric trickle of long-distance trade into several distinct routes, which have remained in constant use ever since. The northerly overland route, also known as the Great Khurāsān Road, led from the western Himalayas, stretched over the length of the Iranian plateau, bypassing its arid interior (the deserts of Dašt-e Kavīr and Dašt-e Lūt), and skirted the southern subtropical shores of the Caspian Sea before descending onto the Mesopotamian lowlands through the Hamadan-Kermanshah-Baghdad corridor (map 1). This route was, in later times, a significant arm of the so-called Silk Roads which connected East Asia with the Mediterranean.

The southerly overland route, from the lower Indus Valley through the modern Iranian provinces of Baluchistan, Sistan, and Kerman, was less favorable to caravans owing to drier and harsher climatic conditions. Alexander the Great's ill-fated return to Babylon from India across the Gedrosian desert (southern Baluchistan) in the late fourth century BCE exposes the perils of this route for large-scale convoys.[5] But this route intersected at various points with the maritime coasting route

2. Reade 2008, 12–13.
3. Boivin, Fuller, and Crowther 2012, 464, 2015, 350–51.
4. Ratnagar 2004, 22–23; van de Mieroop 1999, 30–31.
5. Arr. *Anab.* VI.21.1–26.5.

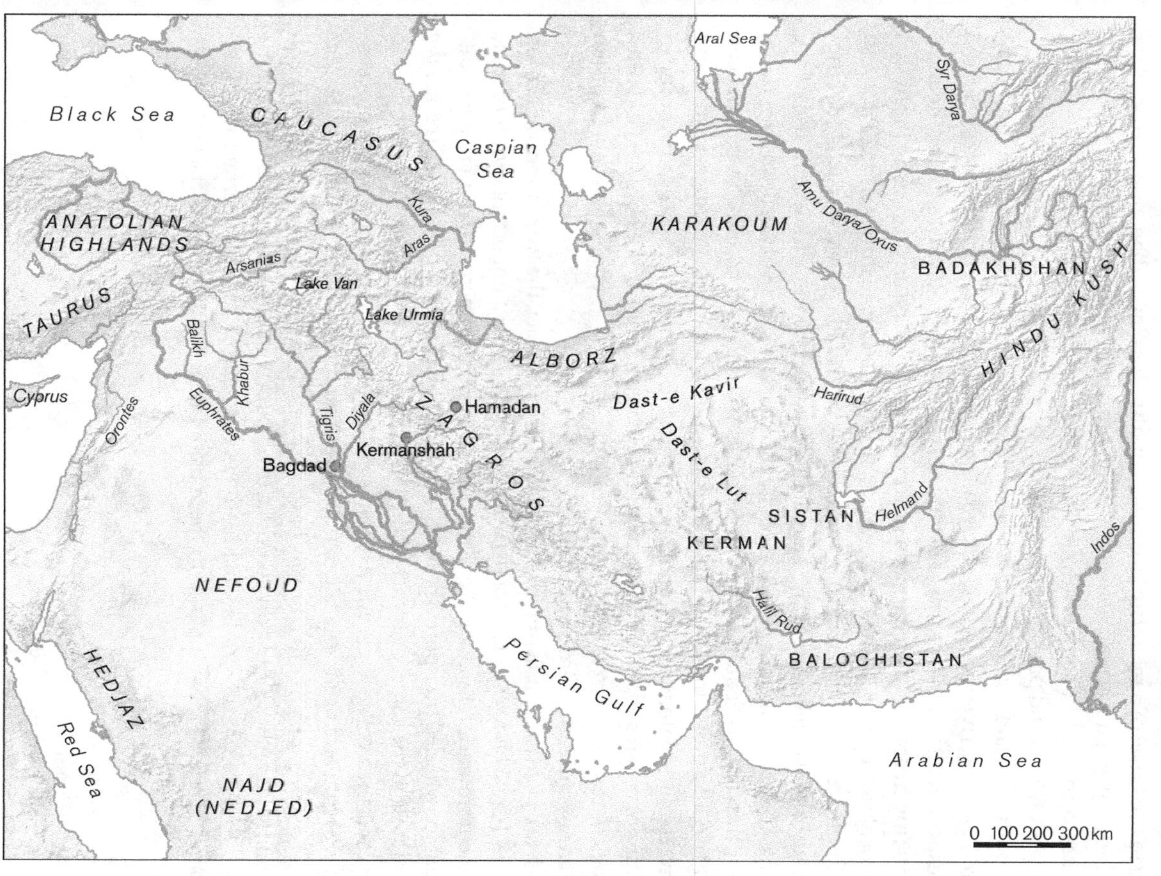

MAP 1. The Middle East, showing key geographic features. © Peter Palm (Berlin).

from northwest India to the Persian Gulf. It was, after all, a Gedrosian navigator named Hydrakes who guided Alexander's fleet to Karmania (the Iranian provinces of Kerman and Hormozgan).[6] A Persepolis Fortification tablet (c. 500 BCE) attesting to the travel of a relatively large party of fifty, possibly Indian, nobles from Susa to India through Karmania also raises the possibility that maritime and overland routes coincided in southern Iran wherever terrestrial travel became too hostile for pedestrians and pack animals alike.[7]

Both overland routes should not, however, be mistaken for single thoroughfares, as lesser riverine routes, side roads, and seasonal pastoral tracks offered the prospect of alternative journeys. Many of these lesser routes remain, for the most part, imperceptible to the modern observer, as premodern roads were determined by local perceptions of the environment, not by physical features alone.[8] Pack animals like zebu cattle, Bactrian camels, donkeys, and horses were crucial for long-distance movement across the Iranian plateau. Before the widespread use of the horse and the donkey, camels and zebu cattle, a domesticate of northwestern India, were the chief pack animals between South Asia and eastern Iran.[9] A pair of zebu cattle could pull a load of up to nine hundred kilograms over rough terrain and traverse forty kilometers in about ten hours.[10] Depictions of zebu cattle occur in Mesopotamian and west Iranian contexts by the late fourth millennium, and osteological remains of zebu cattle are attested as far west as Shahr-i Sokhta in Sistan (Iran) by the early third millennium BCE.[11]

By the late third millennium BCE, if not slightly earlier, the two-humped Bactrian camel (*Camelus bactrianus*) was to become the single most important mover of trade and communications along the plateau routes, especially across terrain that was difficult for wheeled transport. The Persian poet Sa'dī (thirteenth century CE) recalls the hardiness of the camel over that of the horse: "To walk and then sit is better than to run and then collapse. . . . An Arabian steed gallops in haste but a camel proceeds slowly night and day."[12] The late antique Greek author Didymus (fourth or fifth century CE) even remarks that he saw Bactrian camels "competing alongside horses and winning (in races)."[13] Not only could Bactrian camels endure extremes of temperature (−40 to +40 °C), their individual load-bearing abilities over arid terrain (250 kilograms) rendered them superior to the ancient bullock and donkey carts.[14]

6. Arr. *Ind.* 27.1.
7. PFNN 615; Giovinazzo 2000–01, 68–69.
8. Horden and Purcell 2000, 128.
9. Chen et al. 2010.
10. Matthews 2002, 440.
11. Matthews 2002, 443.
12. Sa'dī, *Golestān* IV, trans. Thackston 2008.
13. Didymus ap. *Geoponica* XVI.22, trans. Dalby 2011.
14. Cunliffe 2015, 20–22.

The presence of camel bones, dung, and hair at Shahr-i Sokhta in eastern Iran and camel burials with four-wheeled wagons at Gonur Depe, and miniature clay imitations from Altyn Depe in southern Turkmenistan, indicate that the domesticated variety of the Bactrian camel was a pack animal in these regions by the third millennium BCE.[15] The original habitat and domestication trajectory of the wild Bactrian camel was more northerly, extending from northwestern China through to Mongolia and southern Kazakhstan.[16] Mesopotamian references to the Bactrian camel as the "camel of the road" (Sumerian *amsiharran*), indicating some role in terrestrial transportation, are seen as early as the mid-third millennium BCE.[17] Also, Piotr Steinkeller has convincingly interpreted the references to the *gugur,* a large ungulate from the central Iranian plateau (Šimaški), in Ur III texts from the Puzriš-Dagan archive (2040s BCE) as another word, possibly foreign, for Bactrian camels.[18] The Bactrian camel may have reached the Mediterranean by the early second millennium, since a hematite seal from northern Syria dating between 1800 and 1650 BCE depicts a divine couple sitting on a Bactrian camel (figure 3).[19]

The persistent importance of camels as agents of transport east of the Zagros is also suggested by the efforts of the Assyrian king Aššur-bēl-kala (1074–1057 BCE) to breed them in eleventh-century BCE Assyria: "He sent out merchants and they brought back *burḫiš* oxen, female Bactrian camels and *tešēnu*-animals, he collected the female Bactrian camels, bred (them), and displayed herds of them to the people of his land."[20] And Aššur-bēl-kala may not have been the first Assyrian monarch to import and breed camels, since the bones of nine Bactrian camels dating to the thirteenth or twelfth century BCE were found at the Middle Assyrian settlement of Dūr-Katlimmu (modern Tell Sheikh Hamad) in western Syria.[21] An Aramaic administrative document from the mid-fourth-century BCE archives of the satrap of Achaemenid Bactria (modern Afghanistan) indicates continued royal interest in the breeding of camels in later periods, particularly through tax exemptions granted to Bactrian camel-keepers (Old Persian *uštrapāna*).[22]

15. Bonora 2021, 750–52; Compagnoni and Tosi 1978; Heide 2010, 344–60; Masson and Sarianidi 1972; Sataev 2021, 450–51.

16. Berthon et al. 2020; Bonora 2021, 750–52.

17. Horowitz 2008, 601, 603–05. The earliest mention of the Bactrian camel is in an Early Dynastic list of animals from Šuruppak. While the Sumerian term *amsi* normally denotes an elephant, bilingual Sumerian-Akkadian lexical texts equate the compound form of the Sumerian word *amsi* (e.g., *amsikurra, amsiḫarran*) with the Akkadian for camel (*ibilu*). See also Maaijer and Jagersma 2003/2004, 355; Potts 2004a; Steinkeller 2009.

18. Lafont 2020; Steinkeller 2009. On the region of Šimaški, stretching from the Isfahan region to the south Caspian zone, see Steinkeller 2014, 697–98.

19. Heide 2010, 345; Walters Art Museum, Baltimore, inventory no. 42.804.

20. RIMA II, A.0.89.7, iv 26–28.

21. Becker 2008, 85–86; Fales 2010, 80.

22. Briant 2009, 149; Naveh and Shaked 2012, 68–69 (A1). On camels in other Achaemenid contexts, see Henkelman 2017, 55–63.

FIGURE 3. Hematite cylinder seal with a Bactrian camel ferrying a divine couple, Syria, c. 1800–1650 BCE. Walters Art Museum, Baltimore (public domain).

The breeding of Bactrian camels is, of course, not an end in itself but a vital infrastructural investment in long-distance communications along the plateau routes. The high estimation of camels in eastern Iranian cultures (figure 4) is also intimated by the appearance of camelophoric (*-uštra*) names in the Avesta (c. 1100–500 BCE), the oldest extant Iranian textual corpus and the sacred

FIGURE 4. Arachosian (Greater Kandahar region) tribute-bearers leading a Bactrian camel, northern staircase of the Apadāna, Persepolis, Iran, c. 486–465 BCE. Photograph by Nick Taylor via Wikimedia Commons (CC BY 2.0).

scripture of the Zoroastrian faith, most notably that of the prophet Zaraθuštra himself, whose name translates as either "driver of camels" or "possessor of mature camels."[23]

While the horse was domesticated in the Pontic-Caspian steppe zone by the third millennium BCE, its impact on the Middle East and South Asia was only palpable from the second millennium BCE on.[24] As a pack animal it faced stiff competition from the hardy donkey and the Bactrian camel, whose load-bearing capacity remained superior.[25] But horses offered a speedier mode of transport for a range of other mobile groups capable of moving vegetal commodities, including herders, diplomats, couriers, and soldiers. The fifth-century BCE Greek historian Herodotus esteems the royal Persian horse-mounted couriers as the fastest messengers known

23. Schmitt 2002b. On the Avesta, see Skjaervo 2011, and for camels in the Avesta, see Schwartz 1985, 660. Aššur-bēl-kala's inscription incidentally furnishes the earliest extant use of an Indo-Iranian loanword for the Bactrian camel in Mesopotamia (*udru*, cf. Sanskrit *úṣṭra*).

24. On horse domestication events, see Anthony 2007, 193–224; Anthony and Brown 2011; Cunliffe 2015, 77–80; Librado et al. 2021; Wilkin et al. 2021. On the horse in the Middle East and South Asia, see Anthony 2007, 412–18; Fuller and Madella 2001, 368; Shev 2016; Thapar 2002, 85.

25. On donkeys and their persistence as a pack animal long after the introduction of horses, see Mitchell 2018.

to him.[26] It was in warfare, above all, that horses found exceptional use. The Assyrian Empire's expansion into the Zagros region of western Iran from the late ninth century BCE on was prompted, in particular, by the need to acquire finely bred horses for the army.[27] Unlike other tributaries, who paid in metal, the Zagrosian tribute to Assyria came in the form of horses. In many ways, the horse complemented rather than competed with the older modes of transportation.

Unlike its northern cousin, the dromedary or one-humped camel was domesticated for milk, meat, and traction much later, in late-second-millennium BCE southeastern Arabia. It substantially enhanced connectivity across the *terrae steriles* of the Arabian Peninsula and, by the early first millennium BCE, in the Levant as well.[28] It is around the same time that the Arabian frankincense and myrrh trade takes off spectacularly, and with it the aromatics and precious stones received from the wider Indian Ocean world.[29] Some of the cultivars discussed in this book, like the sacred lotus and taro, could well have arrived in Egypt and the Mediterranean via South Arabian–Red Sea routes, since the evidence for their presence in the Persian Gulf and Mesopotamia is woefully thin.

THE OCEAN

The relative quiet of the Atlantic and the Pacific in early human history presents a dramatic contrast to the constancy, volume, and vibrancy of human activity in the Indian Ocean.[30] The historian Felipe Fernández-Armesto remarks that the "precocity of the Indian Ocean as a zone of long-distance navigation and cultural exchange is one of the glaring facts of history: enormously important and puzzling, when you come to think about it, yet hardly remarked, much less explained, in the existing literature."[31] Trans-oceanic navigation and networking was to a great degree born and perfected in the Indian Ocean theatre. The volume of travel and the distances traveled by peoples, commodities, and ideas over this stretch of water in the prehistoric and early historic periods dwarfed those of other water bodies whose edges were colonized by human settlers.

Monsoonal winds—seasonally reversing airstreams generated by the differential heating of land and sea during summer and winter—were the most important movers of long-distance trade in the Indian Ocean and defined the contours of interactive spaces (map 2). The northeasterly winds prevailing in winter above the equator

26. Hdt. VIII.98. On the Persian couriers, see Briant 2002, 369–71.

27. Radner 2003, 2013.

28. Almathen et al. 2016; Cousin 2020; Rosen and Saidel 2010, 72–76; Sapir-Hen and Ben-Yosef 2013.

29. Altaweel and Squitieri 2018, 166–68.

30. Rangan, Carney, and Denham 2012, 319; Vink 2007, 54.

31. Fernández-Armesto 2001, 382.

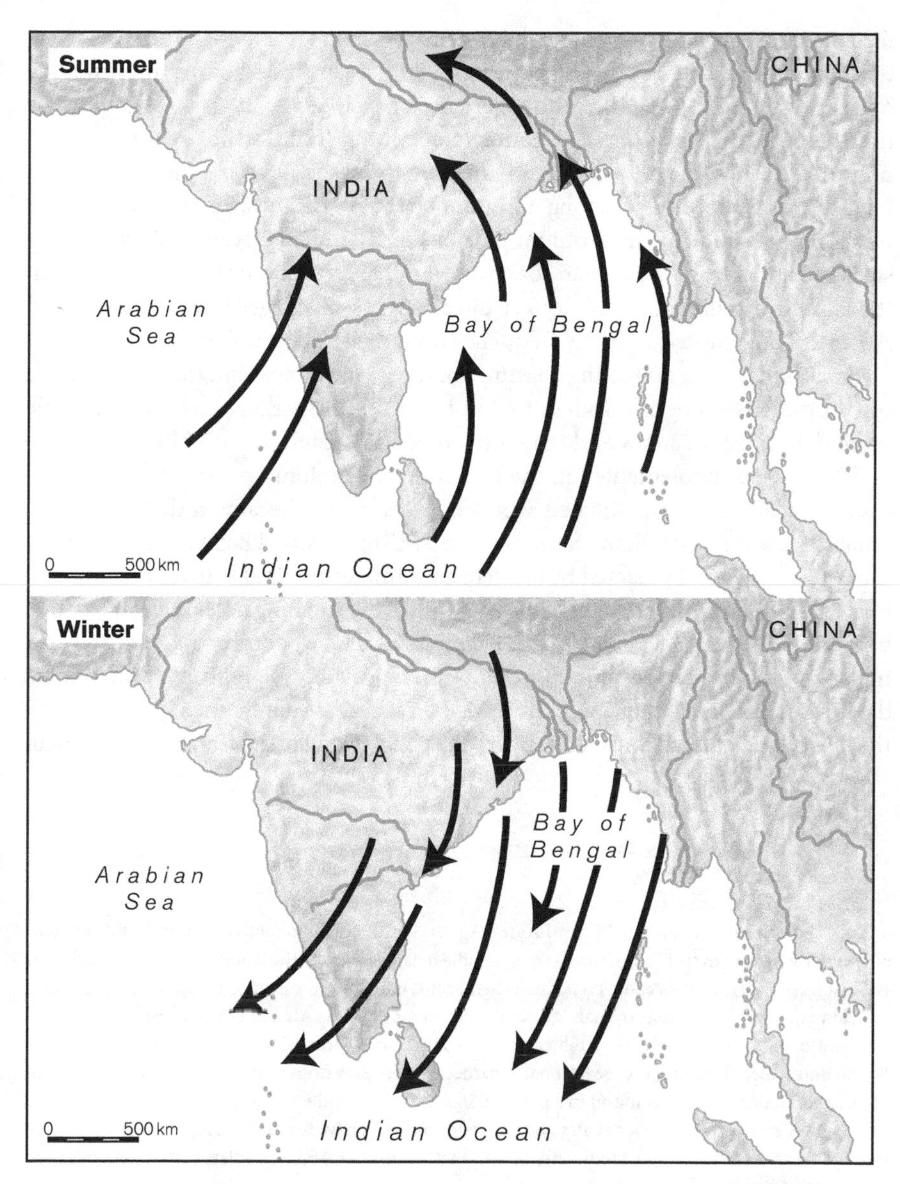

MAP 2. Seasonally reversing monsoonal winds in the Indian Ocean region. © Peter Palm (Berlin).

facilitated east-to-west movement, while journeys in the reverse direction were hastened by the summer southwesterlies.[32] The skill of harnessing monsoon winds for open-sea sailing, buttressed by the experience of deep-sea fishing, was known to the inhabitants of the Harappan civilization in northwest India by the late third millennium BCE, when they took to sailing directly to Mesopotamia.[33] The regularity of long-distance movements along monsoonal routes is, however, a relatively late development (mid-first millennium BCE on) in the Indian Ocean.[34] While the Persian Gulf and the South Arabian coast were familiar territory for Indian sailors from the late third millennium on, longer direct open-sea voyages to the Red Sea and Africa are not attested with any frequency until the late first millennium BCE.[35]

Small-scale interconnecting coasting routes are the typical and the earliest mode of connectivity along the Indian Ocean littoral.[36] The earliest carriers were little more than dugout canoes and seaworthy reed boats caulked with bitumen (figure 5).[37] They were supplemented in the late third millennium by more efficient long-distance seafaring craft: stitched wooden boats, which became a distinctive and ubiquitous feature of Indian Ocean seascapes.[38] The standard Babylonian version of the epic of Gilgameš redacted by Sîn-lēqi-unninni unmistakably invokes the sewn-boat tradition in its description of the plugging of the stitching holes in Utanapištim's boat (XI. 64: *sikkāt mê ina qablīša lū amḫaṣ*).[39] Stitched wooden boats have survived in parts of the Indian Ocean until recent times, providing significant ethnographic data on premodern seafaring craft.[40] Mediterranean seafarers unacquainted with this technique of boat construction were struck with surprise, fear, or even disdain

32. Boivin et al. 2013, 219; Fernández-Armesto 2001, 384–85.

33. Boivin and Fuller 2009, 166.

34. Boivin et al. 2014, 555.

35. The role of East Africa in the Bronze Age Indian Ocean interactive sphere is still unclear, although the spread of tropical African crops to India between the late third and early second millennia BCE suggests early trans-oceanic contacts. Whether this contact was direct or mediated by communities along the South and East Arabian littoral is still a moot point, as African crops transmitted to India have not been recovered in early Arabian archaeological sites (Boivin, Blench, and Fuller 2009, 266; Boivin and Fuller 2009, 114, 165–66). Copal, a hardened tree resin from East Africa (Zanzibar, Mozambique, and Madagascar), was found in a pendant in a mid-third-millennium context at Eshnunna (Tell Asmar) in Mesopotamia (Meyer et al. 1991). African copal may have reached Mesopotamia via Arabian intermediaries, lending support to the hypothesis that peninsular Arabia mediated Mesopotamian and Indian interactions with East Africa.

36. Boivin et al. 2013, 266–67.

37. Boivin and Fuller 2009, 164–65; Cleuziou 2003, 134; Cunliffe 2015, 51–52; Ratnagar 2004, 214–15; Vosmer 2003.

38. Bagg 2017, 132–34; Cleuziou and Tosi 2000, 64; Ratnagar 2004, 215–21; Zarins 2008, 213–14.

39. George 2003, 706–07; Pedersen 2003, 2004. On boat fastenings in the Indian Ocean region, see Ratnagar 2004, 222–25.

40. McGrail et al. 2003; Sheriff 2010; Varadarajan 1993.

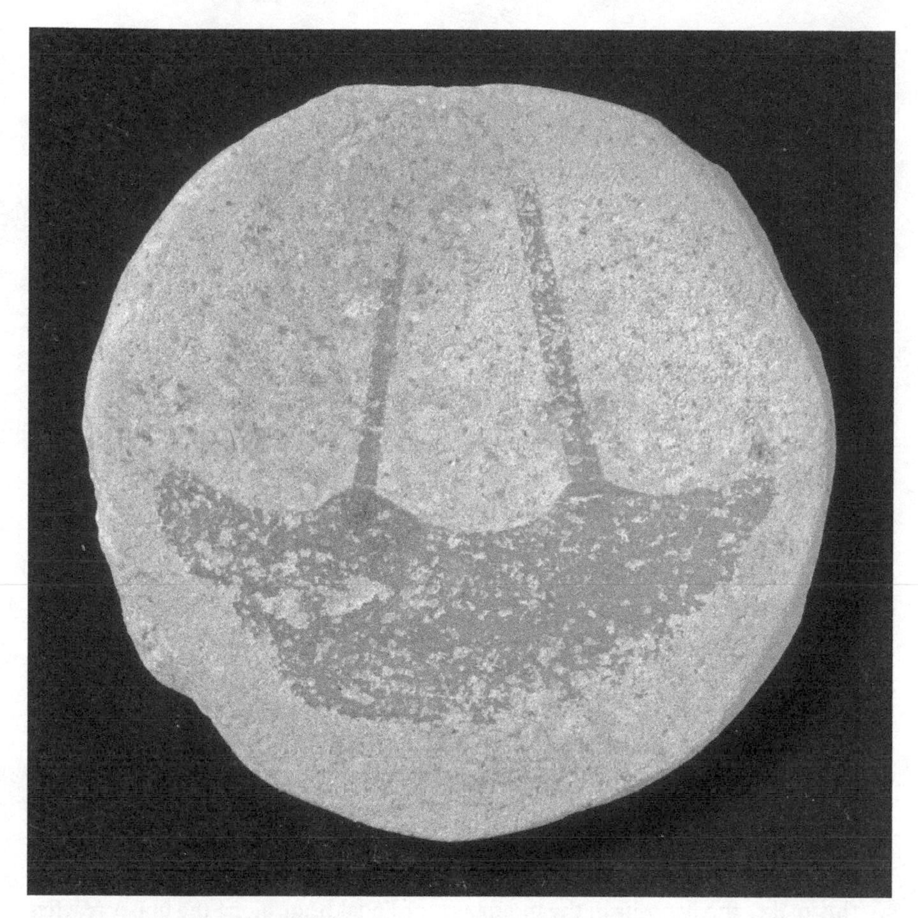

FIGURE 5. Pottery disc with a double-masted (likely reed) boat, al-Sabiyah, Kuwait, c. 5500–5000 BCE (Ubaid period). Courtesy of the British Archaeological Expedition to Kuwait, Dr. Robert Carter and Mr. Mohammed Ali.

for the stitched-boat tradition.[41] Ḥalfon ha-Levi, a twelfth-century CE Jewish merchant crossing the Indian Ocean from Egypt exclaims: "We set sail in a ship with not a single nail of iron, but held together by ropes; may God protect with his shield!"[42]

As a rule of thumb, the volume of waterborne trade was greater in all periods prior to the emergence and widespread application in the nineteenth century of steam and internal combustion engines in land transportation. Trade in bulk, particularly the movement of liquids and their storage jars, timber, metal, grain, and

41. Ray 2003, 59–61.
42. Goitein and Friedman 2008, 11 n27.

other heavy manufactured produce, is best done by waterborne transport. Notwithstanding the perennial risks of shipwreck, freshwater shortage, and piracy, maritime transportation remained superior to terrestrial travel for much of antiquity. A caravan journey from Afghanistan to Mesopotamia could take over three months.[43] In contrast, vessels plying the routes between the Indus and the Straits of Hormuz could make it in less than a month.[44]

THE FOURTH TO EARLY THIRD MILLENNIUM BCE: THE LAPIS LAZULI ROADS

The routes between the Indo-Iranian borderlands and the Middle East might well be dubbed the Lapis Lazuli Roads, as that azure stone, exclusively derived from eastern sources, remains the single most diagnostic feature of the east–west trade in archaeological assemblages in both early and later phases.[45] Owing to the specific geological conditions under which lapis lazuli can be formed, its natural geographical distribution is extremely narrow. Geological surveys have established four main sources for lapis lazuli in the Old World, all of which lie east of Iran: Badakhšan in eastern Afghanistan, Iškašim in the Pamir Mountains of Tajikistan, the Lake Baikal region in Siberia, and the Mogok region of Upper Myanmar (Burma).[46] While Iškašim is only about 130 kilometers northeast of Badakhšan, its lapis lazuli deposits are located on a precipitous cliff face 4,600 meters in elevation, making it extremely unlikely that this site was exploited in antiquity.[47] The deposits in Siberia, bordering a glacier at 5,029 meters in altitude, are equally remote.[48] The famed Mogok Stone Tract in Upper Myanmar, rich in jade, gold, placer diamonds, sapphires, and above all rubies, is also known to produce lapis lazuli, but it is unclear whether lapis lazuli from Mogok was mined and traded westward in antiquity.[49]

This makes the deposits in the mountains of Badakhšan, along the upper reaches of the Kokcha in eastern Afghanistan, the most feasible source of lapis lazuli in antiquity and certainly the main source for regions immediately to the west of Afghanistan. The Badakhšan deposits are themselves hard to reach, with the four known mines (Sar-i Sang, Chilmak, Stromby, and Robat-i Paskaran) ranging in altitude between 1,800 and 5,000 meters.[50] Sar-i Sang (map 3), the most productive and still-active

43. Herrmann 1968, 36 n75.

44. Carter 2013, 590.

45. Sarianidi 1971.

46. Delmas and Casanova 1990; Law 2014, 420; Moorey 1999, 85–92; Zöldföldi and Kasztovsky 2009.

47. Law 2014, 426.

48. Herrmann 1968, 28.

49. Law 2014, 424; Waltham 1999. Burmese rubies are attested much later in Mesopotamia, where they were used as eye inlays for a statuette of a goddess in Parthian Babylonia (Calligaro 2005).

50. Herrmann 1968, 22, 24.

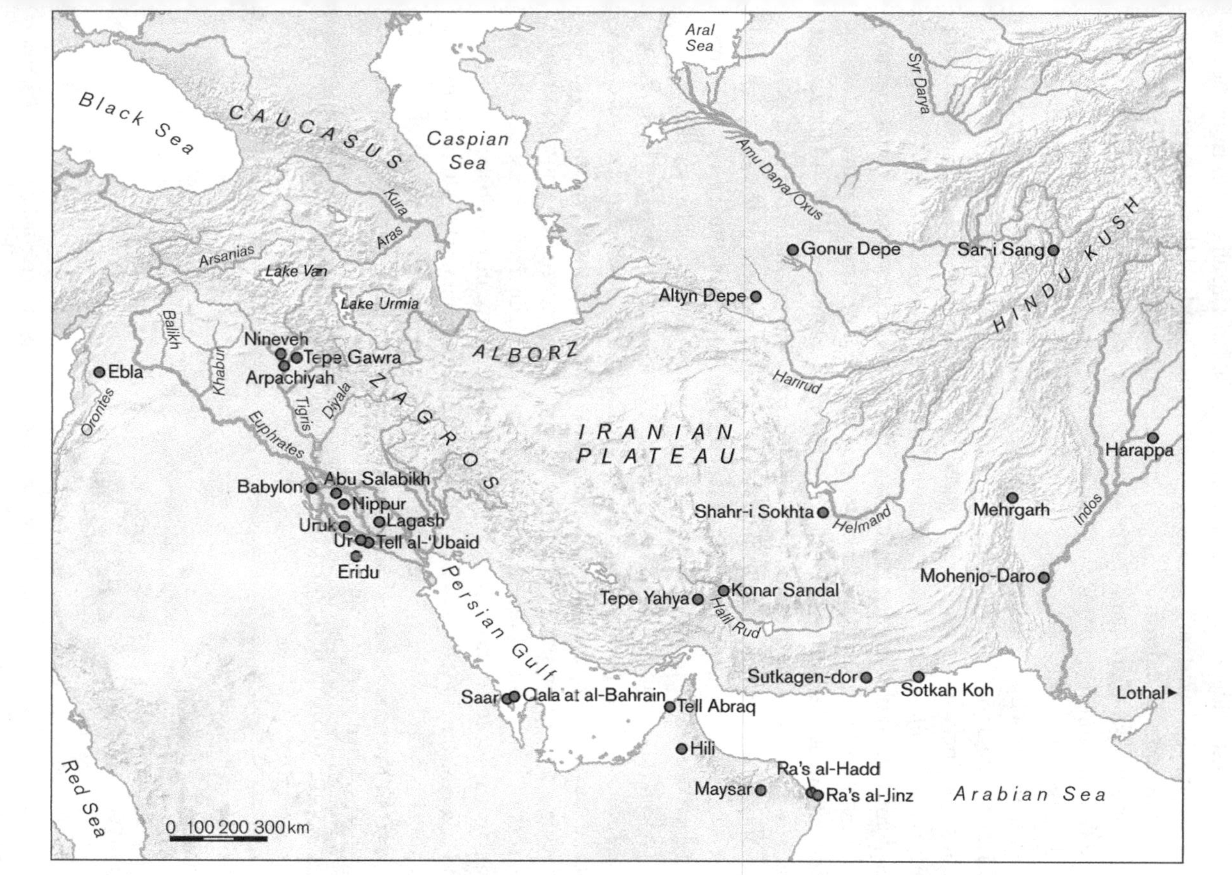

MAP 3. Important fourth- and third-millennium BCE sites mentioned in the text. © Peter Palm (Berlin).

mine, is on a steep mountainside, with the approach having to be renegotiated every spring following the ruin of old pathways by snowfall and fierce winter storms.[51]

The earliest evidence for trade in and the use of lapis lazuli derives from the site of Mehrgarh, in Pakistani Baluchistan, where small beads of the stone were recovered from seventh-millennium BCE graves.[52] In the west, lapis lazuli (Sumerian *zagin* or *gin*, Akkadian *uqnû*) is first attested in northern Mesopotamian sites like Tepe Gawra, Nineveh, and Arpachiyah by the early fourth millennium BCE and as far as Egypt from the mid-fourth millennium on.[53] The magnitude and immense geographical reach of the early trade in lapis lazuli is best reflected in the discovery of over 22 kilograms of unworked lapis lazuli in the artisanal quarter of a late-third-millennium palace in Ebla, Syria.[54] Sumerian textual sources of the late third and early second millennium BCE are replete with references to lapis lazuli.[55] Considered auspicious, the stone found extensive use in the sacred and palatial sphere as a material for cultic objects, cylinder seals, amulets, and ornaments; or as inlays in a variety of objects, including gaming boards, weapons, musical instruments, and statuary; or was simply hoarded as treasure in its raw form.[56] The familiarity of lapis lazuli to Mesopotamian audiences is also palpable in the development of the Sumerian and Akkadian words for lapis lazuli into a color term. In Sumerian, the sky, certain plants, and even the beards of men were described as being the color of lapis lazuli.[57]

Lapis lazuli certainly did not travel alone but was accompanied by a host of other perishable and non-perishable commodities, the former having left scarce archaeological traces. Among the raw materials coming from the east, tin, the most common alloying agent in bronze, was undoubtedly a vital import. The most important sources of tin for the Middle East and the eastern Mediterranean are beyond the Iranian plateau, in southern Central Asia (eastern Kazakhstan, Uzbekistan, Tajikistan, and Afghanistan) and northwest India (the Aravalli mountains).[58] Minor tin deposits are also found closer to Mesopotamia, in the central Zagros region of western Iran (Deh Hossein, Sanandaj-Sirjan zone) and in the Helmand basin of Sistan, in eastern Iran (Chah Kalapi, Shahkuh, and Chah Ruh).[59] Tin-bronzes are first attested in small quantities in Mesopotamia (Tepe Gawra, Kish, Ur, Tell Judeidah) and Iran

51. Herrmann 1968, 24.

52. Law 2014, 419.

53. Aston, Harrell, and Shaw 2000, 39–40; Moorey 1999, 88.

54. Casanova 2001, 158; Pinnock 1995, 150–52.

55. Winter 2010.

56. Winter 2010.

57. Thavapalan 2020, 310–11.

58. Cierny and Wiesgerber 2003; Garner 2015; Helwing 2009; Parzinger and Boroffka 2003; Steinkeller 2014, 698; Stöllner et al. 2011.

59. Nezafati, Momenzadeh, and Pernicka 2008, 9–11. Helwing 2009 disputes the importance of the Zagrosian tin deposits in antiquity.

(Susa, Kalleh Nissar) in the late fourth and early third millennia and become more frequent in the late third millennium.[60] The mastery of large-scale tin-bronze casting by the end of the third millennium is indicated by the boast of Rimuš (2278–2270 BCE), the second king of the Akkadian Empire, that he erected a statue of himself in tin (probably tin-bronze) at the sanctuary of the god Enlil in Nippur.[61]

The constancy of demand which fell to local suppliers of materials like lapis lazuli and tin probably encouraged the development of small-scale professional trade networks by the end of the fourth millennium. This period also marks the eastward expansion of a southwest-Iranian material culture termed Proto-Elamite (c. 3100–2900 BCE). This cultural phenomenon, perhaps linked to the spread of an ethno-linguistic group, is marked by the adoption of distinctive ceramic forms (bevel-rim bowls, nose-lugged bichrome jars), the use of seals, and inscribed tablets in the Proto-Elamite script.[62] Proto-Elamite tablets have been found in Kerman (Tepe Yahya), several sites on the central Iranian plateau (Qazvin, Kashan, the Tehran plains), and as far east as Sistan (Shahr-i Sokhta), while the pottery types (especially bevel-rim bowls) extend further east into Pakistani Baluchistan.[63]

While the exact impetus behind the expansion of Proto-Elamite culture across the Iranian plateau remains unclear, genetic and linguistic studies which provide evidence for wide-ranging east–west population movements in prehistoric and historic periods suggest that a similar population-language-culture dispersal may be at work here.[64] In this regard, one study of DNA sequences from ancient osteological remains in the Middle Euphrates Valley has yielded evidence of a genetic link between Bronze Age populations of the Middle East and those of modern South Asia. The molecular biologist Henryk Witas and colleagues examined mitochondrial DNA sequences extracted from the dental remains of four individuals, two from Early (c. 2650–2450 BCE) and Middle (2200–1900 BCE) Bronze Age Terqa in Syria and two from Roman (200–300 CE) and late antique (500–700 CE) Tell Masaikh in Syria.[65] All four individuals carried mitochondrial DNA haplotypes belonging to the M4b1, M49, and/or M61 haplogroups, which are absent in modern Syria but found in Trans-Himalayan and South Asian populations. Haplogroup M lineages are believed to have diversified in the Indian subcontinent between 5,800 and 2,500 years ago.[66]

The presence of a distinctive South Asian genetic signature in the Middle Euphrates Valley indicates an early (pre-2500 BCE) population movement from the

60. Nezafati, Momenzadeh, and Pernicka 2008, 9–11; Radivojević et al. 2013, 1031.

61. RIME II E2.1.2.18; Foster 2016, 8, 324.

62. Petrie 2013, 15. On the undeciphered Proto-Elamite script and tablet finds, see Dahl, Petrie, and Potts 2013; Desset 2012.

63. Lamberg-Karlovsky 2013, 564; Petrie 2013, 15.

64. Lamberg-Karlovsky 2013, 566.

65. Witas et al. 2013.

66. Witas et al. 2013.

east. The alternative suggestion of Witas and colleagues, that this might represent itinerant merchants, is untenable because direct Indo-Mesopotamian trade relations are only a feature of the late third millennium BCE, and even then it is extremely unlikely that South Asian merchants ventured as far inland as the Middle Euphrates Valley. The genetic evidence for an early migration from the east inevitably implicates itself with the much-disputed origins of the Sumerian-speaking peoples of southern Mesopotamia.[67] Odontometric analyses of individuals from Chalcolithic (Eridu, Tell al-'Ubaid) and Bronze Age (Ur) sites in southern Mesopotamia have previously been used to suggest Sumerian affinities with Indian populations.[68]

In this regard, it has been proposed that the Sumerian language itself is a sort of coastal creole, including, among other contributors, the languages spoken by the peoples of coastal Iran and the Indus Valley.[69] Although there is no scholarly consensus on Sumerian origins or the affiliations of the language, the bioarchaeologist Arkadiusz Sołtysiak notes that the South Asian origin theory for Sumerian-speakers is "relatively better grounded, although no author [has] tested it in [a] proper way and it still remains only a speculation."[70] The cuneiformist Julian Reade, remarking on the unfamiliarity of India to scholars of the ancient Middle East, pithily notes that the "Indian precedence over Early Dynastic Mesopotamia introduced a new and disturbing element into ancient history."[71] The Biblical baggage aside, there remains a strong historiographical bias in approaching the history of the ancient Middle East from the point of view of the modern West, both conceptually and geographically. This has even extended to the appropriation of the pre-Islamic Middle East as the "cradle of Western civilization."[72] The future work of scholars familiar with the archaeological, historical, and linguistic setting of regions to the east and north of the Middle East will hopefully promote more holistic readings of the ancient Middle East, as well as elucidate contentious issues like the Sumerian question and its alleged eastern associations.

Whatever the origins of the Sumerian peoples, the archaeological and genetic data unmistakably indicate strong links between Mesopotamia and populations in the east. The earliest Sumerian polities, in particular the city of Uruk, may have even sponsored long-distance trading and diplomatic ventures in the early third millennium BCE, if later literary materials are to be believed.[73] Sumerian literary texts of the late third and early second millennium speak of Uruk's interactions

67. Reade 1997; Witas et al. 2013.

68. Sołtysiak 2006.

69. Blazek 1999: Høyrup 1992; Reade 1997, 223; Vermaak 2012.

70. Sołtysiak 2006, 151.

71. Reade 2001, 28.

72. Pollock 2005.

73. On Uruk in the early third millennium BCE, see the essays and bibliography in Crüsemann et al. 2013.

with a distant and wealthy eastern polity named Aratta, whose El Doradoesque portrayal and no-show in prosaic economic and administrative texts have led scholars to posit that it belongs to the realm of fabulous fiction rather than reality.[74] Yet it might not be entirely fictional, since an Indus Valley locality named Araṭṭa is attested in Sanskrit sources of the first millennium BCE.[75] Although the Sumerian and Sanskrit references to Aratta are separated by over a millennium, the potential conservatism of toponyms and the easterly location of Aratta means that the homonymity is too much of coincidence to be ignored. While the Mesopotamian references to Aratta are generic and mythic, it is not impossible that they reflect a topographic reality.

THE LATE THIRD TO EARLY SECOND MILLENNIUM: THE MIDDLE ASIAN INTERACTION SPHERE

While the expansion of Proto-Elamite culture, east–west population movements, the widespread use of lapis lazuli along the Nile–Indus corridor, and the Sumerian legends about the eastern country of Aratta already suggest some degree of connectivity across the Iranian plateau in the late fourth and early third millennium, the earliest historically verifiable organized trade between India and the Middle East belongs to the late third millennium BCE. The Harappan civilization of northwest India, stretching from the Makran coast in the west to the upper Ganga-Yamuna Doab in the East and southward into Gujarat; the Sumerian and Akkadian-speaking populations of southern Mesopotamia; and polities in Oman (Magan), eastern Arabia (Dilmun), and Iran (especially Elam in Khuzestan, Marḫaši in Kerman, and Šimaški in the central Iranian plateau) were the key players in the creation of an intensively interactive zone between the late third and early second millennium BCE.[76] In Mesopotamian chronology this extends from the Early Dynastic III to the Old Babylonian period. The Harappan archaeologist Gregory Possehl has fruitfully described this interactive zone as the "Middle Asian interaction sphere," whose frontiers extended from southern Central Asia in the north to the Arabian Peninsula in the south and from the Mediterranean in the west to the Indian subcontinent in the east.[77]

74. Mittermayer 2009; Potts 2004b; Steinkeller 2014, 704; Vanstiphout 2003.

75. See e.g. *Baudhāyanaśrautasūtram* 18.13, 18.44; Parpola 2015, 216; Witzel 2001, 18.

76. Note also other eastern Iranian and South-Central Asian polities mentioned in Mesopotamian sources: Tukriš, which is probably to be identified with the archaeological culture termed the Bactria-Margiana Archaeological Complex, covering the territories of modern-day northern Afghanistan, southern Uzbekistan, and eastern Turkmenistan; and Kupin, which was probably somewhere in Baluchistan (Steinkeller 2014, 693, 701–04).

77. Possehl 1996, 2002.

Harappan artifacts in Mesopotamia, Iran, and the Persian Gulf are the main witnesses for the vibrant trade links between India and the Middle East in this period.[78] Textual references to Meluḫḫa, the Sumerian-Akkadian designation for the Harappan civilization and its peoples in northwest India, represent the other key source for understanding early trade and political links.[79] There is no doubt in the identification of Meluḫḫa with the Harappan realm, since it is described in Mesopotamian texts as the source of unmistakably South Asian articles like peacocks, carnelian, and sissoo wood. Furthermore, the few Meluḫḫan loanwords in Sumerian and Akkadian have some resonances with Indic, particularly Dravidian, languages.[80]

The rarity of Mesopotamian artifacts in India and the extensive distribution of Indus artifacts in the Persian Gulf indicate that Mesopotamian traders were largely uninvolved in sailing to India.[81] The newfound maritime confidence of the Harappans from c. 2500 BCE on expressed itself in the founding of trading enclaves on the southern Iranian coast (Sutkagen-dor, Sotka Koh), in Oman (Ra's al-Jinz, Ra's al-Hadd, Hili, Maysar), in Bahrain and eastern Arabia (Saar, Ras al-Qala), and perhaps even in southern Mesopotamia.[82] The presence of distinctive Indus cooking and serving vessels alongside personal adornments like shell bangles and toys at these "colony" sites suggests that Harappan merchants were accompanied by their womenfolk and children, a rare phenomenon in ancient trade.[83]

Harappan trade commodities included raw and finished beads of semiprecious stones (lapis lazuli, carnelian, chalcedony, heliotrope), shell artifacts (*Turbinella pyrum*, etc.), ivory, animal figurines, gold, and a variety of timbers (sissoo, ebony, teak).[84] Textiles almost certainly formed an important element of exchange, although they are largely invisible in the archaeological record, as are ephemeral agricultural products like wine, dates, ghee, honey, reeds, and cured foods, which are thought to have filled the diagnostic Harappan black-slipped amphora-like jars found across the Persian Gulf.[85] The late-third-millennium horizon also saw the introduction of at least one South Asian cultivar and a few animals to the Middle East along east–west trading routes. Sesame (*Sesamum indicum*), a major oil-

78. Collon 1996, 209–25; Frenez 2018; Possehl 1996, 147–82; Ratnagar 2004, 106–211.

79. Muthukumaran 2021; Possehl 1996, 138–44; Ratnagar 2004, 98–102.

80. Muthukumaran 2021; Mukhopadhyay 2021.

81. Boivin and Fuller 2009, 164; Vidale 2004, 261–62.

82. Cleuziou 1992; Frenez et al. 2016; Parpola 2015, 210–12; Wright 2010, 225–28.

83. Blackman and Méry 1999; Cleuziou 1992; Cleuziou and Méry 2002, 296–98; Frenez 2018, 389, 393; Frenez et al. 2016; Kenoyer 2008, 24–25; Thornton 2013, 609. Compare, for instance, the Old Assyrian trade network in Anatolia, in which Assyrian merchants left their wives in Aššur to manage their households and, on occasion, maintained secondary local wives in Anatolia. A few Assyrian women did, however, accompany their husbands on the trip to Anatolia (Heffron 2017; Michel 2010, 2014).

84. Morello 2014, 541–43; Potts 1993; Ratnagar 2004; Reade 2001, 26–28.

85. On textiles, see Smith 2013. On the black-slipped jars, see Laursen and Steinkeller 2017, 106; Uesugi 2019.

producing crop of Indian origin, was cultivated in the Middle East by the late third millennium BCE.[86] The earliest archaeological evidence for sesame in the Middle East presently derives from the site of Abu Salabikh (c. 2300 BCE).[87] There may be some genetic relationship between the Akkadian elliptical reading for *ellu*, "pure" or "clear," a word frequently used in connection with sesame oil or sometimes straightforwardly denoting sesame oil, and **eḷḷu*, the Proto-Dravidian term for sesame, whose variants are still used in modern South Dravidian languages.[88] The phonological and semantic overlap is unlikely to be mere coincidence.

A few Indian animals, including water buffaloes, zebu cattle, peacocks, and monkeys, were also introduced into Mesopotamia. *The Curse of Agade*, a Sumerian composition of the Ur III or early Old Babylonian period (late third to early second millennium BCE), describes the former capital Agade as a place where the "monkey, mighty elephant, water buffalo (and) beasts of exotic lands rub shoulders in the broad streets."[89] The water buffalo (Sumerian *abzaza*) is well known from seals of the Sargonic period, including those belonging to royalty, like Tar'am-Agade, the daughter of king Naram-Sin, and high administrative officials, like the scribes of the king Šarkališarri and Enheduanna, the high priestess of the moon-god Nanna at Ur.[90] The water buffalo's disappearance in later periods suggests that there were never many buffaloes imported from India to begin with. It is not clear whether the water buffalo was bred for ornamental and ritual purposes or if it had an economic role in Sargonic Mesopotamia, much as it does in the marshlands of modern Iraq and Khuzestan in Iran.

South Asian zebu cattle (*Bos indicus*), on the other hand, were undoubtedly important sources of milk, meat, leather, dung for fuel and fertilizer, and traction. While zebu cattle were already familiar in Mesopotamia and western Iran from eastern Iranian and Indian sources from as early as the late fourth millennium BCE, additional waves of zebu from the east and localized breeding increased zebu stocks in the Middle East from the late third millennium BCE on (figure 6).[91] Although locally domesticated cattle (*Bos taurus*) were available in the Middle East, zebu cattle proved attractive as they are heat tolerant, highly resistant to livestock diseases, and need less water.[92] Genetic (mitochondrial, autosomal, and

86. Bedigian 2003, 22; Fuller 2003c; Zohary, Hopf, and Weiss 2012, 112–13.

87. Fuller 2003c, 132.

88. Southworth 2005, 204, 224, 332. On Akkadian terms for sesame, see Stol 1985.

89. ETCSL 2.1.5, lines 21–22; Foster 2016, 351 (lines 21–22).

90. Buccellati and Kelly-Buccellati 2002, 13; Potts 1997, 257–59. On the domestication of water buffaloes (*Bubalus bubalis*) in India, see Nagarajan, Nimisha, and Kumar 2015. There may have been a vestigial wild water buffalo population in Yemen until the early second millennium BCE, but this was probably not the source of the water buffaloes in Mesopotamia (Potts 2019).

91. Boivin and Fuller 2009, 159; Matthews 2002.

92. Boivin et al. 2014, 569; Matthews 2002, 440.

FIGURE 6. Copper-alloy plate with reclining zebu, eastern Iran, late third to second millennium BCE. Metropolitan Museum of Art, New York (public domain).

Y-chromosomal) data on modern Middle Eastern cattle populations indicate considerable introgression between *Bos taurus* and *Bos indicus,* especially in Iraq.[93] The crossbreeding of zebu and local taurine stocks to produce hardier drought- and disease-resistant cattle probably dates back to the Sargonic period.

The peacock, onomatopoeically named the *ḫaya*-bird in Sumerian, was probably introduced to Mesopotamia in Sargonic times but, like the buffalo, disappears after the Old Babylonian period and only re-emerges in the first millennium BCE

93. Edwards, Baird, and MacHugh 2007.

FIGURE 7. Red calcite figurine of a squatting monkey, possibly a rhesus macaque, Susa, Iran, c. 2400–2100 BCE. Musée du Louvre, Paris. Photograph by the author with the permission of the Agence photographique de la Réunion des Musées Nationaux et du Grand Palais.

as a result of reintroductions from India.[94] Monkeys, frequently found as figurines from the Early Dynastic to the Old Babylonian periods, represent Indian and African imports, as there are no primates native to Mesopotamia or Iran.[95] As the monkey figurines are stylized, it is not possible to identify the species with precision. Some, like a red calcite monkey from late third millennium BCE Susa found alongside other Indus-related objects (figure 7), bear generic similarities to the rhesus macaque (*Macaca mulatta*), which is found abundantly from Afghanistan to mainland Southeast Asia.[96] The likelihood that these figurines mostly represent the rhesus macaque is strengthened by the discovery of a late-third-millennium BCE

94. ETCSL 4.14.3 ("Nanše and the Birds"), Segment A, lines 49–53; ETCSL 1.1.3 ("Enki and the World Order"), line 229; Bodson 1998, 166–77, 1999, 75–81; Meissner 1913.

95. Dunham 1985; Ratnagar 2004, 203–07. Note the Proto-Elamite gypsum statuette of a monkey dated c. 3000 BCE and another grey marble statuette of two monkeys of slightly later date in the collection of the Fogg Museum, Harvard University (nos. 1986.601, 1983.174), which appear to be among the earliest depictions of primates in the region. Besides India and Africa, southwestern Arabia, which has a disjunctive distribution of hamadryas baboons (*Papio hamadryas*) from the late Pleistocene on, is another potential source of monkeys (Kopp et al. 2014).

96. Aruz 1992, 97–98; Dunham 1985, 261.

grave for a pet rhesus macaque at the city of Shahr-i Sokhta, in eastern Iran.[97] A few other exotic animals may have arrived by way of tributary and diplomatic embassies. A (wild?) cat (*gullum*) of Meluḫḫa finds mention in an Old Babylonian bilingual (Sumerian-Akkadian) collection of proverbs, and elsewhere Ur III economic texts refer to models of an unidentified Meluḫḫan bird.[98]

The Harappans were not only involved in the trade of goods from the Indian subcontinent but also profited from the shipping of Persian Gulf resources like copper, timber, and hard stones.[99] While the Harappans dominated the sea lanes along with shippers and traders from coastal Oman (ancient Magan) and Dilmun (Bahrain), the kingdom of Marḫaši (or Paraḫšum in Akkadian sources) in modern-day Kerman acted as a middleman for South and Central Asian goods like tin, gold, and lapis lazuli, which were transmitted westward via the southern overland routes.[100] These commodities were exported to Mesopotamia together with local luxury chlorite stoneware in the so-called Intercultural Style, which drew inspiration from both western and eastern iconography.[101] Apart from Marḫaši, whose eastern boundaries extended to the Bampur Valley in Iranian Baluchistan, late-third-millennium and early-second-millennium polities like Šimaški in the central Iranian plateau and Tukriš in Bactria-Margiana, located along the northerly land route or the Great Khurāsān road, were also responsible for the movement of South Asian commodities to Mesopotamia and beyond. The peoples of Mesopotamia were not passive recipients of foreign trade but were themselves actively involved in regional trading networks. Most of the large seafaring "Magan-boats" attested in Sumerian texts of the Ur III period were not actually from Magan (Oman) but were seacraft made and serviced by Mesopotamian shippers and traders under the aegis of the Ur III state.[102] These state barges carried vast quantities of Mesopotamian produce like textiles, barley, aromatic oils, and leather for consumers across the Persian Gulf.[103]

In Mesopotamia, the earliest distinctively Indian artifacts are encountered at mid-third-millennium (Early Dynastic IIIa) levels in the form of long carnelian

97. Minniti and Sajjadi 2019.

98. ETCSL 6.2.1, N 3395, segment A, line 6; Possehl 1996, 141–43. Steinkeller (2013, 426) suggests that this might be a reference to chickens.

99. Thornton 2013, 601.

100. Steinkeller 1982, 250–52, 2013, 2014, 693. The polity of Marḫaši is also known in archaeological literature as the Jiroft civilization, after the modern region which has yielded the most diagnostic sites for this culture, or the Halil Rud civilization, after the main fluvial body in the area.

101. Perrot and Madjidzadeh 2003; Steinkeller 2013; Vidale and Frenez 2014.

102. Presumably named after a type of seafaring vessel used by Omani traders—or, alternatively, the ships may have been named after their final destination.

103. Laursen and Steinkeller 2017, 106–07; Steinkeller 2013, 418–22.

and etched carnelian beads in the Royal Tombs of Ur.[104] The forms of ornamentation found in the Royal Tombs of Ur, particularly those associated with queen Puabi of the Meskalamdug dynasty and her retainers, bear strong associations with the Indus region. Both the queen and her retainers are equipped with distinctive gold and silver floral headdresses which are otherwise only attested in the iconography of the Harappan civilization. Puabi and another occupant of a tomb at Ur were also dressed in girdles strung with lapis lazuli and carnelian beads of South Asian origin. The wearing of heavy jeweled girdles by women, incidentally still a distinctive feature of Indian costume, is perhaps an Indian contribution to Mesopotamian ornamentation. Either the court of Ur was smitten with Indus fashions, as the art historian Joan Aruz suggests, or the retainers and perhaps Puabi herself may have come from further east.[105] DNA and isotopic analyses hold great potential in unraveling ancient population movements, including perhaps the identity of the occupants of the Royal Tombs of Ur. Kenoyer, Price, and Burton's strontium isotopic analysis of human dental remains from cemeteries in Harappa and Ur has so far been inconclusive, since access to Mesopotamian samples was limited to two human teeth from Ur. But their analysis of Harappan dental samples has suggested population movements from other areas in the Indus valley and beyond.[106]

Contacts between northwestern India and Mesopotamia are much more pronounced in the subsequent Sargonic (or Old Akkadian) and Ur III periods. A coincidence of Harappan mercantile entrepreneurism and the rise of an expansionist imperial polity in Mesopotamia kindled intensive commercial and political contact between the two regions.[107] Sargon (2334–2279 BCE), the eponymous founder of the first historically verifiable Mesopotamian empire, famously claims to have "moored the boats of Meluḫḫa (Harappan India), Magan (Oman) and Dilmun (Bahrain) at the quay of Akkad," his capital somewhere near present-day Baghdad.[108] The presence of Harappans in Sargonic Mesopotamia is vividly confirmed by the seal inscription of one Šu-ilišu, who is described as an interpreter of the Harappan language (*emebal Meluḫḫa*). The Ur III king Šulgi also claimed the ability to speak the language of the Meluḫḫans.[109] Harappan visitors, including merchants, craftsmen, and perhaps other kinds of specialists like puppeteers, are also evidenced by Indus seals and seal impressions at Mesopotamian and western

104. Reade 2001, 26.

105. Aruz 2003, 243; Kenoyer 1997, 274; Tengberg, Potts, and Francfort 2008, 933–34.

106. Kenoyer, Price, and Burton 2013.

107. Steinkeller 2013, 415.

108. RIME II, E2.1.1.11, E2.1.1.12; Foster 2016, 322. Sargon's inscription is preserved in Sumerian and Akkadian manuscript copies from Old Babylonian Nippur.

109. ETCSL 2.4.2.02, lines 211–12; For the identification of Meluḫḫa with the "black land" see Parpola 2015, 218.

Iranian sites, including Ur, Girsu, Umma, Nippur, Kish, Eshnunna, and Susa.[110] Some Sargonic administrative texts provide evidence for the issue of food rations to Harappans and even a Marḫašian for a journey back to Meluḫḫa.[111] It is possible that Harappan traders settled in southern Mesopotamia as early as the Sargonic period. A village in the Girsu-Lagaš District bears the curious name of Village of Mr. Meluḫḫa (Eduru-Meluḫḫa) in fourteen administrative documents of the subsequent Ur III period, and some individuals in contemporary archival documents are toponymously named Meluḫḫa.[112] Perhaps the Mr. Meluḫḫas of Mesopotamia had some kind of genealogical or professional link with the Harappan world.[113]

The first kings of Akkad maintained an aggressive military presence in the Persian Gulf and Iran with the intention of securing control of the region's natural resources (e.g., hard stones and copper) and the trade routes leading to the Indus. But Mesopotamian expansionism soured political relations with polities in the east. Despite close commercial links, Mesopotamian sources indicate that the Harappan polity (Meluḫḫa) was allied to the chief enemies of the Sargonic dynasty in the east: the powerful kings of Marḫaši, in modern-day Kerman. The Harappans came to the aid of Abalgamaš, king of Marḫaši, and his general Sidgau in a battle in western Iran against Rimuš (2278–2270 BCE), the second king of the Sargonic dynasty.[114] The Harappans, led by one king (x)-ibra, again implicated themselves in a revolt against Naram-Sin of Akkad alongside Marḫaši.[115] A later copy of a Sargonic inscription also suggests that the Harappans opposed king Maništūšu (2269–2253 BCE) in a battle fought in Oman.[116] This is not at all unlikely given the archaeological evidence for a strong Harappan settler presence along the Omani coast.[117]

The political relations of the later kings of Akkad and their Ur III successors with Marḫaši and her staunch ally Meluḫḫa become more amicable. Texts from Nippur testify to a marriage between a prince of Akkad, perhaps the future king Šarkališarri (2217–2193 BCE) or his brother Ubil-Ištar, and a Marḫašian princess.[118] The Ur III king Šulgi gave his daughter Liwwir-miṭṭašu in marriage to a king of Marḫaši.[119] The

110. Parpola 2015, 121.

111. See e.g. BdI Adab 102; Foster 1977, 39, 2016, 185; Parpola 2015, 215; Steinkeller 1982, 259 n90.

112. Laursen and Steinkeller 2017, 80–82; Parpola 2015, 218; Parpola, Parpola, and Brunswig 1977; Vidale 2004. On the port of Guabba, see Steinkeller 2013, 417–19.

113. Laursen and Steinkeller 2017, 81.

114. RIME II, E2.1.2.8; Foster 2016, 324; Parpola 2015, 215; Steinkeller 1982, 256, 2014, 693.

115. Parpola 2015, 216.

116. Steinkeller 1982, 258 n80, 1987–99, 334; see also Maništūšu's conquest or raiding of thirty-two settlements across the Persian Gulf in RIME II, E2.1.3.1.

117. Frenez et al. 2016; Parpola 2015, 210–12; Wright 2010, 225–28. On Marḫaši's influence over coastal Oman, see Potts 2005; Steinkeller 2006.

118. Potts 2002, 345, 2004b, 8; Steinkeller 2014, 692, 697.

119. Steinkeller 1982, 259 2014, 692

envoys of the Marḫašian kings were regular features at the court of the Ur III kings.[120] It is quite probable that Marḫaši entertained close matrimonial and diplomatic links with Harappan elites well before the détente with the kings of Akkad and later Ur. But the close relationship between Marḫaši and Meluḫḫa is impossible to reconstruct in detail owing to the absence of insider documents. Consequently, early Irano–Indian ties remain visible only in broad outline through outlier Mesopotamian perspectives. That Marḫaši held close relations with the Harappans even in later times is suggested by Marḫaši's gift of a "speckled dog or feline (*ur*) of Meluḫḫa," perhaps an Indian cheetah or leopard, to Ibbi-Sin, the last king of the Ur III dynasty of southern Mesopotamia.[121] The involvement of the Harappans in the political affairs of Mesopotamia, although mediated by Marḫaši, is a strong testament to a well-connected information network, resembling in some ways the better-known Late Bronze Age "international" system exemplified by the Amarna correspondence (fourteenth century BCE) between the Pharaoh and his Middle Eastern counterparts.[122]

The Bronze Age interaction sphere between Mesopotamia and the Indus was not limited to political and trade contacts but also extended to intercultural exchanges. The case of Harappan aesthetic influences on the elite of Early Dynastic Ur has already been considered. A form of puppetry may have been another Indus contribution to the Middle East. Three ithyphallic terra cotta puppets excavated at Nippur, dating between 2100 and 2000 BCE, have strong parallels with terra cotta puppets found in Harappan sites like Lohumjo-daro, Chanhu-daro, Mohenjo-daro, and Lothal.[123] Stone versions of similar animal and human puppets, also with Indus precedents, have been retrieved from the Marḫašian site of Konar Sandal South in Iran.[124]

Cubical and spherical weights of the Indus type appearing at various Middle Eastern sites like Tell Abraq, Konar Sandal South, Shimal, Susa, and Ur vouch for the influence of Harappan metrology.[125] This system of weights, despite being of Indian origin, was known in Mesopotamian texts as the "standard of Dilmun," as it was widely adopted across the Persian Gulf.[126] The finds of Harappan dice and gaming pieces at sites like Ur, Lagaš, and Barbar, probably associated with a precursor of the *chaupar-pachisi* type of board game, either indicate the popularity of

120. Steinkeller 1982, 260–61 n95.

121. RIME III/2, 1.5.4; Parpola 2015, 216; Potts 2002, 347–51. Cheetahs are presently extinct in South Asia.

122. See Rainey 2015 for a translation and discussion of the Amarna correspondence.

123. Dales 1968, 20–22; Possehl 1994, 179–86.

124. Pittman 2013, 65.

125. Ash 1978, 7; Magee 2014, 113–14; Pittman 2013, 65; Ratnagar 2004, 246–55.

126. Possehl 2002, 226.

an Indian game in the Middle East among local populations, like chess in later times, or represent games played by diasporic Harappan merchant families.[127]

Cultural influences percolated in the opposite direction as well. The famous Near Eastern "contest scene" between man and beast has its equivalents in the iconography of the Indus Valley, although here the local tiger usually stands in for the lion or bull.[128] The Indologist Asko Parpola discusses the case of the distinctive Mesopotamian royal and heroic coiffure in the form of a plaited double-bun, which has Harappan and Vedic parallels as well.[129] The examination of cultural affinities and exchanges between ancient Mesopotamia, Iran, and India is still in its infancy. The chronology and direction of influences, if not independent developments, are unclear for many common cultural features, including deities with horned headdresses, the decoration of eyes with kohl, the profession of snake charmers, and the use of shells as libation devices.[130] Culture, as the art historian James Cuno aptly states in this regard, "has never known political boundaries and has always been mongrel and hybrid, [providing] evidence of contact between peoples and their intertwined history."[131]

THE SECOND TO EARLY FIRST MILLENNIUM BCE: DECLINE AND REVIVAL

With the end of the mature Harappan polity around 1900 BCE, the political crisis engendered by the collapse of the centralized state in Mesopotamia, and the attendant rise of multiple centers of power controlled by Amorite princelings, the vibrant maritime trade of the Persian Gulf faltered and entered a long phase of gradual decline. Yet the end of the mature Harappan polity, and perhaps a centralized state, did not entail civilizational collapse, and outposts of Harappan culture survived, especially in Gujarat and further south, in the Tapti Valley and the upper

127. Hallo 1996, 111–13; Ratnagar 2004, 200. *Chaupar-pachisi* has been simplified into ludo in the modern West.

128. Possehl 2002, 227.

129. Parpola 2015, 227–29, 2020.

130. The profession of the snake charmer (Sumerian *mušlah*, Akkadian *mušlahhu*) is already attested in Mesopotamia by the mid-third millennium BCE. Balul, the earliest named snake charmer, was in the employ of the king Ur-Nanše of Lagaš, who ruled sometime between 2550 and 2400 BCE (RIME I, E1.9.1.2, E1.9.1.4). The ancient Indian snake charmer (Skt. *sarpavid*) and his Mesopotamian counterpart were not the frivolous entertainers familiar to modern audiences but professionals skilled in the lore of snakes (Skt. *sarpavidyā*), including interpretation of snake omens, extraction of venom, and propitiation of snake deities. See Macdonell and Keith (1912, 438) for Vedic references to the "science of snakes" and snake charmers. Note also Potts 2007b on a parallel if not related ritual involving serpent sacrifice in the Persian Gulf and India, for which there is archaeological evidence from mid-first-millennium BCE Bahrain.

131. Cuno 2011, 172.

reaches of the Godavari in Maharashtra (e.g., Daimabad).[132] The early second millennium marked the heyday of the late Harappan port of Kuśasthalī or Dvārakā, successively ruled, according to later Indic mythic-historiographic compendiums, by the Raivata, Puṇyajana, and Yādava dynasties.[133] The surviving outliers of the Harappan civilization continued to maintain maritime contact with the Persian Gulf. Harappan seals, some perhaps from secondary contexts, are attested at Nippur in Mesopotamia and the northern Persian Gulf (Failaka) as late as the Kassite period (c. 1400 BCE).[134]

Direct political and trading relations between South Asia and Mesopotamia, however, effectively came to an end. Polities in the Iranian plateau and the Persian Gulf mediated the "Indian trade" in a manner reminiscent of trading relations prior to the late third millennium BCE. In this regard, the Indian historian Romila Thapar notes that "the more spectacular maritime trade was occasional, but in its interstices there was a steady small-scale contact, often coastal, which involved transporting essential supplies quite apart from luxury items."[135] The durability of small-scale trickle trade networks was not premised on the existence of imperial polities or luxury commodities, since the regular pattern of exchange in the Persian Gulf was one that was sustained by agro-pastoral and fishing communities in the littoral zones.[136]

Early Old Babylonian documents between c. 1900 and 1750 BCE indicate that the peoples of Dilmun (Bahrain and eastern Arabia) ranked foremost among the Persian Gulf middlemen for Indian commodities, including precious and semiprecious stones like lapis lazuli and carnelian, ivory, and timber.[137] Dilmunite merchants traveled as far north as Mari and Šubat-Enlil (map 4), the short-lived capital of a northern Mesopotamian empire created by Šamši-Adad I (c. 1808–1776 BCE).[138] Conversely, Mesopotamian merchants like Ea-Nāṣir (c. 1810 BCE) also ventured to Dilmun to procure regional produce, most significantly copper from Oman.[139] The presence of Late Harappan ceramics at Dilmun sites like Qala'at al-Bahrain and Saar in Bahrain (c. 2000–1850 BCE) indicate that the peoples of the Indus and Gujarat only sailed as far as eastern Arabia and Bahrain in this period.[140]

132. Chakrabarti 2004; Ray 2003, 99; Sarkar et al. 2016.

133. Shastri 2000: 3–4; Thapar 1978, 222–23.

134. Chakrabarti 1999, 199; Possehl 1996, 150.

135. Thapar 1997, 12.

136. Boivin, Blench, and Fuller 2009, 252, 272; Boivin and Fuller 2009, 115, 165; Boivin et al. 2013, 266–67; Fuller and Boivin 2009, 32–33; Fuller et al. 2011, 545–47, 2015.

137. Groneberg 1992; Leemans 1960, 23–56.

138. Eidem and Højlund 1993; Groneberg 1991; Leemans 1960, 141.

139. Leemans 1960, 48–55.

140. Carter 2001. For an overview of the archaeology of Early Bronze Age Bahrain, including connections with Oman and the Indus, see Crawford 1998.

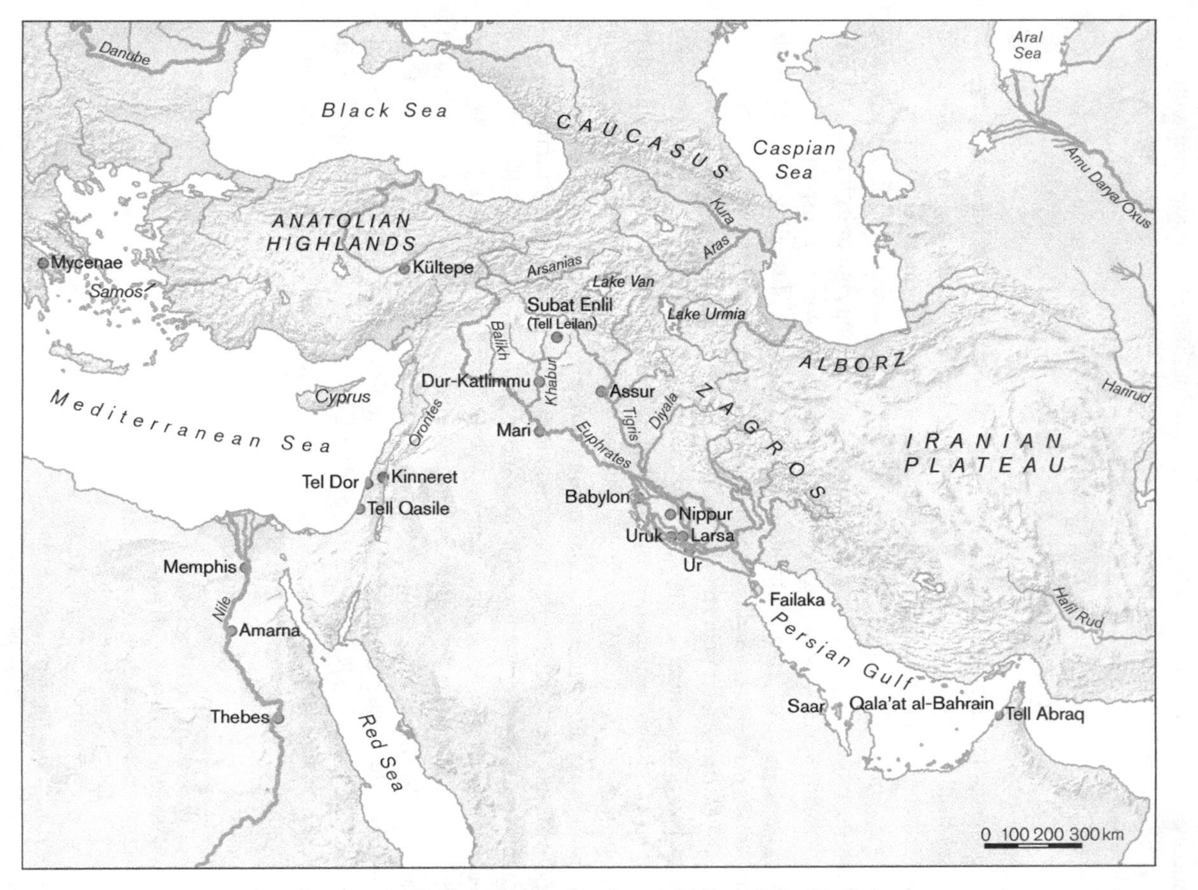

MAP 4. Important second-millennium BCE sites mentioned in the text. © Peter Palm (Berlin).

On the Mesopotamian side, much of the Gulf trade was initially managed through the Ningal temple at Ur, which levied a tax on all incoming trade commodities.[141] For instance, a trader named Milku-dannum dedicated a *šuba*-stone, perhaps agate, from Meluḫḫa to the temple in the reign of Sumu-ilum of Larsa (1894–1866 BCE).[142] The prerogative of regulating and taxing Persian Gulf commodity trade was later transferred to the Larsaite royal court in the reign of Warad-Sin (1854–1825 BCE).[143] But this regulation and taxing seems to have been the sole involvement of the temple and palace in trade in this period. Long-distance trade enterprises were otherwise coordinated and sustained by private entrepreneurs pooling their resources.[144] This early form of profit-oriented joint-stock (Akkadian *naruqqu*, "sack") venture is more clearly outlined by the voluminous Kültepe archives belonging to Assyrian merchants in Anatolia (nineteenth to eighteenth century BCE).[145] Overland trickle trade routes through the Iranian plateau, which fed into Assyrian networks, may have been less affected by developments in the Persian Gulf, although the pace of communications here was invariably slower than on the maritime routes. The lapis lazuli frequently attested in the Kültepe archives of the Assyrian traders in Anatolia probably derived from overland routes through northern Mesopotamia rather than the maritime routes.[146]

Despite the termination of direct contacts, the Indus lands remained familiar in the Mesopotamian literary imagination and continued to appear in contemporary literary compositions like *Enki and the World Order* and *Enki and Ninhursag*, lexica, incantations and ritual treatises (e.g., the Lipšur litanies), and the inscriptions of the kings of Akkad, which were copied as part of Mesopotamian scribal activity.[147] Indus products were still imported via intermediaries on a substantial scale, enough in one case to make a throne of Indian (Meluḫḫan) wood, perhaps teak, as well as a votive lapis lazuli axe and sedan chair for the god Tishpak at his temple in Eshnunna, in the Diyala Valley of central Mesopotamia.[148] Texts from Mari in the Middle Euphrates region also refer to furniture fashioned from Meluḫḫan timber, including a kind of table (Akkadian *kanniškarakku*).[149]

141. Goddeeris 2007, 201.

142. Leemans 1960, 27. The involvement of the temple in long-distance trade activity probably dates back to the late third millennium BCE. Ur-Nammu, the first king of the Ur III dynasty, claims in his inscriptions to have returned the "ships of Magan" (i.e., Oman) to the god Nanna (RIME III/2 1.1.17, 1.1.18).

143. Goddeeris 2007, 201; Laneri 2014, 404.

144. Carter 2013, 590–93; Laneri 2014; Leemans 1960.

145. Barjamovic 2011, 1–34, 2018; Larsen 2015.

146. Larsen 2015, 200; Veenhof and Eidem 2008, 82–84.

147. Possehl 1996, 142–44.

148. Frankfort et al. 1940, 194.

149. ARM 21 (Durand 1983), no. 298.3 (370–71), no. 289.10 (326–27); Kupper 1992, 166.

Animals from the Indian horizon also continued to make their way into the Middle East, most notably the Asian elephant, which was highly valued as a source of ivory throughout the Bronze and early Iron Ages. A fictive royal letter from Ur dating to the Old Babylonian period (c. 1800–1750 BCE) refers to an elephant keeper (*maṣṣār pīrim*) among other professions.[150] A contemporaneous terra cotta plaque from the Diqdiqqah suburb of Ur depicts one such mahout mounting a caparisoned elephant.[151] The Old Babylonian levels of the Merkes quarter of Babylon have incidentally yielded a leg bone belonging to an Asian elephant.[152] Canan Çakırlar and Salima Ikram have convincingly demonstrated that the elephants documented in Middle Eastern archaeological and textual sources between 1800 and 800/700 BCE represent several waves of anthropogenic transmissions from South Asia, not least because there is no credible evidence for continuity between Pleistocene and Holocene elephant populations in the Middle East.[153] At some point, the Bronze Age imports from India abandoned their human captors and were feralized, much like the thriving hippopotamus escapees from the menagerie of the late Columbian drug lord Pablo Escobar. These pachyderms roamed freely in the Middle and Upper Euphrates and Orontes Valleys, where they became hunting trophies for the likes of pharaohs and Assyrian kings. What is far from clear, however, is whether the early-second-millennium BCE elephants were conveyed to the Middle East via terrestrial or maritime routes. The former seems more likely, since there appears to have been little direct maritime traffic between India and Mesopotamia in the Old Babylonian period.

The mid-to-late Old Babylonian period (c. 1750–1595 BCE) marked a nadir in Indo–Middle Eastern trading relations. Textual sources for the Gulf trade also become woefully thin in this period. The exhaustion of surface copper deposits in Oman gradually reoriented the focus of Mesopotamian trade away from the Gulf toward the Mediterranean, so Mesopotamia now received copper from Cyprus, through Syria.[154] Sometime in the same period, the Sumerian-Akkadian toponym Meluḫḫa, formerly denoting the Indus region, came to be predominantly applied to Nubia. This toponymic shift, the result of loss of contact and the similarity of some sub-Saharan African trade products to Indian ones (e.g., exotic woods, ivory, and wild animals like leopards and elephants), emerges most clearly in the Amarna letters, the mid-fourteenth-century BCE correspondence of contemporary Middle Eastern kings found at Amarna, the short-lived Egyptian capital.[155] The decline of Indo–Mesopotamian relations in the mid-second millennium BCE indi-

150. Westenholz 1997, 155 (i.38).
151. Woolley and Mallowan 1976, 182.
152. Reuther 1926, 10.
153. Çakırlar and Ikram 2016.
154. Crawford 1996, 15–17, 20; Heimpel 2003, 38.
155. Balogh 2011, 162 n111; Leemans 1960, 165; Rainey 2015, EA 70:19, 95:40, 108:67, etc.

cates that long-distance connectivity was not an organic and accretional process that is to be taken for granted but the product of well-controlled knowledge networks, deliberate economic design, and particular historical circumstances.

Though long-distance trade across the Persian Gulf was reduced to a trickle, it never completely disappeared, and revival was not far off. The long-lasting Kassite dynasty of Babylon (c. 1595–1155 BCE) oversaw a revitalization of the Gulf trade in the late second millennium BCE. In his letter to the Kassite king Kadašman-Enlil II (1263–1255 BCE), Ḫattušili III, the Hittite king of Anatolia, bemoans the poor quality of lapis lazuli dispatched from the Babylonian court, underscoring the expectation that Babylon was the purveyor of the finest lapis lazuli.[156] The earlier Amarna correspondence (mid-fourteenth century BCE) reveals that the Kassite rulers were cajoled into sending sizeable quantities of lapis lazuli as diplomatic gifts to their counterparts in Egypt.[157] The meager amounts of lapis lazuli sent by Mitannian (Syrian) and Assyrian kings to the Pharaoh suggest that the Kassites received their supplies through maritime or the southerly land routes via Elam rather than the northern overland trading routes.[158]

The maritime hypothesis seems likely in light of Kassite domination over parts of the western Persian Gulf, including the island of Bahrain (Dilmun), where a governor was installed around 1450 BCE.[159] Materials of Indian origin like agate, carnelian, and ivory have been excavated at the Kassite (City III) levels of Qala'at al-Bahrain (mid-fifteenth to late thirteenth century BCE), the most important settlement and seat of the Kassite governor in Bahrain.[160] The increased use of chalcedony types (agate, onyx, sardonyx) for seals, amulets, and jewelry, including the distinctive chalcedony "eyestones" in Kassite Babylonia, was probably the result of vibrant trading activity with India, which was a major source of chalcedony.[161] The memory of an easterly Meluḫḫa was also not altogether eclipsed. The early Kassite (sixteenth or fifteenth century BCE) Agum-kakrime inscription, surviving in seventh-century BCE manuscripts from Nineveh, refers to the eyestones of Meluḫḫa, almost certainly Indian chalcedony, in a fragmentary passage detailing dedications to the god Marduk and his consort Zarpanītum in Babylon.[162] The last we hear of an eastern Meluḫḫa in extant Mesopotamian records is in the titulary of the Assyrian king Tukultī-Ninurta I (1233–1197 BCE), who claims symbolic lordship of Meluḫḫa alongside Dilmun (Bahrain) following his conquest of Kassite

156. Beckman 1999, 143 (no. 23).

157. Rainey 2015, EA 2, 7, 8, 9, 10, 11, 13 (dowry of a Babylonian princess).

158. Brinkman 2017, 8; Moorey 1999, 90; Olijdam 1997.

159. Hoyland 2001, 16; Potts 2006.

160. Olijdam 1997.

161. Clayden 1989, 150, 2009, 40–41.

162. Stein 2000, 157, col. 3:47. On the authenticity of this inscription, previously thought to be a forgery, see Paulus 2018.

Babylonia.[163] Tukultī-Ninurta I's title may not simply be an archaizing vaunt but probably reflects Assyrian estimations of Kassite networks in the Persian Gulf and beyond. The similarity of thirteenth- and twelfth-century BCE ceramics from Shimal in Oman and Tell Abraq in the UAE to those found in contemporary Pirak in Pakistan suggests that the Persian Gulf region was still sustaining contacts with regions further east.[164]

The transition from the Bronze Age to the Iron Age in the Middle East and the Mediterranean, associated with migratory movements like those of the "Sea Peoples" and Aramaeans, has usually been perceived as disruptive to long-distance trade connections.[165] But residue analysis of Phoenician flasks from early Iron Age sites in Israel hints that the Indian spice trade might extend back to the early Iron Age. Ten Phoenician flasks from Tel Dor, Tell Qasile, and Kinneret dating between the eleventh and late tenth centuries BCE have yielded traces of cinnamaldehyde, a distinctive but not exclusive chemical signature of *Cinnamomum* species which are native to India and Southeast Asia.[166] The flasks may have contained the essential oil of either cinnamon (*Cinnamomum verum*) or cassia (*Cinnamomum cassia*), the two most commercially significant *Cinnamomum* species. South India and Sri Lanka were the closest sources of cinnamon, while cassia is a native of Indochina.[167] The finds of *Cinnamomum* essential oils in early Iron Age Phoenician flasks are perhaps not completely anomalous, since a single cassia blossom (a source of aromatic oil) was previously identified among the seventh-century BCE botanical remains at the Heraion of Samos.[168] But cinnamaldehyde is also found in species like myrrh which can be sourced from regions closer to the Levant, so these identifications remain inconclusive.[169] In the case of the Samos find, a species-specific identification on the basis of a single floral specimen also makes for rather shaky evidence.

Still, Iron Age textual data for cinnamon and cassia may lend support to these archaeochemical analyses. In the Book of Exodus (30:23–4), the Israelite god instructs Moses to gather fragrant cinnamon (Heb. *ḳinnemōn beśem*) and cassia (*ḳiddāh*), among other aromatics, to make the holy oil used to anoint the Taber-

163. RIMA I, A.0.78.1: 15.

164. Olijdam 1997.

165. Kuhrt 1995, 385–401.

166. Gilboa and Namdar 2015; Namdar et al. 2013. Haw 2017 disagrees with Gilboa and Namdar's conclusions.

167. Namdar et al. 2013, 14.

168. Kučan 1995, 52–53. Note also alabastra from Hellenistic Babylonia and Egypt with Greek labels attesting to the presence of aromatics like cinnamon, spikenard, and sweet marjoram (Finkel and Reade 2002). Residue analysis of Mesopotamian and Egyptian alabastra and ceramic lugged jars which contained aromatic oils holds great promise in elucidating the early spice trade. Finkel and Reade (2002, 38) note that "small Babylonian ceramic jars can still scent the air when washed."

169. Haw 2017.

nacle and the Aaronite priests. While this tradition may belong to an earlier Iron Age phase, the extant passage in Genesis is a redaction belonging to the so-called Priestly Source, dated between the sixth and fifth centuries BCE.[170] The importance of cinnamon in the Levant is also recalled by Herodotus, who remarks that the Greeks learned "from the Phoenicians to call it cinnamon."[171] The outlandish story recounted by Herodotus of cinnamon collected from the broken nests of giant Arabian birds was probably a Phoenician fabrication to enhance the product's status and keep other Mediterranean traders, especially those of Greek origin, in the dark about the origins of cinnamon.[172] This sort of fanciful advertising has more recent ethnographic parallels. The inhabitants of the Siassi islands in New Guinea, for instance, sell ceramics to non-pottery-producing tribes as the shells of deep-sea mollusks harvested with great difficulty.[173]

In Mesopotamia, too, the earliest extant references in cuneiform records to both cassia and cinnamon date to the sixth century BCE.[174] The Akkadian word for cinnamon, *šalīḫātu,* is a borrowing from the Sabaic *slḫt,* surviving in modern Arabic as *salīḫa,* suggesting that the Arabs may have been the main purveyors of this spice, much as they were in later times. Pliny the Elder preserves this Semitic name for cinnamon as *serichatum.*[175] It remains to be seen, however, whether the ancient cinnamon and cassia are to be straightforwardly equated with the Asian *Cinnamomum verum* and *Cinnamomum cassia.* Leaving aside the correspondence of the ancient Akkadian and Sabaic terms for cinnamon with Arabic, Theophrastus's careful testimony in the late fourth century BCE lends support to this identification, since he explicitly states that the best aromatics, including cinnamon and cassia, were derived from Asia, in particular Arabia and India.[176]

The growing, but often tenuous, evidence for cinnamon provides some context for the previously isolated discovery of Indian peppercorns in the abdomen and nostrils of the mummy of the pharaoh Ramesses II (1279–1213 BCE).[177] The Late Bronze Age also witnesses the earliest appearance of rice, citrons, and possibly an Indian cucurbit in Mediterranean–Middle Eastern textual and archaeological records (see chapters 3 to 5). Taken cumulatively, this evidence, although disputed in part, suggests more intensive connections with South Asia in the Late Bronze Age and Early Iron Age horizons than was previously assumed. The arrival of these

170. Davies 2001, 18–20. On cinnamon and cassia in the Bible, see Zohary 1982, 202–03.

171. Hdt. III.111.

172. Hdt. III.111. Pliny the Elder (*HN* XII.85) also dismisses Herodotus's story as a fabulous tale invented to boost the price of cinnamon.

173. Sherratt 1999, 20.

174. Jursa 2009, 161–65.

175. *HN* XII.99.

176. Theophr. *Hist. pl.* IX.7.2–3.

177. Lichtenberg and Thuilliez 1981; Plu 1985.

new spices and cultivars should perhaps be associated with the expansion of Kassite political and commercial networks in the Persian Gulf. For the moment, however, the data pertaining to trade in botanical commodities and crop transfers from South Asia to the Middle East and the Mediterranean are much stronger for the first millennium BCE.

THE FIRST MILLENNIUM BCE: THE AGE OF EMPIRE

From the ninth century BCE on, political and economic conditions in the Middle East were increasingly ripe for long-distance exchanges. The rise of large, stable imperial polities in the Middle East throughout the first millennium BCE (the Assyrian, Neo-Babylonian, Achaemenid Persian, and Seleukid Empires) made long-distance travel safer. A secure agrarian base encouraged urban and demographic growth, and sustained demand for long-distance produce. Similar socioeconomic developments, especially the rise of an urban leisured class with interest in long-distance trading, can be discerned in contemporary northern India.[178]

Textual and archaeological sources from the late Assyrian and Neo-Babylonian horizon (eighth to sixth century BCE) yield evidence for previously unattested Indian commodities, including cotton, ginger, bdellium, and an unidentified "Indian" wood, perhaps teak or sandalwood.[179] The Nimrud and Tell Tayinat manuscripts of the "succession treaty" imposed on the officials and vassals of the Assyrian king Esarhaddon (680–669 BCE) cite ginger (Akkadian *zinzaru'u*), a rhizome native to tropical Asia, as a medicament in a curse clause.[180] The earliest references to ginger (*zingíberi*) in Greek sources are also in medical contexts.[181] The paucity of references to ginger and its exclusive role in medicine suggest the use of imported dried rhizomes rather than fresh, locally grown ginger.[182] Andreas of Karystos, a physician of

178. On the second (i.e., post-Harappan) urbanization of northern India, see Allchin et al. 1995; Erdosy 1988.

179. See Jursa 2009 on new eastern spices and aromatics in first-millennium BCE Mesopotamia.

180. See SAA 2 6: 643 and Lauinger (2012, 111) for Esarhaddon's "succession treaty." *Zinzaru'u,* a *hapax legomenon* in Akkadian, is identified as ginger on the basis of lexical cognates in Semitic and Indo-European languages, all of which are direct or indirect borrowings from the Indo-Aryan *singivera* or a Dravidian source (Tamil *inji-vēr*). The medical applications of *zinzaru'u* in Esarhaddon's succession treaty also support the identification with ginger (Watanabe 1987, 208).

181. Andreas of Karystos and Menestheus of Stratonikeia (second century BCE) ap. Galen *Gloss.* s.v. *Indikon* (19.105K); Pseudo-Orpheus (third to second century BCE) ap. Aet. Amid. 1.139 (CMG 8.1 p. 70); Kleophantos (first century BCE) ap. Galen *Antid.* 2.1 (14.108–109 K.); Diophantus of Lycia (late first century BCE) ap. Galen *Comp. Med. sec. Loc.* 9.4, 13.281 K.; Arbinas of Indos (second to first century BCE) ap. Galen *Antid.* 2.1 (14.109–111 K.); Cornelius (first century BCE) ap. *Antidotarium Brux.* 40 (Theodorus Priscianus, 374–75 R.); Dioscorides (first century CE) *Mat. Med.* II.160.

182. Ginger was grown in later times in some parts of the Middle East. See Varisco (2002, 347) for ginger in medieval Yemen.

Ptolemy IV (late third century BCE), and Menestheus of Stratonikeia (second century BCE), the earliest extant Greek authorities on ginger, referred to it as the "Indian" (*indikón*) spice.[183] Neither the Assyrian nor the Neo-Babylonian state, however, sustained direct links with South Asia, unlike their Persian and Hellenistic successor states. It is nonetheless likely that Middle Eastern traders, scholars, and elites in the early first millennium BCE had some vague hearsay knowledge of the region. The mention of the eponymous *sindû* wood, perhaps teak, in the annals of the Assyrian king Sennacherib (704–681 BCE) is the first unambiguous reference to the Indus region in Middle Eastern records of the first millennium BCE.[184] It is unclear, however, whether the Assyrians knew *sindû* ("Indus," from the Sanskrit *sindhu*) to be a country and not simply the name of an exotic wood.

The Assyrian Empire, the largest and most significant polity in the Middle East in the early first millennium BCE, had access to South Asian commodities through maritime and overland routes. The introduction of cotton and sissoo cultivation in Assyria was probably mediated by southern Mesopotamia (see chapters 2 and 8). The ivory, elephant hides, and timber (from species like ebony and *ellūtu*) which the Assyrians received as tribute from southern Babylonian potentates undoubtedly came from South Asia, Oman, and the eastern Iranian coast.[185] A few other South Asian commodities in Assyria were probably acquired through the transit trade along Iranian plateau routes. The inscriptions of Tiglath-pileser III (744–727 BCE) refer to a trans-Zagrosian locality called Māt-tarlugallē or Land of the Roosters, suggesting the transmission of the domesticated chicken (*Gallus gallus domesticus*) from South Asia along these routes.[186]

Lapis lazuli is predictably a prominent commodity along the plateau routes and is well attested in both textual and archaeological sources of the early first millennium BCE. The Assyrian king Esarhaddon (680–669 BCE) describes the tribute of "blocks of lapis lazuli" provided by "distant-dwelling Medes" (*madāya ša ašaršunu ruqu*) in whose territory Bikni, the mountain of lapis lazuli, was to be found.[187] The latter is probably to be identified with Mt. Damāvand, south of the Caspian Sea.[188] The Assyrian blanket term Medes (*Madāya*) for Iranian-speaking ethnic groups in and beyond the Zagros could have included merchants coming from as far east as

183. Ap. Galen *Gloss.* s.v. *indikón* 19.105K. The earliest archaeological finds of ginger are presently in the form of starch grains recovered from charred residues on cooking vessels in late-third-millennium BCE Harappan Farmana (Kashyap and Weber 2013).

184. RINAP 3, 17.vi.14b, 17.vi.23, 17.vii.31, 44.41b, 44.63b, 46.123b, 46.148b, 49.20b; Parpola 1975.

185. See e.g. Shalmaneser III (858–824 BCE): RIMA III, A.0.102.5: vi 7, A.0.102.8 27'-29', A.0.102.6 ii 51–4; Šamšī-Adad V (823–811 BCE): RIMA III, A.0.103.2, iv 11–29; Tiglath-Pileser III (744–727 BCE): RINAP 1, III.47, 26, r. 23, III 51.18.

186. RINAP 1, 17.1, 41.4b, 47.29, 47.37b.

187. RINAP 4, 1.iv.32, 1.iv.46, 2.iii.53, 2.iv.1, etc.

188. Radner 2003, 59.

Bactria (modern Afghanistan). The Greek physician Ctesias, who worked at the court of the Persian king Artaxerxes II in the early fourth century BCE, refers to a Bactrian merchant dealing in Indian gemstones, suggesting an important, albeit unstressed, intermediary role for Bactrians and other eastern Iranians along the plateau routes.[189] A banded agate eyestone inscribed with the Assyrian king Esarhaddon's dedication to the Babylonian deity Marduk was acquired in Kabul sometime in the late 1970s, but it is not clear whether this was the result of some ancient transaction or more recent loot.[190]

Apart from lapis lazuli, a number of other precious stones filtered into Mesopotamia from South Asia, via either overland or maritime routes. Margaret Sax's mineralogical analysis of 361 Assyrian and Babylonian cylinder seals in the British Museum's holdings yielded evidence for the use of green microcrystalline grossular garnet and brown-and-white agate of Indian origin.[191] Mineralogical analysis of seals and jewelry promises to be of further use in uncovering Assyrian trade links with the east. Brown-and-white agate, for instance, finds extensive use in diadems, necklaces, and other ornaments found in the eighth-century tombs of the Assyrian queens in Kalḫu (Nimrud), as does garnet in jewelry from Neo-Babylonian Uruk (map 5).[192]

The Achaemenid Persian period (late sixth to fourth century BCE) saw the introduction of further varieties of Indian gemstones, including rubies and beryls sourced from South India, Sri Lanka, and Myanmar. Phylarchus, a Greek historian of the third century BCE, remarks that the "gold plane-trees and the gold grapevine beneath which the Persian kings commonly sat to conduct their business" were wrought with "grapes made of emeralds as well as of Indian rubies and extremely expensive jewels."[193] Posidippus of Pella, a third-century BCE poet celebrating Achaemenid heirlooms acquired by the Ptolemies in Alexandria, also makes mention of rubies, beryls, and other Indian stones.[194] Finely crafted metalware also ranked among the costlier Indian exports in this period, as indicated by one Pseudo-Aristotelian treatise on mirabilia (fourth to third century BCE):

> They also say that among the Indians the copper is so bright, pure and free from rust that it cannot be distinguished in color from gold; moreover that among the cups of Darius there are certain goblets and these not inconsiderable in number, as to which, except by their smell, one could not otherwise decide whether they are of copper or gold.[195]

189. Ctesias *Indica* F45.6.

190. RINAP 4, 142; cf. an inscribed agate eyestone of an earlier Assyrian king Adad-nārārī I (1307–1275 BCE) found in a tomb at Khodjali in Azerbaijan: RIMA I, A.0.76.46.

191. Sax 1991, 112–13.

192. Collon 2008, 107–110; Moorey 1999, 83.

193. Ath. XII.539d.

194. Posidippus of Pella, *Lithiká* (Austin and Bastianini 2002), nos. 1–3, 6, 8; Kuttner 2005, 151–56.

195. *On Marvelous Things Heard* 49, trans. Hett 1936.

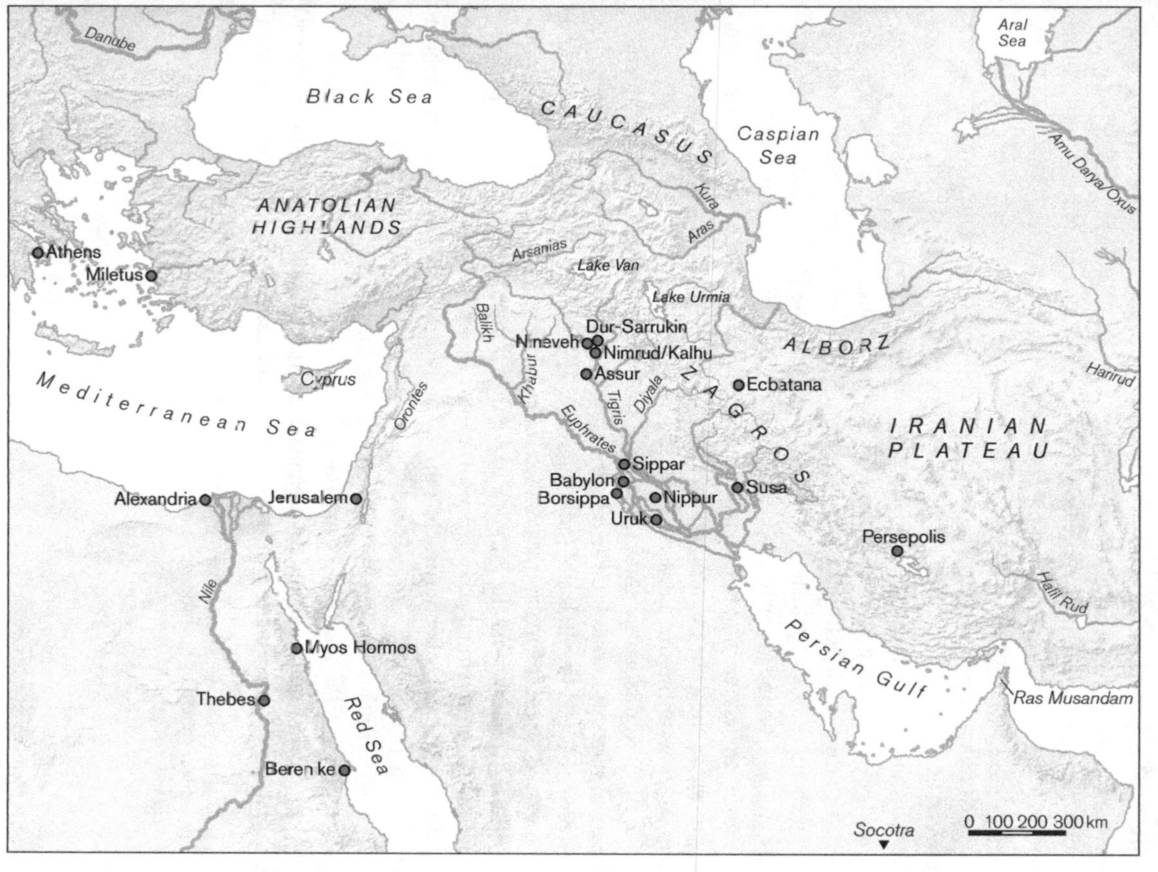

MAP 5. Important first-millennium BCE sites mentioned in the text. © Peter Palm (Berlin).

This author almost certainly refers to brass (Sanskrit *pittala*), an alloy of copper and zinc, which is gold-like in appearance when polished and (like copper and iron) produces a distinctive metallic smell when it comes into contact with sweaty hands.[196] The gold-like character of brass is also borne out by the Sanskrit synonyms *dīptaloha* ("shining metal") and *pītala* or *pītaka* ("yellow metal").[197]

The Achaemenid period represents a new maximum of political and trading relations between South Asia and the Middle East in the first millennium BCE. By this period, we can speak of systemic connectivity—that is, structured and repetitive exchanges of peoples, goods, and services—and dense knowledge networks between South Asia and the Middle East. Cyrus the Great's eastern territorial conquests in the late sixth century BCE included the Indian polities of Thataguš (Sattagydia), in the Bannu region of the Middle Indus, and Gandhāra, which corresponds to the region between the lower Kabul Valley and the upper reaches of the Indus, in the Punjab (Pakistan).[198] Darius I (r. 522–486 BCE) would later extend Persian rule into the lower Indus Valley (Old Persian *Hinduš*).[199] The political integration of the lands lying between southern Bulgaria and the lower Indus under the banner of the Achaemenid Persian kings is perhaps the most spectacular testament to transcontinental Eurasian connectivity.

Regularized diplomatic and bureaucratic intercourse between the Persian court and its Indian satrapies is indicated by the frequent travel of officials and diplomats between the Persian capitals and India (figure 8). The itineraries of these travelers and the imperial rations issued for part of their journey are documented in the Persepolis Fortification archive from the middle years of the reign of Darius I (509–493 BCE).[200] Some of these individuals traveled with a sizeable entourage. The caravan of one Gandhāran named Zakurra (probably from the Indic *Cakra*) included 290 men, 31 mules, and 12 camels.[201] Ctesias (early fourth century BCE) also refers to Indian envoys who regularly brought gifts to the Persian king.[202] While redacted at a later age, the *Matsya-Purāṇa* (early centuries CE), an encyclopedic Sanskrit text with genealogical, cosmological, and mythological elements, dimly preserves the memory of Indian diplomatic missions to the Achaemenid court in Susa. In this text, Susa

196. This is caused by the production of carbonyl compounds when skin comes into contact with some types of metals (Glindemann et al. 2006).

197. On the history of brass production in India, see Biswas 1993.

198. On Thataguš, see Fleming 1982, 102–12. For Gandhāra, see Kent DB 16–17; Xen. *Cyr.* I.1.4; Arr. *Ind.* I.3; Pliny *HN* VI.92.

199. Kent DPe 17–18, DPh 7, DSf 44, DSe 24, DNa 25, DSm 10. On the archaeology of Achaemenid India, see Fleming 1993; Magee and Petrie 2010. For Achaemenid and contemporary Greek sources on northwest India, see Badian 1998; Giovinazzo 2000–01; Karttunen 1989; Potts 2007a; Vogelsang 1990.

200. PF 785, 1317, 1318, 1358, 1383, 1397, 1410, 1425, 1437, 1440, 1511, 1524, 1525, 1529, 1548, 1550, 1552, 1556, 1558, 1572, 1601, 2057.

201. PFNN 0431 (possibly the same individual called Zakarna in PF 1139). See Henkelman 2017, 187–90.

202. Ctesias *Indica* F45.19, 39, F45m, F45dβ, F45pγ.

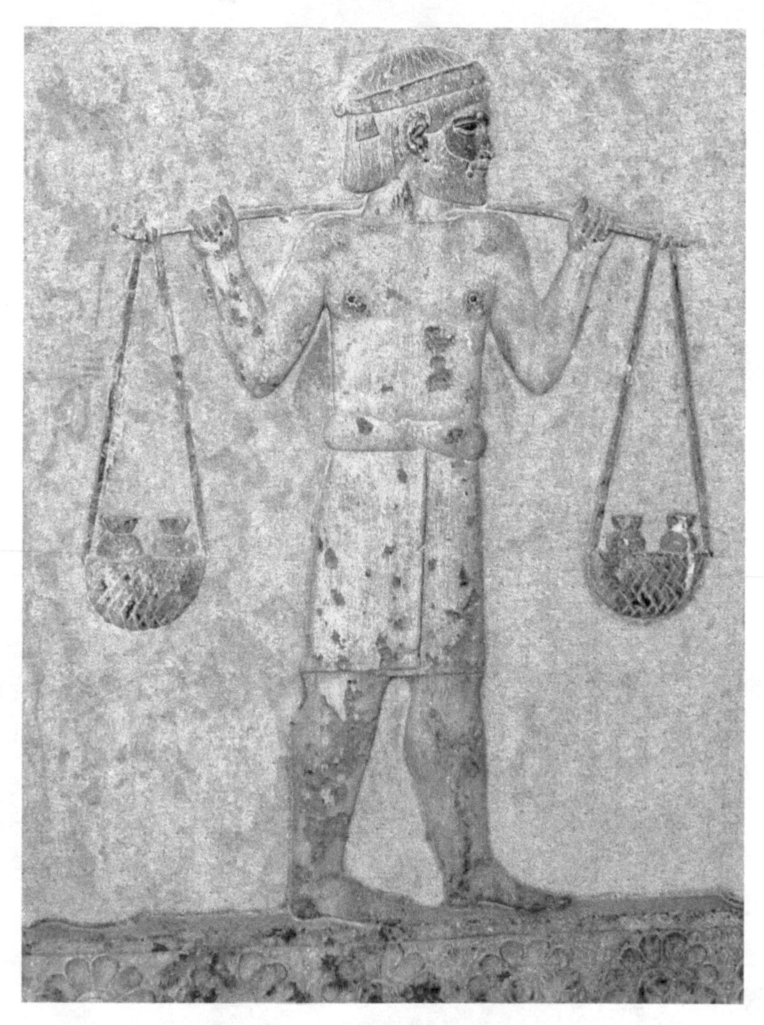

FIGURE 8. Indian tribute-bearer on the eastern staircase of the Apadāna, Persepolis, Iran, c. 522–486 BCE. Photograph by A. Davey via Wikimedia Commons (CC BY 2.0).

acquires mythical dimensions and is described as the beautiful city of wise Varuṇa, the god of the ocean (124.22: *suṣā nāma purī ramyā varuṇasyāpi dhīmataḥ*).[203]

It is possible that Indian embassies came not only from the officials and petty princelings of the northwest who acknowledged Achaemenid suzerainty but also from the powerful neighboring kingdoms of Kosala and Magadha in the eastern

203. On the dating and the utility of the *Purāṇas* as historical documents, see Rocher 1986.

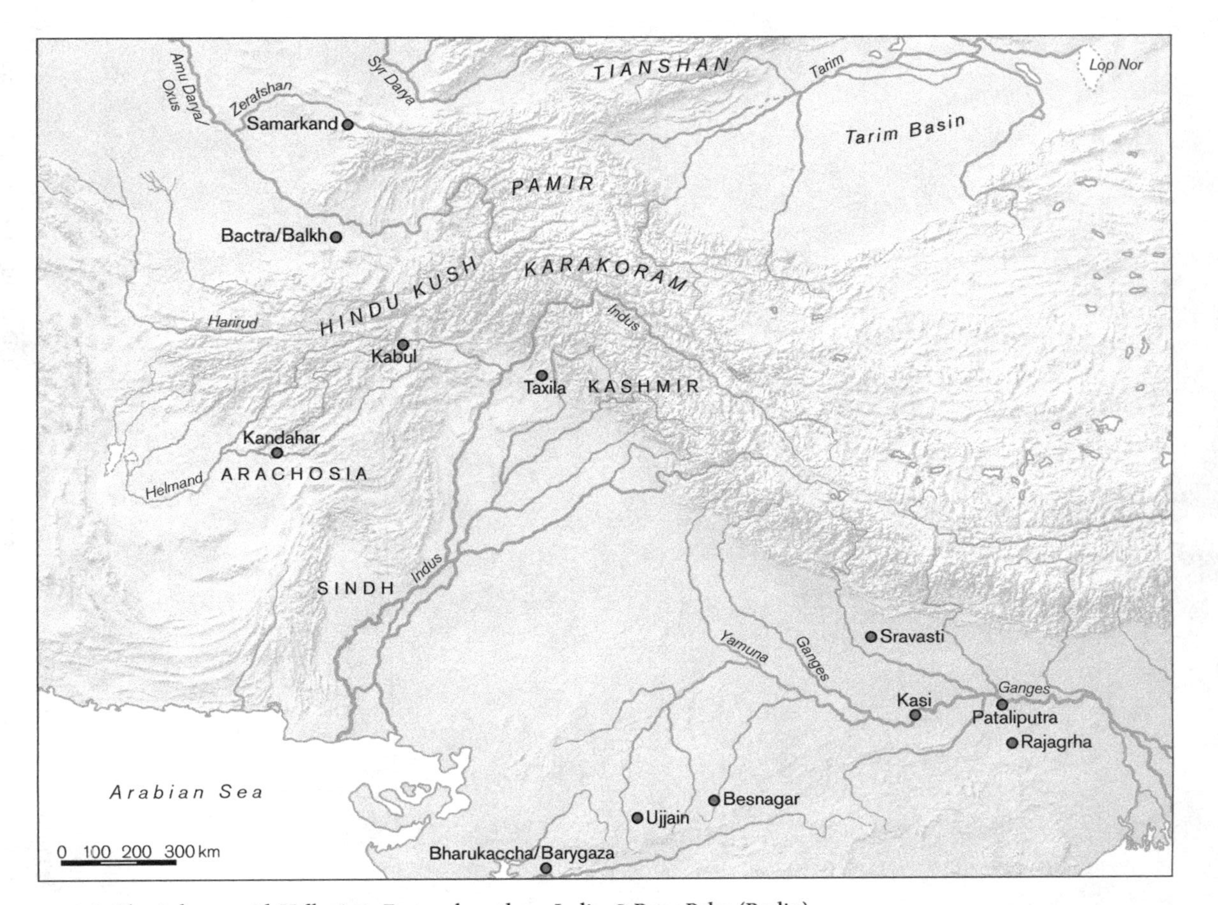

MAP 6. The Achaemenid-Hellenistic East and northern India. © Peter Palm (Berlin).

Gangetic Plain (map 6).[204] By the mid-fourth century BCE, the polity of Magadha under the bellicose Nanda dynasty had expanded across the Indo-Gangetic Plain to meet the Achaemenid frontier in the northwest. The founder of this dynasty, Mahāpadma Nanda, is flamboyantly described in the *Viṣṇu-Purāṇa*, among other Indian texts, as the exterminator of kings, the sole sovereign, and one whose commands were not transgressed.[205] While Achaemenid–Magadhan interactions are poorly documented, the Nanda dynasty of Magadha figures as a significant element in narratives of Alexander's incursion into India.[206] Reports of the military might of the Nandas, in Greco-Roman sources the kingdom of the Gandaridae or Gangaridae, allegedly incited the mutiny of Alexander's army on the banks of the Vipāśā River, much to the Macedonian king's chagrin.[207]

Indians figure in various vocations across the Middle East in the multilingual textual sources of the Achaemenid period. The Babylonian Murašû archive (late fifth century BCE) provides evidence for the billeting of numerous foreign military colonists, including Indians (*Indumāja*), in the rural hinterland of Babylonian cities like Nippur in exchange for their military service.[208] The Achaemenid rulers made extensive use of Indian contingents in their campaigns and, by the fourth century BCE, had adopted a few Indian war elephants, which were to become a major feature of Hellenistic armies.[209] While some Indians fought for Achaemenid armies, a few were also seized by them. A Gandhāran female slave (*Gandaruitu*) bearing the name Nanā-silim, perhaps obtained as a war captive during Cyrus's campaign in Gandhāra, appears in the Egibi archive from Babylon during the reign of Darius I (508/7 BCE).[210]

Soldiers aside, Ctesias (early fourth century BCE) saw an Indian mahout in Babylon, and Chares of Mytilene (late fourth century BCE) observed Indian magicians in the spectacles organized for the mass wedding celebrations of Alexander's officers at Susa, suggesting the presence of Indian entertainers at the Persian court.[211] Since Ctesias remarks that he tasted Indian cheese (*turós*) at the Persian

204. Kulke and Rothermund 2004, 57–60.

205. *Viṣṇu-Purāṇa* IV.24.3–4. The *Viṣṇu-Purāṇa* (early centuries CE) belongs to the same class of texts as the *Matsya-Purāṇa*, whose encyclopedic concerns include royal genealogies, cosmologies, legal codes, ritual, pilgrimage, and, above all, stories concerning the gods (Rocher 1986).

206. Fauconnier 2015.

207. Diod. Sic. XVII.93–94; Plutarch, *Life of Alexander* 62; Curtius IX.2.3; *ME* 68; Arr. *Anab.* V.25.1–2.

208. Zadok 1977, 125; Dandamaev 1992, 54, 59, 63, 144, 165. Kessler 2002 suggests that the toponym Sittacene, *Sattagū* in Babylonian texts, in the east Tigridian region derives from the settlement of colonists from Sattagydia, one of the Indian countries conquered by Cyrus.

209. Hdt. VII.65–66, 86, VIII.113, IX.31; Strabo XV.1.6; Diod. Sic. XIV.22.2, XVII.59.4; Arr. *Anab.* III.11.5, III.8.3–6; Ctesias *Indica* F45.7. On the Asian elephant in the late Persian and Greco-Roman periods, see Charles 2008, 2010; Scarborough 1985; Scullard 1974; van Oppen 2019.

210. Dar. 379:44; Zadok 1977, 124. See Pliny *HN* VI.92 for Cyrus's campaign in Gandhāra.

211. Ctesias *Indica* F45bα; Ath. XII. 538e; Aelian *VH* VIII.7.

court, it seems likely that Indian (or at the very least Indian-influenced) cooks were present in the Persian king's kitchens: Indian cheese (Sanskrit *kilāṭa*, Hindi *panīr*), unlike its Western counterparts, is produced fresh by the acidification of heated milk and cannot be stored for export over long distances.[212] Finally, some personal names (e.g., Indukka and Hindukka) etymologically linked to the toponym India (Hinduš) in Babylonian and Elamite texts may represent hypocorisms and thus offer further evidence for the settlement of Indians in the central provinces of the Achaemenid Empire.[213]

The mid-first millennium BCE also witnesses an intensification of direct, long-distance, profit-oriented maritime trade across the Indian Ocean, with Indians and Arabs as its chief architects. This phenomenon is perhaps tangential to the development of empires in the Middle East and more a testament to the renewed maritime confidence, mobility, and technological capabilities of Indian and Arab coastal communities.[214] Kāludāyin, an early Buddhist monk and contemporary of the Buddha, candidly spells out the motivation of merchants who crossed the ocean: the desire for wealth (*āsāya vāṇijā yanti samuddaṃ dhanahārakā*, in *Theragāthā* 530).

Open-sea sailing with the aid of monsoonal winds became more common in this period.[215] In one of his discourses, the Buddha casually refers to seafaring merchants who employed a land-sighting bird on the open seas.[216] The *Saṅkha-Jātaka*, a story from a didactic compendium in the Pāli Buddhist canon (mid-first millennium BCE), explicitly refers to ships powered by monsoonal winds.[217] The various seas (Khuramāla, Aggimāla, Dadhimāla, Nilavaṇṇakusamāla, Nalamāla) distinguished by color and attribute in the *Suppāraka-Jātaka*, another didactic Buddhist tale, probably represent corruptions of actual Indian Ocean localities traversed by merchants, but their identity is uncertain. The Khuramāla sea is said to be distinguished by the presence of deep-sea billfish (marlin, sailfish, swordfish), which are described as "men-like fish with razor-sharp noses."[218] Perhaps this refers to the East African coast, where billfish are more commonly sighted, rather than the Persian Gulf.[219] The *Nāyādhammakahāo*, a cycle of Jain didactic stories dating between the late centuries BCE and the early centuries CE, refers to a distant land called Kāliyadīva where Indian seafaring merchants observed and even captured "aston-

212. Ctesias *Indica* 48; Asher 2015, 115–17; Kumar et al. 2014.
213. Dandamaev 1992, 40, 86, 130; Henkelman, Jones, and Stolper 2006, 8–9; Tavernier 2007, 202.
214. Boivin, Blench, and Fuller 2009, 265.
215. Boivin, Blench, and Fuller 2009, 265.
216. *Dīghanikāya* XI.85.
217. *Jātaka* 442.
218. *Jātaka* 463.
219. Howard and Starck 1975, 23–25.

ishing" (*accherā*) wild horses in "brownish, grey and black hues."[220] As horses, donkeys, and wild onagers were well known in South Asia, Dinendra Jain has suggested that this may be a secondary description of an encounter with zebras in an East African locality like Zanzibar.[221]

Within the Persian Gulf, Indian merchants appear to have deposited their wares in eastern Arabia and left the local dispersal of Indian commodities to Arab merchants. This accounts for the persistent attribution of Indian aromatics like cinnamon, cassia, and nard to Arabia in early Greek sources.[222] Yet Theophrastus, a contemporary of Alexander, accurately remarks that these aromatics derive "from India, whence they are sent over sea."[223] Nearchus in the late fourth century BCE states that "the Assyrians (i.e. Mesopotamians) imported cinnamon and other spices" from Maceta, the promontory of Oman (Ras Musandam), indicating that Indian ships typically sailed up to this point.[224] Some Indian merchants may have ventured further into the Gulf on occasion. The *Bāveru-Jātaka*, an edifying canonical Buddhist story of the mid-first millennium BCE, narrates the passage of Indian merchants to Babylon (*Bāveru*, from Old Persian *Bābiru*), where they peddle a crow and a dancing peacock to astonished Babylonian citizens.[225]

More typically, it was Arab merchants who undertook the shipment of Indian commodities to Mesopotamia. Greek sources credit the Neo-Babylonian king Nebuchadnezzar (r. 605–562 BCE) with the (re)foundation of the trading emporium of Teredon or Diridotis at the head of the Gulf.[226] This port settlement was perhaps somewhere in the vicinity of the ancient city of Eridu, although cuneiform texts do not offer confirmation.[227] Nearchus notes that merchants brought to Teredon or Diridotis "frankincense from the land of Gerrha (eastern Arabia) and all the other sweet-smelling spices Arabia produces."[228] Arab merchants appear to have gone beyond Teredon, since Aristobulus observes in the fourth century BCE that the Gerrhaeans in northeastern Arabia "import most of their cargoes on rafts to Babylonia, and thence sail up the Euphrates with them, and then convey them by land to all parts of the country."[229] This situation undoubtedly applied to the sixth century BCE as well, when the riverine distribution of exotic commodities is observed in

220. *Nāyādhammakahāo* 17.9–24.

221. Jain 1980, 73.

222. Hdt. III.107, 111; Diod. Sic.II.49.3; Strabo XV.1.22, XVI.4.19; Arr. *Anab.* VII.20.2; cf. Pliny *HN* 12.14.

223. Theophr. *Hist. pl.* IX.7.2; see also Aristobulus ap. Strabo XV.1.22.

224. Arr. *Ind.* 32.7.

225. *Jātaka* 339.

226. Potts 1990, 85 (e.g., Berossus ap. Eusebius *Praeparatio Evangelica* XLI).

227. Jursa 2010, 110.

228. Arr. *Ind.* 41.6–7; cf. Strabo XV.3.5; XVI.3.2; Pliny *HN* VI.32; Amm. Marc. XXIII.6.11.

229. Aristobulus ap. Strabo XVI.3.3.

Neo-Babylonian archival texts. Cinnamon and cassia appear, for instance, among the merchandise procured by one Nabû-dūr-paniya for the Ebabbar temple in Sippar.[230] Nabû-dūr-paniya did not venture far to purchase these exotic commodities, since he is elsewhere said to have sold oil to the temple from the local riverine port (*muhhi kāri*).[231]

The last quarter of the first millennium BCE saw an intensification of trends already observed in the Achaemenid period, with the spatial extent of Indian diplomatic contacts now encompassing the Mediterranean as well.[232] The regularity of long-distance trade flows, reflecting extensive knowledge networks and mutual trust in commercial exchange, also continued unabated. Trade commodities in this period include previously unattested South and Southeast Asian spices and aromatics like turmeric, amomum, cardamom, costus, Indian bay leaf or malabathron, cloves, spikenard, and camphor.[233]

The far-flung diplomatic dealings of the Mauryan Empire of India are perhaps most famously represented in the second and thirteenth rock edicts of the Mauryan king Aśoka (third century BCE), which invoke the kingdoms of Antiochos II, Ptolemy Philadelphos, Magas of Cyrene, Antigonus Gonatas of Macedonia, and Alexander of Epirus.[234] Envoys of the Seleukid (Megasthenes, Daimachos) and Ptolemaic (Dionysios) kings resided at the Mauryan capital of Pāṭaliputra (modern Patna) for extended periods.[235] The kings of the Seleukid Middle East and Mauryan India clearly corresponded, although modern historians are not privy to these exchanges beyond anecdotes like the Mauryan king Chandragupta's gift of aphrodisiacs to Seleukos I (r. 311–281 BCE), which were claimed to make men "as randy as birds."[236]

The Greeks, and more broadly the Mediterranean world, also learned in this period of the existence of lands beyond India through the mediation of Indian informants. Megasthenes, the Seleukid or more properly Arachosian envoy to the Mauryan court, is the earliest known Greek author to describe Southeast Asia as the land of gold (*khrusē̃ khṓra*). This is evidently a calque of the Indic Suvarṇadvīpa or Suvarṇabhūmi (Golden Land), which refers perhaps to the

230. MacGinnis 1996, no.16 (BM 67001).

231. Jursa 2009, 168 n103 (Nbn. 821).

232. On Hellenistic contacts with India, see Karttunen 1997; Muthukumaran 2012; Raschke 1975; Salles 1996.

233. Casson 1984; Dalby 2003; Greppin 1999; Jursa 2009; Miller 1969; Muthukumaran 2012.

234. Bloch 2007, 93–95, 125–32; Thapar 2012, 377–84.

235. Kosmin 2014, 31–58, 261–72; Roller 2016. Megasthenes was more properly dispatched to Mauryan India by Sibyrtios, the satrap of Arachosia stationed in Kandahar.

236. Ath. I.18e. A similar Indian aphrodisiac is mentioned in Theophr. *Hist. pl.* IX.18.9.

auriferous regions of Sumatra, western Borneo, and the Malay Peninsula—or more likely to the fortunes gained from trading in that region.[237]

The diplomatic missions of Aśoka to the west probably included Buddhist proselytisers as well.[238] Jain tradition recalls the dispatch of proselytizer-diplomats by Aśoka's grandson Samprati to barbarian countries, presumably including the regions his grandfather had been in contact with.[239] The memory of Aśokan diplomatic contacts with Ptolemaic Egypt may be preserved in the "Swallow and the Sea," a Demotic Egyptian narrative of the Roman period, where Aśoka is corrupted as Ausky, the "prince of Arabia," who writes to Pharaoh.[240] The Indian Ocean connections of this Demotic text are also suggested by the unmistakable parallel with the Indian story of the *ṭiṭṭibha*-birds (sandpipers or plovers) and the ocean (*ṭiṭṭibhadampatīkathā*) preserved in the Sanskrit *Pañcatantra*, a compendium of didactic stories which worked its way in later times into La Fontaine's *Fables* via the Arabo-Persian *Kalīla wa-Dimna*.[241]

The multicultural and cosmopolitan nature of "Middle Asian" societies at the nexus of trade and communications routes is best reflected in Hellenistic Bactria and northwestern India. The transmutability and flexibility of ethnic identities in these liminal zones is especially suggested by the case of Sophytos, a well-traveled Indian merchant from Alexandria-in-Arachosia (Kandahar) of the second century BCE, whose loquacious autobiographical inscription in Greek, an acrostic no less, is replete with Homeric diction and fabulously rare words conjuring the rarefied intellectual climate of Callimachian Alexandria.[242] The roughly contemporary Heliodorus, an ambassador of the Indo-Greek king Antialkidas to the Indian Śuṅga king Bhāgabhadra, left an inscription in Prakrit on a freestanding temple pillar in Vidiśā (Besnagar) which describes him as a Bhāgavata—that is, a worshipper of the Indian deity Vāsudeva-Kṛṣṇa.[243] While the Indian Sophytos parades his Homeric erudition, the Greek Heliodorus's inscription quotes didactic precepts which have echoes of a passage in the *Mahābhārata* (11.7.19: *damas tyāgo 'pramādaś ca te trayo brahmaṇo hayāḥ*).[244] Elsewhere, at the Greek city of Miletos on the Anatolian coast, a Greek astrometeorological inscription dating to the late second century BCE cites the Indian scholar Kallaneus as an authority on stellar movements

237. Megasthenes ap. Solinus 52. 6–17; Pliny *HN* VI.80; Pomponius Mela I.70; Seldeslachts 1998, 282–83; Wheatley 1961, 177–84.

238. Karttunen 2014.

239. Hemacandra, *Sthavirāvalīcarita* XI.89–90.

240. Betro 1999. See Collombert 2002 for text, translation, and commentary.

241. Viṣṇuśarman, *Pañcatantra* I.8. While the *Pañcatantra* was committed to writing c. 200–400 CE, the stories are much older. On the sources of the *Pañcatantra*, see Falk 1978.

242. Bernard, Pinault, and Rougemont 2004; Rougemont 2012, 173–82 (no. 84).

243. Salomon 1998, 265–67.

244. For a detailed study of both personalities, see Mairs 2014, 102–45.

and meteorological phenomena.[245] David Pingree, a historian of the exact sciences, points out that Kallaneus (probably the Greek version of the Sanskrit *Kalyāna*) was using a Greek rather than Indian method of prediction, given that lunar conjunctions with fixed stars were used as indicators of meteorological phenomena in India, in contrast to the heliacal movements cited in the Milesian parapegmata.[246] The mingling of Mediterranean–Middle Eastern and South Asian cultures is further suggested by the case of Dhammarakkhita, an Indo-Greek (Yona) monk who preached Buddhism in western India in the third century BCE.[247]

On the maritime front, Indians and Arabs continued to dominate Indian Ocean trading networks in the Hellenistic period. Iamboulos, the protagonist of a late-Hellenistic utopian travel narrative set in the Indian Ocean world, bears a name of Arab extraction.[248] The frequency of direct Indian voyages to the African coast also increases in the Hellenistic period. Agatharchides of Knidos (second century BCE) remarks that merchant vessels from Patala, on the lower Indus, frequently visited the island of Suquṭrā (Socotra), which lay 230 kilometers east of Cape Guardafui on the Horn of Africa.[249] This now-remote island was in antiquity a major trading site attracting Indian, Arab, and Greek settlers.[250] A shipwrecked Indian rescued by the Ptolemaic coastguard off the Red Sea later acted as guide for the voyage to India of Eudoxus of Cyzicus (late second century BCE) on behalf of his Ptolemaic patrons, providing evidence for Indian vessels plying the seas up to Egypt.[251] A passage in the *Mahāniddesa*, a canonical Buddhist exegetical text in Pāli dating to the reign of Aśoka if not earlier, confirms the expansive trading world of Indian merchants, which stretched from Southeast Asia to "the country of the Greeks, the country of the distant Greeks (and) Alexandria (in Egypt)."[252] This no doubt refers to the neighboring Seleukid Empire and Ptolemaic Egypt, whose capital Alexandria (*Allasanda*) is explicitly named. Like Eudoxus, Greek-speaking merchant mariners from Seleukid Asia and Ptolemaic Egypt also set sail to India, although in far fewer numbers than the Arabs and Indians who traveled westward. Among

245. Lehoux 2005, 130–33, 136: Miletus II 456A Col. I.6, II.5, 456D Col. I.8–9, 456N 10.

246. Pingree 1976, 143–44.

247. *Mahāvaṃsa* XII.4–5.

248. The adventures of Iamboulos in fictionalized Indian Ocean islands and his visit to the Mauryan capital of Pāṭaliputra are reproduced in garbled form by Diodorus Siculus: II.55.2; II.60.1–3. On Iamboulos's name, see Altheim and Stiehl 1964, 83–84.

249. Agatharchides of Knidos, *On the Erythraean Sea*, Book 5, 105a-b; Diod. Sic. 3.47.8–9; Burstein 2012.

250. Cosmas Indicopleustes III.65; *Periplus Maris Erythraei* 30; al-Masʿūdī 1966–1979 para 879; Yāqūt 1866–1870 III.102. See also Strauch 2012 on the inscriptions left by Indian mariners and merchants on Sūqutrā between the first century BCE and early centuries CE.

251. Strabo II.3.4.

252. *Mahāniddesa* PTS 154–5, 415. On the dating of this text, see Norman (1983, 84–87) and Sarkar (1981, 297–301).

those who have left faint traces in documentary sources is one Sosandros, who authored a now-lost Indian periplus.[253]

There is little need to discuss *in extenso* the beginnings of Indo–Roman trade, which represents an inheritance and amplification of Hellenistic contacts with India.[254] While the Roman interaction with India is today the best-known long-distance trading relationship of antiquity, the Romans were in fact newcomers to an interconnected world over three millennia in the making. Notwithstanding the unaccounted number of caravans traveling up the Euphrates into Syria, by 26 or 25 BCE some 120 ships made their way to India from the Roman Red Sea annually, according to the reckoning of the Greco-Roman geographer Strabo.[255] A similar if not larger figure should be expected for the Persian Gulf region, which is geographically and historically more integrated with the west coast of India.

Black pepper (*Piper nigrum*), native to South India, was the most prominent commodity in Greco-Roman exchanges with India.[256] It is first attested in the Mediterranean as a medicament, particularly for ocular conditions, in Hippocratic pharmacopoeia (fifth to fourth century BCE).[257] The earliest author to explicitly cite pepper in the context of cooking is unsurprisingly a physician, Diphilos of Siphnos, who provides a recipe for sautéed scallops with pepper.[258] The extreme popularity of pepper in the Mediterranean, purchased with gold no less, baffled Pliny the Elder:[259]

> It is quite surprising that the use of pepper has come so much into fashion, seeing that in other substances which we use, it is sometimes their sweetness, and sometimes their appearance that has attracted our notice; whereas, pepper has nothing in it that can plead as a recommendation to either fruit or berry, its only desirable quality being a certain pungency; and yet it is for this that we import it all the way from India! Who was the first to make trial of it as an article of food?[260]

Remarkably, during the Roman general Sulla's siege of Athens between 87 and 86 BCE, pepper was said to be in abundant supply even though the city's olive oil and

253. BNJ 714. There is no good reason to assume that Sosandros is the author of the *Periplus Maris Erythraei* of the first century CE.

254. On the Roman trade with India, see especially Begley and Puma 1991; Cimino 1994; Cobb 2015, 2018a, 2018b; De Romanis 2020; De Romanis and Tchernia 1997; Fitzpatrick 2011; Parker 2002, 2008; Raschke 1978; Seland 2010; Sidebotham 1986, 2011; Tomber 2008; van der Veen 2011, 5–8.

255. Strabo, *Geographia* 2.5.12.

256. Asouti and Fuller 2008, 47.

257. See e.g. Hippoc. *Morb. Mul.* I. 81, II.158; Hippoc. *Acut. Sp.* 34; Hippoc. *Epid.* V.67, VII.64; Hippoc. *Nat. Mul.* 32. Pepper is cited seventeen times in the Hippocratic corpus (Totelin 2009, 191, 194).

258. Ath. III.90f.

259. Pliny *HN* VI.101, XII.84; cf. *Akanānūru* 149.

260. Pliny *HN* XII.29.

wheat had been exhausted.[261] Almost half a millennium later, Alaric, the leader of the Visigoths, who would sack Rome in 410 CE, was pacified by the Roman Senate in the years preceding the conquest with a payment of 3,000 pounds of pepper, along with gold, silver, and silk.[262] Pepper was hoarded much like gold, silver, and staples, but it was not prohibitively expensive, and its consumption cut across social classes. Gambax, a solider of humble rank, could purchase Indian pepper for just two denarii in Vindolanda, on the periphery of Roman Britain.[263] Black pepper has also been recovered from several Roman sites in northern Europe, including a latrine in a legionary camp near Oberaden in Germany.[264] At Rome itself, the *horrea piperataria* (pepper warehouse), a specially designated commercial area in the porticoes of the Neronian Sacra Via, was the main trading site for pepper and a host of other imported spices and aromatics.[265]

Rome's entry into the Indian Ocean world at the end of the first millennium BCE also coincided with the emergence of new pan-oceanic carriers of trade, most notably the Tamils of the far south of India and northern Sri Lanka, whose dispatch of embassies to Rome in the age of Augustus suggests an earlier involvement in the western Indian Ocean world.[266] Archaeological and epigraphic data in the Tamil-Brahmi script from the early centuries CE indicate that Tamil mercantile communities operated from the Straits of Malacca in Southeast Asia to the Red Sea frontier of Egypt, a trading network spread over 7,000 kilometers—a scale hitherto unknown.[267] Like other mobile communities, Tamil traders and settlers moved botanical produce as trade commodities and for their own consumption. The macro-remains of coconut, rice, pepper, and mung bean at Roman Berenike and Myos Hormos on the Red Sea coast (first century CE), and the mung bean and

261. Plutarch, *Life of Sulla* 13.3.

262. Zosimus V.41.4.

263. Tab. Vindol. II.184.

264. Cappers 1998, 313; Kučan 1992; van der Veen 2011, 44; Vandorpe 2010, 39, 49, 89.

265. The construction of the *horrea piperataria* is frequently attributed to the Emperor Domitian (81–96 CE), but he probably restored a pre-existing structure (Richardson 1992, 194–95; Rickman 1971, 104–07).

266. Two contemporary Tamil dynasties (Pāṇṭiya, Cēra) are explicitly named in the Indian embassies to Augustus (Strabo XV.1.4; Suet. Aug. 21.3; Florus II.34.62; Orosius VI.21.19). The Pāṇṭiyas ruled in the basin of the Vaikai and the Porunai (Tāmiraparaṇi), with a capital at Maturai and a major port and pearl fishery at Koṟkai, while the Cēras, a trans-Western Ghats power, were centered on the inland capital of Karūr (Vañci), with ports at Muciṟi and Tonṭi on the west coast (Subbarayalu 2014).

267. Note especially Tamil-Brahmi epigraphs at Phu Kao Thong (second century CE) and Khuan Luk Pat (third century CE) in southern Thailand, at Khor Rori in Oman (first century CE) and Berenike and Quseir al-Qadim on the Egyptian Red Sea coast (first century CE). South Indian ceramics have been found from the Red Sea region to Indonesia from the late centuries BCE to the early centuries CE (Bellina et al. 2014; Chaisuwan 2011; Mahadevan 2003, 49; Muthukumaran 2019; Rajan 2011; Subramanian 2012).

horsegram at Khao Sam Kaeo in southern Thailand (c. 300 BCE), have been interpreted as foodstuffs moved by Tamil seafarers and traders for their own consumption.[268] While the impetus for this massive spatial stride is poorly investigated in modern scholarly literature, it was certainly no overnight development but one that came on the heels of long-established small-scale regional networks, combined with maritime daring in the fashion of early Phoenician and Greek traders in the Mediterranean.

The survey of communications and commerce between South Asia, the Middle East, and the Mediterranean from the fourth millennium BCE on highlights the problem with speaking of crop introductions as singular events or "packages," since multiple, sometimes competing, networks, agencies, and routes often operated simultaneously. The importance and intensity of a specific exchange network was predicated on variables like political stability, material prosperity, social complexity, infrastructural investments, technological innovations, risk management, economic opportunism, and knowledge networks. Some chronological horizons, namely the late third millennium BCE and the late first millennium BCE, are characterized by direct, intensive, and regularized contact, but otherwise transit trade managed by intermediaries in the Persian Gulf, southern Arabia, and eastern Iran appears to be the default mode of contact between South Asia, the Middle East, and the Mediterranean.

The next few chapters will continue to examine Afro-Eurasian connectivity through the lens of individual tropical and subtropical botanical transfers from South and Southeast Asia to the Middle East and the Mediterranean. I will draw attention first to the cultivation status of the crop in South Asia, which is either the place of origin or the clearinghouse so to speak, before turning to the spread of the crop to the Middle East and the Mediterranean. Each chapter also assesses the peculiar environmental and cultural constraints faced by the incoming crops.

268. Cappers 1998, 313–17, 2006, 78–79, 104–05, 132–33; Castillo, Bellina, and Fuller 2016; Castillo and Fuller 2010; Fuller et al. 2015; van der Veen 2011, 46–50; van der Veen et al. 2018; cf. evidence for Indian cotton in Thailand in the late first millennium BCE (Cameron 2010, 2015).

2

Wool from Trees: Cotton

Cotton fiber is the white unicellular epidermal seed trichome or root hair derived from four domesticated woody shrub species of the genus *Gossypium*: *G. hirsutum* (upland cotton), *G. barbadense* (sea island cotton), *G. herbaceum* (Levant cotton), and *G. arboreum* (tree cotton). The trichomes or hairs on domesticated cottons are of two types. The long seed trichomes (2.2 to 6 centimeters), also known as lint, on the ovule epidermis number between 13,000 and 21,000, depending on the type of cotton.[1] These long hairs, which are a distinctive feature of domesticated cottons, are not only a major source of yarn for textile manufacture but are also used in the production of a variety of fibrous equipment, like lamp wicks, ropes, and fishing nets. The shorter hairs (2 to 6 millimeters), known as fuzz or linters, are morphologically similar to the hairs produced by wild cotton species and remain stuck to the seed after ginning. In more recent times, this fuzz, which only starts to develop five to seven days after the growth of lint fibers, is harvested for its cellulose content, which is used in the manufacture of a variety of commodities, including plastics, paper, and explosives.

In addition to fibers, cotton seeds are a source of comestible oil used as lamp fuel, soap, lubricant, and emollient. But these uses are less common, owing to the rapid rancidification of cotton oil. Following oil extraction, the residual protein-rich cottonseed cake is also used as livestock fodder, while the plant stalks function as efficient fuel sources in tree-deficient ecosystems like Egypt.[2] The seeds, flow-

1. Bewley, Black, and Halmer 2006, 105.
2. Samuel 2001, 435–36.

ers, leaves, and root of the cotton plant have also been ascribed healing properties in South Asian and Middle Eastern medical traditions.[3] The pharmaceutical potential of cotton continues to elicit interest in modern medicine. An unfortunate bout of infertility across Chinese villages consuming cold-pressed cottonseed oil at the height of the Cultural Revolution (1966–76) led to the revelation that gossypol, a yellow polyphenolic compound in cottonseeds, functioned as a non-hormonal male contraceptive.[4] Cotton is presently not only the preeminent fiber crop but also ranks alongside tobacco (*Nicotiana tabacum*) as the most widely grown non-subsistence crop worldwide.[5]

Of these economically valuable taxa, *G. hirsutum* and *G. barbadense* are New World species which were unknown in the Old World until post-Columbian times, and now account for the bulk of global cotton fiber production. The other two species are of Afro-Asiatic origin and produce less lint. The New World cottons are tetraploid (they have four sets of chromosomes), which accounts for their greater lint production than the Old World cottons, which are diploid (with two sets of chromosomes). Old World cottons are consequently of negligible value in industrial production today and largely survive as relict crops. Nonetheless, the receptivity of the early modern textile industry to New World cottons owes entirely to the long-standing familiarity with Old World cotton species, whose anthropogenic diffusion to the Middle East and the Mediterranean from tropical and subtropical Asia in antiquity is our present concern.

SUB-SAHARAN COTTON: *GOSSYPIUM HERBACEUM L.*

Before we examine the spread of cotton cultivation to the Middle East and the Mediterranean, it is necessary to clarify the primary center of cotton domestication, since Old World cottons feature two genetically related but geographically disparate species (*G. herbaceum* and *G. arboreum*). *G. herbaceum* is a native of sub-Saharan Africa, and well-documented wild populations (*G. herbaceum* L. subsp. *africanum)* still exist in the far south of the continent.[6] Despite the presence of a native cotton species, the earliest secure archaeological and textual evidence for the exploitation of cotton for textile production in Africa

3. The Buddha, for instance, prescribes cotton leaves (*kappâsika*) to give an appetizing flavor to the food of the sick (*Mahāvagga* VI.5.5). For the medical uses of cotton in the medieval Islamic Middle East, see Watson 1983, 31, 163.

4. Segal 2003, 116–21.

5. Brite and Marston 2013, 41.

6. Fuller 2008, 3; Zohary, Hopf, and Weiss 2012, 108.

is rather late and coincides with Roman rule in North Africa in the first century BCE.[7]

The earliest cultivated cottons in North Africa are archaeologically attested in the form of bolls and seeds at Qasr Ibrim in Lower Nubia (presently in southern Egypt) in the first century CE.[8] Qasr Ibrim has also yielded a sizeable quantity of cotton textile fragments, presumably of local manufacture, which constitute 80 percent of all textile finds dating to the first four centuries CE.[9] Cotton textile fragments have also been recovered from several Nubian sites of that period and later, including Karanog, Meroë, Aksha, Qustul, Ballana, and Saï (southern Egypt and Sudan).[10] Vergil's casual reference to Nubian cotton (literally "wool") groves in a passage of the *Georgics* (II.120) celebrating the characteristic botanical produce of various exotic lands suggests that the cultivation of cotton was already commonplace in Nubia by the late first century BCE.[11]

Evidence for cotton cultivation in Egypt proper emerges most distinctively from Greek papyri, ostraca, and wooden boards dating between the second and fourth centuries CE from sites in the Dakhla (Kellis) and Kharga (Kysis/Dush) oases in the Western Desert. The same oasis sites have also yielded cotton in the form of seeds, bolls, and textile fragments, confirming the testimony of contemporary written documents. Also, at least one site in the Eastern desert, the Roman garrison of Maximianon (modern al-Zarqa), has yielded cotton seeds dating between the first and third centuries CE.[12] Pliny the Elder was also aware of cotton cultivation in

7. The reports of finds of cotton at late-fourth-millennium Nubian sites by Indian archaeologists (Chowdhury and Buth 1971, 2005) are unacceptable on methodological grounds and undoubtedly represent intrusive material from late Meroitic or post-Meroitic (formerly X-group) periods (Fuller 2002; Zohary, Hopf, and Weiss 2012, 108–09). The first reference to cotton in an Egyptian context concerns the Saite pharaoh Amasis's gift of an embroidered cotton corselet to the Spartans (Hdt. III.47). The claims for the presence of cotton in Egypt before the first millennium BCE are not convincing. Palynological analysis by French scholars who examined the mummy of Ramesses II in 1976 found traces of cotton pollen in the cavity of the king's mummy (Leroi-Gourhan 1985). The results of the study were controversial, as traces of tobacco were also identified in the mummy. The cotton pollen, like the tobacco, must be the result of post-excavation contamination (Buckland and Panagiotakopulu 2001). Cotton fibers interpreted as "pillow stuffing" have also been recovered from Tell el-Amarna, but these are not contemporary with its 18th-Dynasty founder Akhenaten (c. 1353–1335 BCE) and are thought to be either remains from a Roman or Coptic settlement or perhaps more recent material used by nesting gerbils (Thomas 1987).

8. Bouchaud et al. 2018, 393–94; Clapham 2019, 93–94; Clapham and Rowley-Conwy 2006, 2007, 2009; Pelling 2005, 2007; Rowley-Conwy 1989; Wild and Wild 2006; Wild, Wild, and Clapham 2008, 145.

9. Wild and Clapham 2007, 17.

10. Bouchaud et al. 2018, 387–88, 402; Yvanez and Wozniak 2019; Wild 1997, 289.

11. Vergil *G*. II.120.

12. Dakhla oasis (Kellis): Thanheiser 1999, 2002, 304, 308. Kharga oasis (Dush/Kysis, El Deir, Bagawat, Umm el-Dabadib): Jones and Oldfield 2006; Letellier-Willemin and Moulherat 2006; Pelling 2005, 406; Wild and Clapham 2007, 16. Maximianon: Letellier-Willemin 2019; Bouchaud, Tengberg, and Prà 2011, 406, 408. On the textual documents, see Bagnall 2008; Bouchaud et al. 2018, 403–05; Decker 2009, 200; Wagner 1987, 291–93; Wild 1997, 289; Wild, Wild, and Clapham 2008, 143–44.

Upper Egypt and Nubia and even remarks that the priests of Egypt favored cotton garments.[13] The scale of cotton cultivation outside Nubia and the Egyptian oasis settlements remains uncertain. A private letter of the second century CE, probably belonging to the Oxyrhynchite or Arsinoite nome, appears to suggest that cotton was common in parts of the Nile valley by the second century CE:

> Arethousa to Herakles: By all means send me through this shipment twenty drachmai's worth of good cotton thread. See that you do not neglect it, since your brothers have no outer garments, now that their cotton ones are worn out, and they need them, as you know, inasmuch as they spend all their time in the field. [14]

Cotton cultivation was also attempted further west: cotton seeds have been found in third-century CE Garamantine oasis settlements of the Fazzān region of southwest Libya.[15] It seems most likely that the exploitation of the native sub-Saharan species, *G. herbaceum,* only took place under the influence of imported cottons from the east.[16] The archaeobotanist Dorian Fuller even raises the possibility that the Kushite and Upper Egyptian cotton industry could have been exploiting the imported Indian variety of cotton, *G. arboreum,* rather than the local *G. herbaceum.*[17] The seeds and textile products of *G. herbaceum* and *G. arboreum* are virtually indistinguishable in archaeological contexts, so the question of the origin of the cultivated cotton in northeast Africa cannot be settled by morphological analysis alone.[18] DNA sequencing of cotton seeds recovered from a fifth-century CE context in Qasr Ibrim has plainly indicated that the locally used cotton was indeed of the African variety.[19] But this does not rule out earlier imports of Indian cotton fibers.

13. Pliny *HN* XIX.14, *HN* XIII.90; cf. Pollux *Onomasticon* VII.75.

14. SB 6.9026 (P. Mich. 1648), in Winter and Youtie 1944, 257–58.

15. Pelling 2005, 402–06, 2008, 50, 56, 58–59. AMS dating of one seed recovered from Jarma yielded c. 130–300 CE (Pelling 2013).

16. While the textual sources for cotton imports in Egypt provide a relatively early date (sixth century BCE), the earliest archaeologically verifiable cotton textile imports, found at the Red Sea ports of Berenike and Myos Hormos (Quseir al-Qadim), only date to the Roman period. The 1,060 fragments of cotton textiles found at Roman Berenike emanate from two distinct chronological horizons: a small sampling dates to the pre-Flavian or early Flavian phase (first century CE), while the bulk of the finds date between the late fourth and fifth centuries CE. The Indian provenance of these textiles is suggested by the Z-spun threads, the patterning and dyeing methods (e.g., blue check and dot rosette motifs on resist-dyed textiles), and the presence of Z-spun cotton threads on other South Asian imports, like Sri Lankan and South Indian beads strung on a Z-spun cotton string. Both sites have also yielded evidence for Z-spun Indian cotton sails (first to early second century CE), which were found in association with reinforcement bands and brailing rings. Also, six cotton mat fragments (first century CE) with Ghiordes knots, found at Berenike, probably represent remnants of sleeping mats used by Indian sailors. See Bouchaud et al. 2018, 414–15; Wild 1997; Wild and Wild 2001, 2008, 2014a, 211–27, 2014b, 100–04.

17. Fuller 2008, 19.

18. Fuller 2008, 3.

19. Palmer et al. 2012.

In any case, the earliest Greek sources on cotton, dating between the fifth and fourth centuries BCE, consistently associate the textile and plant with India and/or the Persian Gulf. Moreover, the loanwords for cotton in several Mediterranean and Middle Eastern languages (e.g., *pambakís, kárpasos, karpas*) point to an easterly mode of transmission. Even in the medieval period, when cotton was well known in the Mediterranean region, it remained, at the ideational level, an "eastern" textile *par excellence,* as the remark of the eleventh-century Arabic author al-Thaʿalabī demonstrates: "People know that cotton belongs to Khurāsān (northeast Persia up to the Oxus) and linen to Egypt".[20] We can thus confidently discount the likelihood that *G. herbaceum* and production centers in northwestern Africa had any role whatsoever in the dissemination of cotton cultivation to the Middle East and the Mediterranean prior to the first century BCE.

COTTON IN SOUTH ASIA

The other Old World species in question, *Gossypium arboreum* (figure 9), more commonly called tree cotton, is native to South Asia, and wild forms have been observed in the lower Indus (southern Sindh) and the hilly tracts of the central Deccan.[21] The original range of *G. arboreum* is hard to determine, since feral varieties appear to have spread across the Indian subcontinent together with the domesticated crop.[22] Dorian Fuller suggests that the distribution of wild cotton during the wet phase of the early to mid-Holocene might have encompassed parts of the Persian Gulf, including the littoral zone of Arabia, which would have acted as a natural bridge for the wild progenitors of *G. herbaceum* and *G. arboreum.*[23] This might explain the early and rather perplexing presence of cotton fibers and the impression of a Z-spun fabric in plaster at Dhuweila in eastern Jordan c. 4400 BCE.[24] But the cotton fibers at Dhuweila are not isolated, since cultivated cotton is already attested further east, at Mehrgarh, in the Pakistani province of Baluchistan, c. 5000 BCE. The Dhuweila find might therefore be a product of long-distance transactions of the Neolithic.[25]

20. Lamm 1937, 198.

21. Fuller 2008, 3; Hutchinson and Ghose 1937; Santhanam and Hutchinson 1974.

22. Fuller 2008, 3; Wendel 1995; Zohary, Hopf, and Weiss 2012, 108.

23. Fuller 2008, 3–4.

24. Betts et al. 1994. Spun cotton threads typically fall into two distinct traditions of spinning, that is, the process by which fibers are twisted into a sturdy continuous thread: Z-spun (right or clockwise spin), which is associated with India and the Persian Gulf region; and S-spun (left or anticlockwise spin), which is chiefly associated with Egypt and the Levant, where the spinning direction of cotton fibers mimics the spinning of flax, which has a natural S-twist. See Barber 1991, 67; Bouchaud, Tengberg, and Prà 2011, 415; Decker 2009, 199; Wild 1997, 289.

25. Boivin and Fuller 2009, 128.

FIGURE 9. Tree cotton (*Gossypium arboreum*), illustration from E. S. Ayensu, *Medicinal Plants of West Africa* (Algonac, 1978). Department of Botany Collections, Smithsonian National Museum of Natural History, Washington, DC (public domain).

A cotton-like fiber is also reported from the fourth-millennium burials (c. 3700–3200 BCE) of the Majkop culture at Novosvobodnaya, in the North Caucasus region.[26] Microscopic and histochemical analysis has demonstrated that three textile fragments from the Novosvobodnaya burials, now preserved in the State Hermitage Museum, have a "ribbon-like cellulosic structure" which is compatible with cotton. But, while it is clear that these fragments derive from a plant fiber, the textile specialists admit that further investigation is needed to confirm its precise identity.[27]

An additional note on textile taphonomy is pertinent at this juncture. Archaeological textiles, like all ephemeral organic matter, are subject to aggressive processes of deterioration. The spatial range of archaeological textiles is uneven, with a bias toward desert or alpine climatic zones. In other geographical zones, textiles that are carbonized, waterlogged, or mineralized through contact with metals like copper have the best chance of survival.[28] If differential survival were not enough, the identification of excavated textiles in varying states of decay remains challenging, and a great number of textile finds from the ancient Eurasian zone remain unidentified.[29]

The earliest archaeobotanical evidence for the use of cotton in South Asia is available in the form of mineralized thread in copper beads and uncharred seeds from Late (Ceramic) Neolithic or Early Chalcolithic Mehrgarh (c. 6000–4500 BCE) in Baluchistan, Pakistan.[30] Textile impressions of an unidentified fiber, perhaps cotton, have also been reported from contemporary levels at Mehrgarh.[31] And a fourth-millennium BCE grave at the site of Shahi Tump, in southern Baluchistan, has yielded a cotton string preserved in a carnelian bead.[32] These early finds from Baluchistan provide compelling evidence for the domestication of cotton in the pre-Harappan period.[33] Cotton is subsequently amply attested at various Mature to Late Harappan sites (2600–1700 BCE): as pseudomorphs or mineralized fibers at Mohenjo-daro; as pollen at Balakot (Sindh); and as seeds at Hulas, Harappa, Kunal, Banawali, Sanghol (Punjab), and Kanmer (Kutch).[34] From the early second millennium BCE on, cotton begins to appear at sites outside the Indus alluvial plains, particularly in Saurashtra, Rajasthan, and the Upper Ganges Valley.[35]

26. Shishlina et al. 2003.
27. Shishlina et al. 2003, 339.
28. Strand et al. 2010, 151–52.
29. Margariti, Protopapas, and Orphanou 2010, 522; Strand et al. 2010, 152.
30. Costantini and Biasini 1985, 24; Fuller 2002; Moulherat et al. 2002.
31. Lechevallier and Quivron 1981, 80.
32. Moulherat et al. 2002, 1399.
33. Tengberg and Moulhérat 2008, 6–7.
34. See Fuller (2008, 4) for references.
35. Fuller 2008, 4, 10.

Remarkably, although the archaeological evidence suggests the widespread use of cotton across northwestern India by the early second millennium BCE, cotton goes unnoticed in the early Vedic textual corpus. It seems that the Brāhmaṇa priestly elite of the Indo-Aryan-speaking community did not take an immediate interest in the use of cotton in ritual contexts. The material is strikingly absent from the earliest Vedic texts and only appears in the *Kalpasūtras*, late Vedic prose manuals on ritual practice.[36] The earliest appearance of cotton in Indian literature (Sanskrit *kārpāsa*, Prakrit *kappāsa*) appears to be in the *Āśvalāyanaśrautasūtram* (IX.4.17), the *Lāṭyāyanaśrautasūtram* (II. 6.1, IX. 2.14), and the *Gobhilagṛhyasūtram* (II.10.10) of the early first millennium BCE. The *kar-* prefix of the Sanskrit word *kārpāsa* points to a likely Austroasiatic origin for the word.[37] Meanwhile, *tūla*, another ubiquitous word for cotton in classical Sanskrit, is probably of Dravidian origin and derives from a word which originally described feathers (Proto-South Dravidian **tuu-*, cf. Tamil *tūval* "feather," *tuy* "cotton").[38] Both terms indicate that the pre-Indo-Aryan-speaking inhabitants of the Indian subcontinent were familiar with cotton, a fact confirmed by the archaeobotanical findings. Later Sanskrit lexicographers offer further terms for cotton. The most prominent of these are *tuṇḍikerī* ("possessing a beaked fruit"), *samudrāntā* ("reaching to the sea"), and *bhāradvājī* ("relating to the clan of *Bharadvāja*"), the latter only denoting wild cotton shrubs.[39]

COTTON IN ASSYRIA

Our examination of the westward dissemination of cotton will adopt a geographical approach, moving from Mesopotamia and Iran to the Mediterranean, roughly corresponding to the chronology of attestations for this crop. Any account of the spread of cotton cultivation from South Asia to the Middle East must begin with the well-known inscriptional records of the Assyrian king Sennacherib (704–681 BCE), whose efforts to enrich the gardens of his dazzling "palace without a rival" at Nineveh with exotic flora give us the earliest textual evidence for the cultivation of cotton in Mesopotamia:

> I planted alongside it (the palace) a botanical garden, a replica of Mount Amanus, which has all kinds of aromatic plants and fruit trees, trees that are the mainstay of the mountains and Chaldea (i.e. swamps of southern Iraq), together with cotton trees (lit. "trees bearing wool"), collected inside it.[40]

36. See Gonda 1977 for the content and dating of these texts.
37. Ayyar and Aithal 1964, 7–9; Fuller 2008, 16; Southworth 2005, 197, 223; Witzel 1999b, 36–37.
38. Fuller 2008, 15; Southworth 2005, 223.
39. Amarasiṃha, *Nāmaliṅgānuśāsanam* II.7.6.
40. RINAP 3/1 16 vii.17–21. Parallel passage: RINAP 3/1 17 vii.53–57.

> I created a marsh to moderate the flow of those waters and planted a canebrake in it. I let loose in it herons, wild boars and roe deer.[41] By divine will, vines, all kinds of fruit trees, olive trees and aromatic trees flourished greatly in (those) gardens (planted) on newly tilled soil. Cypress trees, sissoo trees and all kinds of trees grew tall and sent out shoots. The marshes thrived greatly. Birds of the heavens, herons whose homes are far away, made nests and wild boars and roe deer gave birth in abundance. I cut down sissoo trees and cypress trees grown in the orchards and marsh reeds from the swamps and I used them in the work required (to build) my lordly palatial halls. They picked cotton (lit. "trees bearing wool") and wove it into clothing.[42]

In spite of the rather laconic references to cotton in Sennacherib's inscriptions, several points can be adduced in regard to the origins, status, and extent of cultivation of cotton in ancient Assyria—present-day northern Iraq. Like the later Herodotean "wool from the tree" (*eíria apò xúlou,* III.47), cotton is named periphrastically in Assyrian records as *iṣṣū nāš šīpāti* or "wool-bearing trees." This highlights the most notable feature of this cultivar but offers little information on the routes or peoples through which cotton reached Assyria. But the reference to cotton in tandem with "trees that are the mainstay of the mountains and Chaldea" points to the likely geographic origin of the cotton shrubs in the Ninevite "botanical garden" (*kirimaḫḫu*). Since cotton could not be procured from the mountainous zones of the Assyrian Empire, this implies that Chaldea, the Babylonian swamp region in the extreme south of modern Iraq also called the Sealand (*māt tâmti*), was its source.[43] The second reference to cotton occurs in tandem with Sennacherib's attempts to recreate southern-Babylonian marshscapes (*agammu*) in Nineveh, reinforcing the connection of the cotton trees with this region.

Elsewhere in his inscriptions, Sennacherib claims that he "planted in great number all types of mountain vine, every type of fruit tree from all over the world, including spice and olive trees."[44] Since the Assyrian "world" (*adnātu*) at that time included several other territories to the east and southwest which are known to have grown cotton well before the Hellenistic period, we cannot entirely discount the Persian Gulf territories, particularly western Iran and peninsular Arabia, as potential sources for the cotton plant. Yet the case for a Babylonian intermediary seems more compelling when we examine the cuneiform sources and archaeological findings from Babylonia and the Persian Gulf region, which indicate a long

41. Literally the "bull of the forest." It could be a water buffalo or some other swamp-dwelling cervid species.

42. RINAP 3/1 17 viii.46–64. Parallel passage: RINAP 3/1 16 viii.29–51.

43. Zawadzki 2006, 27.

44. RINAP 3/1 17 viii.20–21; cf. RINAP 3/2 223 18b-21. For Assyrian gardens and collections of flora and fauna, see Dalley 2013; Foster 1998; Thomason 2005; 169–99; Wiseman 1983.

tradition of local cotton cultivation beginning sometime in the early first millennium BCE.

It appears that cotton growing in the Assyrian heartland was an experiment limited to the palace, since we do not hear of cotton cultivation outside of the royal botanical garden. Sennacherib's inscriptions from 696 and 694 BCE contain the most detailed accounts of Nineveh's reconstruction, and both mention the king's attempts at cultivating cotton.[45] This suggests that cotton was introduced into Nineveh sometime early in the reign of Sennacherib, perhaps after his first campaign in southern Babylonia (703 BCE) against Marduk-apla-iddina, a Chaldaean contender for the throne of Babylon. But there is archaeological evidence for cotton in Assyria before the reign of Sennacherib. The royal burials at the Northwest Palace of Kalḫu, associated with Yabâ, the wife of Tiglath-pileser III (744–727), and Atalia, the wife of Sargon II (721–705), have yielded a single fragment of a cotton textile (map 7). The textile was identified as cotton through optical microscopic examination of the fiber's cross-sectional profile and epidermal structure.[46] But the discovery of a cotton textile among funerary paraphernalia does not prove cotton cultivation, as it could well have been an import. The presence of foreign prestige goods in royal Assyrian burials is vividly confirmed by a fragmentary seventh-century text from Nineveh describing a royal funeral, perhaps that of Esarhaddon, equipped with exotic grave goods like a Babylonian black-fringed carpet and a gold Elamite headdress.[47]

Still, there remains a faint possibility that the cotton cultivated in Sennacherib's palace in Nineveh simply represents a continuation of a palace cotton industry that already existed in Kalḫu, the erstwhile imperial capital. In this respect, Pauline Albenda has reasonably suggested that the floral garland framing a row of winged genii in the eighth- or early-seventh-century BCE paintings recovered from Room 25 of the Assyrian residence at Til Barsip (Tell Ahmar) may represent the side profile of a split cotton pod.[48] And it is interesting that Sennacherib does not claim to be the first Assyrian king to procure cotton, though elsewhere in his inscriptions he readily acclaims and takes credit for other innovations.[49]

The longevity of the Assyrian palace cotton industry is unclear, since the extant Assyrian textual sources outside of Sennacherib's inscriptions do not refer to cotton. The Ninevite cottons have consequently been perceived, following an *argumentum ex silentio*, as a short-lived experiment.[50] Eckart Frahm suggests that

45. RINAP 3/1 16, 17.

46. Toray Industries 1996, 199–200.

47. Kwasman 2015; MacGinnis 1987.

48. Albenda 2005, 58–59. Irene Winter (2007, 378) notes that some of Albenda's interpretations represent "overdetermination based on single or limited samples."

49. E.g., RINAP 3/1 17 vi.89–vii.8.

50. Brite and Marston 2013, 43; Watson 1983, 38.

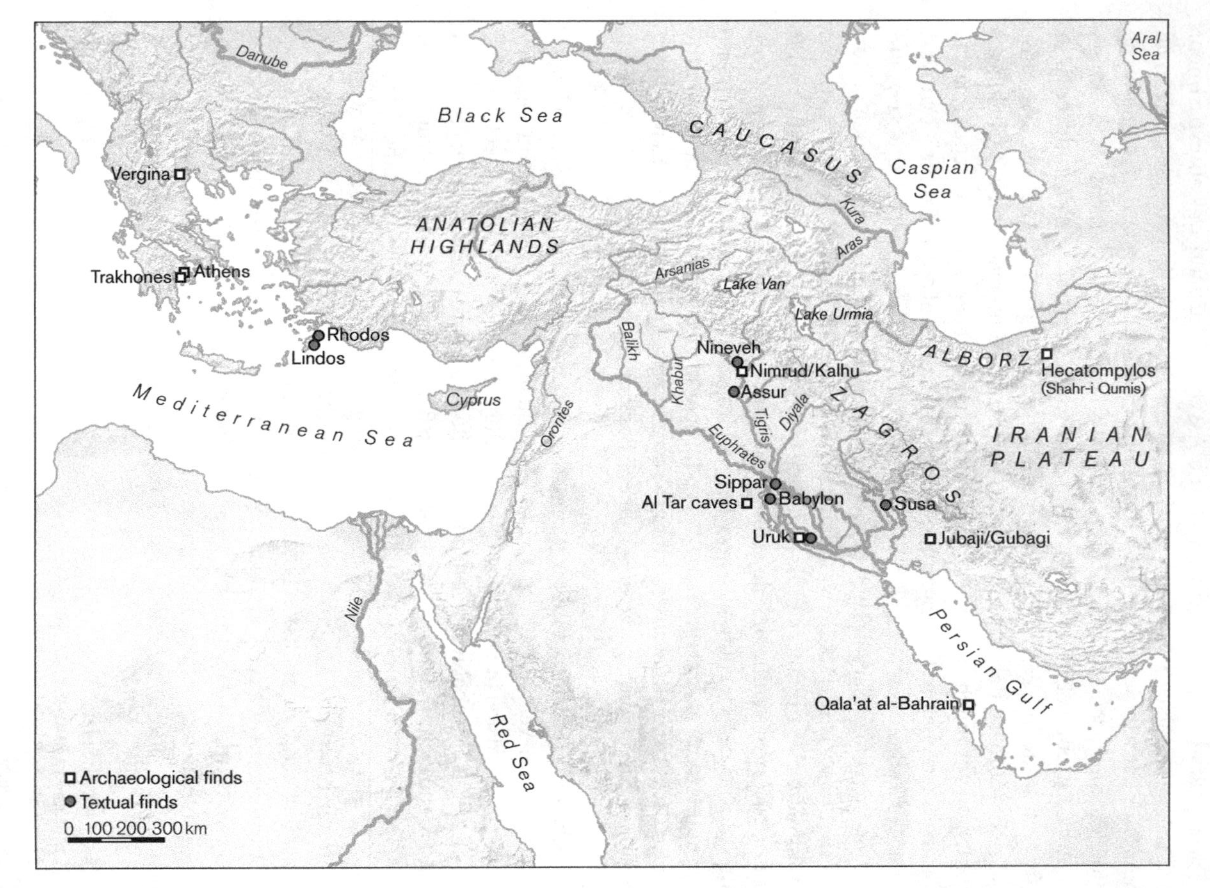

MAP 7. Ancient sites which have yielded evidence for cotton (*Gossypium arboreum*). © Peter Palm (Berlin).

Sennacherib's cultivation of cotton was abortive, since a later inscription, composed in the king's fourteenth regnal year (691 BCE), substitutes grain for cotton in the same passage.[51] But this section of the inscription is highly fragmentary, and the building reports concerning Nineveh become increasingly terse in Sennacherib's extant inscriptions after 694 BCE as other subject matters are added to the royal annals.

It appears unlikely that cotton cultivation would have been abandoned, since a complex network of irrigation canals and aqueducts, engineered by the king for his new capital, enabled a constant and ready supply of water for even the thirstiest of crops.[52] This was also true of Kalḫu, the former capital, and the Assyrian heartland as a whole, which benefited from intensive state-sponsored irrigation projects.[53] Sennacherib boasts that he was able to grow "every type of fruit" in Nineveh "thanks to the waters of the canals that I caused to be dug."[54] Significantly, the cotton plants in Nineveh are described as "trees" (*iṣṣū*) readily bearing fruit, presupposing a few years of good growth. The perennial *G. arboreum* rapidly grows to the height of a small tree (about two meters), so the description of cotton as a tree in the Ninevite gardens is not perplexing. Remarking on cotton in Arabia, Abū Ḥanīfa al-Dīnawarī, a ninth-century CE Persian writer in Arabic, notes that "the cotton trees (in the lands of the Kalb tribe) grow high until they look like apricot trees, and last twenty years."[55] But the wet and cold winters of northern Mesopotamia are a significant impediment to the large-scale cultivation of perennial cotton, which is not frost tolerant.[56] Consequently, it could be that the term *iṣu* in Sennacherib's inscriptions is used loosely, to describe a large annually planted summer crop. Overall, there appears to be insufficient evidence to assume that the limited cultivation of cotton in Assyrian palatial contexts was abandoned after the reign of Sennacherib.

Cotton was certainly deemed eminently suitable for Assyrian royal textiles already in the eighth century BCE, as evidenced by the presence of cotton among the funerary garments of the Assyrian queens buried at Kalḫu.[57] The unspecified workers in Sennacherib's Ninevite inscriptions, who are said to have sheared and woven cotton into textiles (*ibqumū imḫaṣū ṣubātiš*), are almost certainly the female weavers attached to the royal household (*ušpārāti ša šarri*), who exercised stringent

51. RINAP 3/1 18 viii 23; Frahm 1997, 277–78.

52. See e.g. RINAP 3/2 226 1–9 (Jerwan inscription B).

53. On Sennacherib's canal projects, see Fales and del Fabbro 2014; Ponchia 2014, 387–89; Reade 1978; Ur 2005.

54. RINAP 3/2 223 18b-21 (the "Bavian inscription").

55. Abū Ḥanīfa al-Dīnawarī, *Kitāb al-nabāt,* ed. Hamidullah 1973, 217.

56. Brite and Marston 2013, 43.

57. Toray Industries 1996, 199–200.

quality control in the procurement of textile raw materials.[58] To understand the prestige cotton accrued in the Assyrian heartland, let us examine the textual sources from southern Mesopotamia on the arrival and use of cotton.

COTTON IN BABYLONIA

The Identification of Kidinnû *with Cotton*

While the identification of cotton in Sennacherib's inscriptions is unambiguous, the interpretation of the evidence in Babylonian textual sources is less straightforward. The term *kidinnû,* appearing in some fifty Akkadian texts, has only recently been convincingly identified with cotton, and its semantic range encompasses the plant and its fibers, as well as woven textiles.[59] Despite unresolved problems in the etymology of *kidinnû,* there is now general agreement among Neo-Babylonian specialists that this term represents cotton.[60] In addition to *kidinnû,* an inventory of textiles from the Hellenistic Bīt Rēš temple in Uruk (253/2 BCE) uniquely attests to the use of the Sanskrit loanword *karpassu* (< *kārpāsa*).[61] The same Sanskrit word is also the source of the Hebrew *karpas* in the biblical book of Esther (I.6).

Before a thorough discussion of cotton cultivation in Babylonia and the potential trade pathways through which cotton was introduced there, I summarize and supplement the arguments adduced for the identification of *kidinnû* with cotton, as older scholarship has regularly confused this term with *kitû,* a word for linen or finished linen products.[62] While linen is of great antiquity in the Middle East, having been used there from at least the seventh millennium BCE, *kidinnû* is only attested in textual sources of the first millennium BCE.[63] It appears primarily as fabric for the garments of cultic statues in Babylonian temples and is a relatively valuable and rare commodity at least until the middle of the sixth century BCE,

58. RINAP 3/1 17 viii.46–64; Gaspa 2013, 230–32.

59. Muthukumaran 2016; Quillien 2019; Zawadzki 2006, 25–26. Louise Quillien (2019, 5) has convincingly established that the spelling of this word should be *kidinnû* and not *kiṭinnu* or *kitinnu.*

60. Graslin-Thomé 2009, 208–10; Kleber 2011; Payne 2011; Quillien 2019. Near Eastern and Mediterranean archaeologists have also been receptive to the identification (Sauvage 2014, 218; Völling 2008, 66–67). Francis Joannès (2014, 460) has omitted Zawadzki's identification and proposes that it is a woolen garment owing to the use of the determinative SÍG (wool). This identification is, however, untenable, as one text (CT 55, 834) unambiguously indicates that *kidinnû* was used a wool substitute and therefore cannot itself be wool.

61. NCBT 1244, 8, 11, 23; Beaulieu 1989, 69–72.

62. CAD, s.v. *kiṭinnû*; Löw 1924, 241; Oppenheim 1967, 251; Thompson 1949, 113.

63. Linen textile fragments dating to the seventh millennium BCE have been recovered from Nahal Hemar and Jarmo in Israel. Çatal Hüyük and Çayönü in Anatolia have yielded evidence for the use of linen textiles from a sixth-millennium BCE context. See Gleba 2008, 65; McCorriston 1997, 519; Potts 1997, 66–67, 117–19; Völling 2008, 66.

when its use becomes more widespread during the Neo-Babylonian period.[64] Most importantly, there are several texts which unmistakably discriminate between *kidinnû* and *kitû*, suggesting that the former was some other kind of material. The text on the clay case covering the so-called Sun Tablet of the ninth-century BCE Babylonian king Nabû-apla-iddina, surviving in a sixth-century BCE copy, enumerates several garment offerings used in the clothing ceremony (*lubuštu*) for the sun-god Šamaš in Sippar.[65] Among the textiles, the *ṣibtu* garment of Šamaš is followed by the word *kidinnû* in a position where one would expect an indication of the material. Other varieties of textiles, like *šalḫu*, *ḫullānu*, and *mēzeḫu*, are either preceded or followed by the determinative GADA (Akkadian *kitû*), which unmistakably indicates that they were linen products.[66]

Two archival texts from the Ebabbar temple of Šamaš in Sippar, dating to 560 BCE and 502 BCE respectively, and a Hellenistic marriage agreement from Babylon dating to 281 BCE, provide evidence for the word *kidinnû* prefixed with the determinative SÍG (*šīpātu*) for wool.[67] It is clear, however, that this is a case of association by resemblance, much like Sennacherib's description of cotton plants as "trees bearing wool" (*iṣṣū nāš šīpāti*).[68] One of the Sippar tablets (CT 55, 834) explicitly states that *kidinnû* recycled from the *lubāru*-garment of Šamaš was provided, instead of wool, to a weaver named Sūqaya for the manufacture of the *ṣibtu* garment for the bed of Šamaš.[69] Both the use of the determinative SíG and the substitution of *kidinnû* for wool indicate that the quality of the textile crafted from *kidinnû* was held to be similar to wool rather than linen. The similarity to wool is also borne out in other tablets from Sippar, where *kidinnû* is issued alongside red wool (*tabarru*) for the weaving of *ṣibtu* garments for the goddess Anunītu.[70] In fact, archival texts from both the Ebabbar temple of Sippar and the Eanna temple of Uruk indicate that the heavy *ṣibtu* garments, which were used to dress the deities and occasionally as a bedspread for the ritual bed, were only ever made from *kidinnû* or wool.[71] *Kidinnû* is also traded in bulk, as is typical of raw cotton or

64. Zawadzki 2006, 25.

65. BBS 36 = Zawadzki 2013, no. 175; Woods 2004.

66. Zawadzki 2006, 25.

67. Sippar: CT 55, 834 (Zawadzki 2013, no. 582); CT 55, 753. Babylon: CT 49, 165: 8.

68. The comparison of cotton with wool is perfectly logical. The fibers of cotton are short and resemble tufts of wool. The cotton textiles of antiquity had a coarser texture than those of today and even occasionally looked hairy like wool (Schmidt-Colinet et al. 2000, 8; Stauffer 2000, 249). A brocaded cotton weave from Palmyra was initially thought by textile experts to be a woolen garment (Schmidt-Colinet et al. 2000, 8 n36; Stauffer 2000, 249 n5).

69. Zawadzki 2013, no. 582.

70. CT 56, 5; Nbn 879.

71. BM 64060; BM 49188; IBK 165; Zawadzki 2006, 26.

wool, whereas unprocessed linen fibers (*kitû*) are unsurprisingly described as packaged and sold in bundles or hanks (*qātu*, lit. "a hand of").[72]

Although the cuneiform references to *kidinnû* increase in the sixth to fifth centuries BCE, judging by the much more frequent attestations for wool and linen in hundreds of documents, *kidinnû* represents only a small fraction of the textile production in southern Mesopotamia. This is strongly suggestive of rarity and novelty.[73] A few documents provide prices for *kidinnû*, and the extant evidence suggests that it more or less corresponded with the price of wool (at an average of four minas per shekel of silver), although it could also be more expensive in some cases (e.g., BM 79603 has two minas per shekel of silver, and Cam 250 has three minas per shekel of silver).[74]

The Etymology of Kidinnû

The Assyriologist Elizabeth Payne rightly notes that "etymology is not central" to the identification of *kidinnû* with cotton, but it is certainly worthwhile to explore the possible connections of this word with the better-known Arabic *quṭn* or *quṭun* (vulg. *qoṭon*), from which most European terms for cotton are derived.[75] The Arabic word *quṭn* is distinct from, but not altogether unrelated to, the Arabic term for flax/linen (*kattān*), which has cognates in Biblical Hebrew *ketoneṭ*/*kuttōneṭ* (linen tunic), Ugaritic *ktn*, Aramaic *kettān*/*kittān*, and Akkadian *kitû*. The West Semitic forms of this word were in turn borrowed into Mycenaean Greek as *kito* (classical Greek *khitṓn*).

Paul Pelliot notes that both flax and cotton lack proper Semitic etymologies and the terms relating to both categories have been regularly confused.[76] Following Theodor Nöldeke, he suggests that both words might derive from the same unidentified foreign base but were adopted into Arabic from different sources at varying dates.[77] It is not difficult, therefore, to envision the Babylonian *kidinnû* as the product of a similar process. While Semitic philologists generally agree that the Arabic *quṭn* and Akkadian *kidinnû* are of foreign origin, it has proven difficult to find lexical cognates resembling *quṭn* or *kidinnû* in any of the ancient recorded

72. E.g., CAD, s.v. *kitû*; BE 9 65: 20; PBS 2/1 150: 20; Quillien 2014, 277; Zawadzki 2006, 108–09. One text also describes raw linen as thick (*ka-b[a-ri]*: Zawadzki 2013, no. 367).

73. There are over fifty references to *kidinnû* in published texts, most deriving from the archives of the Ebabbar temple in Sippar and of the Eanna in Uruk (sixth to fourth centuries BCE) (Quillien 2019). It is likely that more references will be found as the study of the voluminous corpus of Neo-Babylonian archival texts, numbering in the tens of thousands, progresses.

74. Kleber 2011, 88; Payne 2011. The price of wool fluctuates throughout the Neo-Babylonian period, but not too drastically (Kleber 2010).

75. Payne 2011, 250 n2.

76. Pelliot 1959, 426.

77. Löw 1924, 242; Pelliot 1959, 426.

languages. Pelliot and Vollers have suggested that the Arabic *quṭn* is either of Egyptian or Indian origin.[78] Siegmund Fraenkel associated *quṭn* with the Coptic *kontion.*[79] Although South Asian vocabulary has few lexical items resembling *quṭn* or *kidinnû,* several terms in South Dravidian languages denoting a kind of coarse cotton cloth might provide a lead (Ta. *kiṇṭaṉ*, Ka. *giṇṭa,* Te. *giṇṭemu*).[80]

In addition to the Sanskrit loanwords *kirbās* and *kursuf* (< Sanskrit *kārpāsa*), several other words are used by medieval Arab authors for cotton, including *ʿuṭub* and *ṭūṭ.*[81] The forms *ʿoṭb* and *ajās* are also known from the Yemenite dialect. The diverse names for the cotton plant and its fiber in Arabic do not straightforwardly correlate with any known Indian lexical items. The early lexical diversity pertaining to cotton has led Andrew Watson to speculate that cotton must have been familiar in the Arabian Peninsula from a very early date.[82]

To summarize, the identification of the Babylonian *kidinnû* with cotton can be proposed on the basis of analysis of the relevant textual evidence, whose chronological distribution matches the archaeological attestations of the material in the wider Middle East. The identification of *kidinnû* with cotton does not rely on the equation with the Arabic *quṭn*, although it is entirely possible that both terms derive from the same non-Semitic, perhaps Dravidian, word.

The Origins and Scale of Babylonian Cotton Cultivation

Since cotton is rarely attested in a fiscal or agricultural context in Babylonian texts, it may be doubted as to whether the fabrics and raw fibers mentioned in the cuneiform texts are the result of local cultivation. In this respect, a list of taxes levied on local agricultural produce (*šibšu*) from the vicinity of Babylon (501–500 BCE) cites raw cotton using dry capacity measures alongside barley and emmer.[83] This economic document provides the strongest evidence for Babylonian cotton production rather than consumption of imported fibers. It also follows that cotton must have been growing in Babylonia well before Sennacherib introduces the plant in his palace gardens in Nineveh. The earliest mention of *kidinnû* is in the ninth-century BCE cultic regulations of the Babylonian king Nabû-apla-iddina, inscribed on the clay cover of the so-called Sun Tablet, which survives in a Neo-Babylonian copy.[84] Given the antiquarian fervor of the Neo-Babylonian kings and the conservative nature of temple ritual in this period, it is highly unlikely that Nabû-apla-iddina's

78. Löw 1924, 241; Pelliot 1959, 426.
79. Fraenkel 1886, 42.
80. DEDR 1540.
81. Ducène 2019, 4.
82. Watson 1983, 162 n3.
83. "4 gur (900 liters) of *kidinnû: šibšu* tax" (Dar. 533, 34).
84. BBS 36 = Zawadzki 2013, no. 175.

regulations were tampered with.[85] While cotton may have been familiar in Babylonia from as early as the ninth century BCE, the specifics of cotton's introduction to this region are unfortunately invisible in the textual record. Polities in the Persian Gulf probably mediated cotton's arrival in Babylonia from South Asia, and there is ample reason to suspect that the island of Bahrain was involved in this transmission (see below).

Assyrian kings who visited Babylonian temples in the course of their military campaigns in the ninth and eight centuries BCE may have first encountered cotton textiles draped on cultic statues in Babylonian temples.[86] The use of a strikingly new material, one that was whiter than wool or linen, is unlikely to have escaped Assyrian attention. The Assyrian interest in the textiles of divinities outside of Assyria bears out in at least one, albeit hostile, context. The booty enumerated in Sargon II's plunder of the immensely wealthy temple of Ḫaldi (the tutelary deity of the rival state of Urartu) in Muṣaṣir included nine *lubāru*-garments of the god, richly decorated with golden discs (*nipḫu*) and rosettes (*aiaru*).[87]

The close interactions of Assyrian kings of the ninth and eighth centuries BCE with Babylonian temples where cotton was already used in sacred garments reinforces the suggestion made earlier that Sennacherib's cotton plants in Nineveh, which bear a distinct association with southern Babylonia in his inscriptions, were not an innovation of the seventh century but had probably already featured in the palace gardens of earlier Assyrian rulers. The references to cotton in Mesopotamian records before the sixth century BCE are, regrettably, meager. Apart from the texts of Nabû-apla-iddina and Sennacherib, cotton is not found in the extant corpus of Mesopotamian texts until late in the seventh century, when an archival text from Sippar dating to the accession year of the Assyrian king Sîn-šar-iškun (c. 627 BCE) refers to the use of cotton for the *ṣibtu*-garments of the goddess Anunītu.[88] The dearth of early references to cotton in Babylonia owes entirely to the lacunose archival documentation of the early first millennium BCE. Rich archival documentation, especially that associated with the temples in Sippar and Uruk, starts only in the late seventh century BCE, on the eve of the Assyrian Empire's demise.

Beginning with the Neo-Babylonian period (626–539 BCE), cotton, which was previously only attested in the cultic and royal domain, appears to be readily available to a wider section of Mesopotamian society. A few texts from the early sixth

85. Beaulieu 1994, 2013.

86. Relations between the royal houses of Babylon and Assyria were particularly close in the ninth century, beginning with the reign of Shalmaneser III (858–824 BCE), who visited Babylonia and made offerings in Babylon, Borsippa, and Kutha (RIMA III A.0.102.5: v 3b–vi 5; A.0.102.6: ii 45–54; A.0.102.8: 24'b–29'a; A.0.102.14: 77–84; A.0.102.16: 50b–65'; A.0.102.58 etc.); cf. Adad-nārārī (810–783 BCE): RIMA III A.0.104.8.

87. Sargon's Letter to Aššur, ed. Mayer 2013, 136–37, line 386.

88. Zawadzki 2013, no. 556.

century BCE mention cotton, but the archival documentation for cotton only becomes substantial from the reign of Nabonidus (555–539 BCE) on. The archival texts, chiefly hailing from the Ebbabar temple in Sippar and the Eanna temple in Uruk, attest to the issue of cotton as income to various temple functionaries, including officials in charge of the rations of the king (*ša kurummati šarri*), brewers (*sirāšû*), an alphabet scribe (*sēpiru*), a measurer of staples (*mandidu*), and the overseer of the bakers (*šāpir nuḫatimmē*). Even a boatman (*malāḫu*) who coordinated the visit of the cultic statue of the god Šamaš to Babylon in 521 BCE was paid with cotton for his services.[89]

As private production of garments in households is not documented by cuneiform texts in this period, the extent of cotton's use outside the temple domain is unclear. A dowry of a woman named Amat-Nanâ from Babylon dating to 281 BCE contained a cotton textile (variant spelling: *kidinnītu*) among other valuables, including furniture and silver.[90] It is described as "desirable" (*ḫišiḫtu*) and assessed to be worth twenty-five shekels of silver alongside another garment.[91] The high valuation of the cotton textile might suggest that it was still a luxury well into the Hellenistic period. Unfortunately, it is not clear what kind of textiles the inventory documents. Martha Roth notes that not all the references to textiles in dowry lists denote clothes for the use of the betrothed but probably represent other kinds of goods like cushions, blankets, and curtains.[92] The valuation of textiles could also be affected by their workmanship and quality. The cotton textile from Amat-Nanâ's dowry is therefore a poor yardstick for the use of cotton in private households.

The presence of cotton textile fragments in a Neo-Babylonian (sixth-century BCE) jar burial from Uruk could suggest that it was no longer a luxury material.[93] This would accord well with the moderate prices listed for cotton in Neo-Babylonian texts. But the number of comparable burials with textile remains is small, so the finds are not a good indication of how widespread cotton was.[94] Also, ordinary denizens may choose to be buried with goods considered valuable.

Although the mechanics of cotton production in Babylonia are invisible in the textual record, we may safely assume, owing to the large capital investments, that

89. Zawadzki 2005 (BM 64557).

90. CT 49, 165:8.

91. Joannès 2014, 460; Roth 1989/1990, 31.

92. Roth 1989/1990, 30.

93. Van Ess and Pedde 1992, 257.

94. The cotton from Neo-Babylonian Uruk is the earliest extant archaeological attestation of the material in southern Mesopotamia. Otherwise, cotton is not encountered in the Iraqi archaeological record until the Parthian period (second century BCE). A large cache of fragmentary cotton textiles have been retrieved from the Partho-Sasanian burials dating from the second century BCE through to the fifth century CE at the artificially dug caves of aṭ-Ṭār, west of Karbala in the Iraqi southwestern desert (Fujii 1983/4, 177; Wild 1997, 287).

cotton cultivation was not the prerogative of the average householder but belonged to the institutional economy managed by large landholders like temples and the ruling elites. Cotton was probably grown in tandem with fruit trees and date palms in irrigated groves, just as it was in parts of the Persian Gulf until recent times. Ethnographic data from Oman indicate that cotton was frequently grown in open fields at the periphery of irrigated date palm gardens.[95] Even in India, cotton was grown either in open fields or alongside date palm gardens. Theophrastus notes in this respect that the Indians "plant them (cotton) in the plains in rows, wherefore, when seen from a distance, they look like vines. Some parts also have many date-palms."[96]

Foreign Cotton for the Babylonian Market?

The surge in references to cotton in Neo-Babylonian temple archives from Sippar and Uruk from the reign of Nabonidus on may not be entirely due to the vagaries of source survival. The extension of Babylonian political control to the Persian Gulf and Nabonidus's conquest of the north Arabian kingdom of Tayma could have provided greater access to cotton grown along the Persian Gulf littoral. An administrative document from the archives of the Ebabbar temple in Sippar dating to 547 BCE uniquely attests to two merchants, Nāṣir and Šulā, paying for the rent of property owned by the temple in cotton instead of silver.[97] The extremely low rate of exchange, over nine minas of cotton for one shekel of silver, suggests that the merchants were short on silver credit but had a large supply of cotton to pay in lieu.[98] While the merchants could have traded in locally produced cotton, it is also possible that they had access to supplies from foreign sources, notably localities in and around the Persian Gulf, like Elam in southwestern Iran, Dilmun (Bahrain), and perhaps even the Indus region. Even for linen, local supplies were augmented by foreign imports. For instance, one text from Sippar dating to 503 BCE refers to linen from Egypt, while another undated text of the Achaemenid period from Uruk refers to a linen textile from the Indian satrapy of Gandhāra (*gandarasanu*).[99] The use of both Egyptian and Indian linen (*hinduyin*) in the same context is incidentally recounted later in the Mishnah's provisions for the High Priest's garments on Yom Kippur.[100]

95. Richardson and Dorr 2003.

96. Theophr. *Hist. pl.* IV.4.8. Fuller suggests that the earliest cotton in the pre-Harappan Indus valley was cultivated alongside perennial fruit crops like vines and dates (Fuller 2008, 4; Fuller and Madella 2001).

97. Bongenaar 1997, 285; Graslin-Thomé 2009, 393–94.

98. BM 75584; Zawadzki 2013, no. 570.

99. Egyptian linen: CT 2, 2: 8; Joannès 1992: 182–3; see Quillien 2014, 275–76 on linen imports. Gandhāran linen: GCCI 2, 361: 8, 20; presumably a type of linen which natives of Gandhāra like the grammarian Pāṇini refer to as *umā* (Pāṇini, *Aṣṭādhyāyī*, IV.3.158).

100. Mishnah, *Yoma* 3.7.

COTTON IN IRAN

In light of cotton cultivation in Babylonia, it would seem likely that it also grew in the lowlands of Khuzestan, a well-irrigated region of southwestern Iran which is ecologically similar to the contiguous lower Mesopotamian plain. With a few exceptions, textiles from Iranian archaeological sites have not been methodically scrutinized with contemporary scientific methods, and the data for cotton are murky. Twelve fragments of textiles along with an undetermined number of carbonized fragments were recovered from a royal burial at Arǧān (Arjān), in eastern Khūzestān, dated to the late seventh or early sixth century BCE. English-language publications reporting this find have identified the textiles as cotton, but Susan Motaghed's original laboratory report in Farsi recognizes them as linen on the basis of microscopic examination.[101] An undetermined number of purportedly cotton textiles with gold appliqués have also been reported from the rich neo-Elamite burials found at the village of Gubagi (also spelled Jubaji), near Rāmhormoz, in Khuzestan Province.[102] The tombs belonged to Aninuma and Lārna, Elamite royal women related to the king Šutur-Nahhunte, who is dated to the early sixth century BCE.[103] Further north, the site of Hasanlu (Gilzanu) near Lake Urmia has yielded charred textiles dating between 1100 and 800 BCE which are predominantly woolen. But some of them consist of unidentified bast and vegetal fibers which may include cotton.[104] Elsewhere, five cotton textile fragments of uncertain dating were reportedly found in a bronze bathtub coffin, presumably of the Amlash culture (early to mid-first millennium BCE), in Gīlān Province of northwestern Iran.[105] A surer, albeit later, find of cotton in Iranian territory is represented by a fragment of a child's woolen garment lined with cotton which was recovered from Hecatompylos (Shahr-i Qumis) and dated to the first half of the

101. Alvarez-Mon 2010, 30–32, 2015. This is probably a case of mistranslation, as Susan Motaghed's (1990) report in Farsi identifies the textile as linen and not cotton. The cross-sectional and longitudinal profiles of the textile fibers published in this report correspond to linen (84). The Arǧān tomb textiles were found to be S-spun fabrics (Alvarez-Mon 2010, 30–32), and flax is typically spun counter-clockwise (S-spun).

102. Ahmadinia and Shishegar 2019, 164; Shishegar 2015, 67, 73, 243, 246; Wicks 2012, 27, 69, 2015, 27–8. Roonak Ahmadinia, a researcher at the Islamic Azad University's Department of Archaeology in Tehran, has informed me that the Gubagi (Jubaji) textiles were identified as cotton by the Research Center for Conservation of Cultural Relics, which operates under the wing of the Iranian Ministry of Cultural Heritage, Tourism, and Handicrafts. The most important book-length treatment of the Gubagi finds, including its textiles, is Arman Shishegar's Farsi excavation report published in 2015, which I have had no access to. None of the extant reports in English have specified the scientific methods used to identify the textiles as cotton.

103. On the dating, see Alvarez-Mon 2010, 49–50.

104. Kawami 1992, 17; Rubinson 1990.

105. Alvarez-Mon 2010, 33 n58.

first century BCE by a coin found alongside it.[106] In general, the early archaeological evidence for cotton in Iran is suspect and needs further study. And none of these finds would straightforwardly indicate local cultivation of cotton.

Turning to the textual evidence, while there is relatively abundant documentation for cotton in texts from Babylonia, the same is not true of neighboring western Iran. The textile terminology of the native Elamite language of southwestern Iran is poorly understood, and no lexical item corresponding to cotton has been identified in this language. Meanwhile, sources in Old Persian are largely restricted to formulaic royal inscriptions, and terms for cotton there have to be inferred from lexical borrowings in other ancient languages. In the case of cotton, its use in Persian contexts has to be reconstructed from sources in Greek and Hebrew. Old Persian appears to have had more than one term for cotton. The Iranian loanword *pambakís* in late Hellenistic Greek indicates the use of the restored **pampaka/i/u* in Old Persian for cotton, which eventually gave rise to the Middle Persian *pambag* and New Persian *pambah.*[107] The Hebrew *karpas,* on the other hand, reveals that Old Persian was also using Indic-derived terms for cotton, in this case the Sanskrit *kārpāsa.*

The use of cotton to manufacture soft furnishings like hangings, carpets, cushions, and mattresses appears to have begun in the Achaemenid period. Greek moralizers commenting on the wanton luxury of the Persian elite lifestyle frequently credit the Persians with all kinds of novelties, including lavish textiles.[108] The book of Esther, best described as a late Hellenistic historical novella on the Jewish experience in the Persian Empire, describes the Persian king's palace-garden in Susa as festooned with hangings of white (linen), cotton, and blue-colored textiles (Esther 1.6).[109] Charred textiles have indeed been recovered from the Achaemenid palaces in Persepolis and Susa, but the materials have yet to be analyzed.[110]

Nearchus (fourth century BCE) casually reports that the Macedonians used cotton for "mattresses and the stuffing of saddles."[111] This is most likely a case of Macedonians adopting Persian customs. It is unclear, however, whether he refers to a pre- or post-conquest custom. Since cotton has been identified as a constituent of a funerary pyre textile from the tomb of Philip II of Macedon (Tomb II) in Vergina, the former interpretation seems likely.[112] The Persian influences on Macedonian material culture, including a taste for cotton, probably date back to the early

106. Kawami 1992, 14; Met Museum 69.24.30.

107. Albiani 2006; Bailey 1979, 323–24; Brust 2005, 155; Hemmerdinger 1970, 64.

108. E.g., Aristoxenus, *Life of Archytas,* ap. Ath. XII. 545e.

109. On the date of the book of Esther, see Baumgarten, Sperling, and Sabar 2007, 215–20.

110. Vogelsang-Eastwood 2006, 226–28. On the subject of curtains, it should be noted that a kind of temple drapery or curtain made of cotton is also attested in an inventory of textiles owned by the Bīt Rēš temple in Hellenistic Uruk (NCBT 1244: 23; 253/2 BCE). See Beaulieu 1989, 69–72.

111. Nearchus, ap. Strabo XV.1.20.

112. Moraitou 2007.

fifth century BCE, when the royal house of Macedon accepted Persian suzerainty and contracted matrimonial ties with Persian nobility.[113]

Persian palaces stored immense quantities of luxury textiles and dyes, which were extracted as tribute and hence formed a part of state revenues.[114] Plutarch reports that during Alexander's conquest the treasury at Susa contained five thousand talents (over 125,000 kilograms) of purple dye from the Peloponnesian port of Hermione, which had been stored there for 190 years.[115] Cotton textiles, both imported and locally produced, almost certainly formed a part of the immense textile cache of Persian treasuries. Ctesias, the fourth-century Knidian physician at the Persian court, describes brilliant lac-dyed, perhaps cotton, textiles presented to the Persian king by the Indians:

> In India there is a large creature something like a scarab and red in color. At first sight it looks like a moth. It has huge feet and is soft to the touch. It is born on the amber-bearing trees and lives off its fruit. The Indians hunt this animal, crush it and extract from it a dye for purple textiles, their clothes and anything they want to dye or color in this way. They bring this dyed material to the Persian king. And the Persians find its appearance absolutely marvelous. Compared with the material produced locally in Persia, it is much better. According to Ctesias, it is so impressive because it is even brighter and more brilliant than the famous colored clothes of Sardis.[116]

Ctesias is incidentally the first author outside India to describe the red lac dye extracted from the resinous secretions of the parasitic lac insect (*Kerria lacca*), which is distributed from India through Southeast Asia. Lac (Sanskrit *lākṣā, jatu, alakta, raktā*) has been used in India from antiquity as a cosmetic, a dye, and a constituent of medicaments.[117] It is not clear what material the lac-dyed textiles Ctesias refers to were made of, but dyed cotton textiles were undoubtedly present in Achaemenid Iran. The widespread appeal of red-dyed cottons in antiquity is borne out in an archival text from Sippar dating to 531 BCE which attests to the dyeing of a cotton garment for the goddess Anunītu with madder (Akkadian *ḫūratu*).[118] No examples of such textiles have survived within the Persian heartland, but a remarkably well-preserved red-dyed cotton chemise from the Scythian burials of the fourth to third centuries BCE at Pazyryk, in the Altai region of

113. Gygaea, the daughter of the Macedonian king Amyntas I, was given in marriage to Bubares, a Persian nobleman (Hdt. V.21; VIII 136.1; Justinus, *Epitome of the Philippic History of Pompeius Trogus* VII, 4.1–2). See Paspalas 2006 for Persian influences on Macedonian material culture.

114. Curtius V.6.3.

115. Plut. *Alex.* 36.1–2.

116. Ctesias, *Indica,* F45py, ap. Aelian *NA* IV.46 (trans. Kuhrt 2007); cf. *Indica* 38–39.

117. See Lienhard 2007 for the use of lac in ancient India.

118. Madder was imported to Mesopotamia from the north Syrian and/or Anatolian region. A religious text from Aššur refers, for instance, to madder from Hatti (KAR 60, r. 9, cf. RAcc p. 4 : 24). See Zawadzki (2006, 41–44) for dyes in Babylonian texts and Gaspa (2013, 226) on dyes in Assyrian texts.

southern Russia, could reflect the kinds of cotton commodities in circulation throughout and beyond the Persian Empire.[119] Incidentally, madder- and lac-dyed cotton textiles, dating between 400 BCE and 50 CE, have also been recovered from remote cave burials in the Muktinath Valley of the Himalayas, in western Nepal.[120] The appearance of cotton textiles in this mountainous region bordering the Tibetan plateau corroborates Ctesias's otherwise outlandish remark that the Indians were also trading cotton with the cynocephalic (dog-headed) peoples living to the north.[121] Note should also be made of the finely preserved red woolen sleeve with white cotton bands recovered from a burial at the site of Djoumboulak Koum, a fortified settlement in the Keriya Valley of the southern Tarim basin (Xinjiang), dating between the fifth and third centuries BCE.[122] In all three cases, dry climatic conditions allowed better preservation of the textiles than southerly climes would.

COTTON IN ARABIA

While the archaeological finds of cotton in the Persian Gulf postdate the literary references to cotton in Babylonia, the linguistic and contextual evidence logically implies that cotton reached the latter region through the Persian Gulf. Cotton certainly reached the Middle East through maritime routes, since the beginnings of cotton cultivation in Central Asia appear to be rather late. The earliest archaeologically attested cotton seeds in Central Asia only date to the Hellenistic period. Soviet archaeological expeditions in Central Asia, which are still not satisfactorily published, reported the presence of cotton seeds dating to the early Hellenistic period (third to second centuries BCE) in the valleys of the Amu Darya, Syr Darya, Murgab, and Zeravshan Rivers (ancient Bactria-Sogdiana).[123]

The reconnaissance expedition along the Arabian side of the Persian Gulf commissioned by Alexander the Great in the winter of 325/4 or 324/3 BCE brought back reports of extensive cotton cultivation on the islands of Bahrain and Muharraq, and in eastern Arabia. Androsthenes of Thasos, who led the expedition, committed his observations to a now-lost treatise, "Voyage along the Indian Coast."[124] This text was subsequently consulted and quoted by his contemporary Theophrastus, who reports as follows:

119. Barkova and Polosmak 2005, 44; Good 2011, 147.

120. Alt et al. 2003, 1531.

121. Ctesias *Indica* 41. On the so-called dog-headed peoples of India reported in Greek ethnographic discourse, see White 1991.

122. Desrosiers 2000, 146.

123. Brite and Marston 2013, 44.

124. Ath. III.93b.

They say that the island (of Túlos, i.e. Bahrain) also produces the wool-bearing tree in abundance. This has a leaf like that of the vine, but small, and bears no fruit; but the vessel in which the wool is contained is as large as a spring apple, and closed, but when it is ripe, it unfolds and puts forth the wool, of which they weave their fabrics, some of which are cheap and some very expensive. This tree is also found, as we said, in India as well as in Arabia.[125]

Pliny provides a similar account of the cotton trees of Bahrain (Tylos), which clearly owes much to Theophrastus but is supplemented by a later account of Arabia by Juba of Mauretania (48 BCE to 23 CE):

On a more elevated plateau of the same island, we find trees that bear wool, but of a different nature from those of the Seres; as in these trees the leaves produce nothing at all, and, indeed, might very readily be taken for those of the vine, were it not that they are of smaller size. They bear a kind of gourd, about the size of a quince; which, when arrived at maturity, bursts asunder and discloses a ball of down, from which a costly kind of linen cloth is made. This tree is known by the name of *gossypinus*: the smaller island of Tylos (Muharraq), which is ten miles distant from the larger one, produces it in even greater abundance. Juba states, that about a certain shrub there grows a woolly down, from which a fabric is manufactured, preferable even to those of India. He adds, too, that certain trees of Arabia, from which vestments are made, are called cynæ, and that they have a leaf similar to that of the palm. Thus do their very trees afford clothing for the people of India.[126]

Androsthenes of Thasos crucially notes that cotton was used to manufacture both expensive high-quality fabrics and cheaper varieties on Bahrain, suggesting that it was easily accessible to a wide spectrum of the local population.[127] Juba's account in Pliny, which draws on mercantile sources, hints that local merchants thought the cotton of Arabia to be superior to that of India.[128]

It is not surprising that Bahrain (Akkadian Dilmun, Greek Túlos) and the neighboring island of Muharraq were major cotton production centers. Bahrain functioned for millennia as the natural point of exchange between Mesopotamia, Oman, and the Indus owing to its abundant supply of freshwater and favorable port facilities. Onomastic scrutiny of Dilmunite royal names (*Upēri, Hundaru*) appearing in Assyrian royal inscriptions of the late eighth and seventh centuries indicates close links with the Iranian coast as well, since the names are of Elamite affiliation.[129]

125. Theophr. *Hist. pl.* IV.7.7–8.
126. Pliny *HN* XII.38–39.
127. Ap. Theophr. *Hist. pl.* IV.7.7.
128. Roller 2008.
129. Potts 1990, 337, 2006.

Cotton from the Indus region must have been introduced into the groundwater-irrigated oasis settlements of Bahrain, and perhaps eastern Arabia, well before the ninth century BCE, when it first surfaces in Babylonian records. Yet the earliest archaeological traces of cotton cultivation only date to the period of Achaemenid rule, between the sixth and fourth centuries BCE. Seven carbonized seeds of cotton were recovered from a domestic dump of the Achaemenid period at the important Iron Age site of Qal'at al-Bahrain.[130] Cotton textile fragments were also found in a contemporary "bathtub" coffin from the same site.[131] The presence of small and light spindles at the Hellenistic levels of Qal'at al-Bahrain also suggests the spinning of delicate fibers like cotton.[132]

Beyond Bahrain, the archaeological evidence for cotton cultivation and use on the Arabian mainland thins out. The earliest definite evidence for cotton cultivation on the mainland comes not from eastern Arabia, where it would be expected, but from the Nabataean site of Hegra (Madâ'in Sâlih), in northwestern Saudi Arabia, on the overland incense trade route connecting South Arabia with Jordan. Some two hundred charred cotton seeds dating to the first century CE, presumably byproducts of ginning, were recovered from domestic contexts, and at least nine cotton textile fragments of Z-spun threads of varying quality were found in Hegra's monumental tombs dating between the first and third centuries CE.[133]

In light of the earlier textual and archaeological records for cotton in Bahrain, it seems probable that cotton reached northwestern Arabia from the Persian Gulf region. There may well have been other centers of cotton production in the groundwater-irrigated oasis settlements of mainland eastern and southern Arabia. Theophrastus's account certainly indicates so, as does the lexical diversity of terms relating to cotton in Arabic, which suggests a long-standing familiarity with cotton.[134] A fortified building at the site of Mleiha, in the UAE, has yielded cotton seeds and fibers dating to the second or third century CE, but the strontium isotopic signature of these finds suggests they may have derived from northwestern India rather than being local cultivars.[135] Future archaeological investigations in eastern and southern Arabia are likely to uncover more evidence for early cotton in this region.

130. Bouchaud, Tengberg, and Prà 2011, 410.
131. Bouchaud, Tengberg, and Prà 2011, 411; Højlund and Andersen 1994, 415.
132. Bouchaud, Yvanez, and Wild 2019, 19; Lombard 1999, 178–79.
133. Bouchaud, Tengberg, and Prà 2011, 412.
134. Theophr. *Hist. pl.* IV.7.7; cf. Pliny *HN* XIII.28.
135. Ryan et al. 2021.

COTTON IN THE MEDITERRANEAN

Egypt and the Levant

The earliest reference to cotton in an Egyptian context is provided by the Greek historian Herodotus rather than native Egyptian sources. Herodotus reports that the Saite pharaoh Amasis (570–526 BCE) gave a linen corselet (*thórax*) with gold and cotton figural embroidery to the Spartans and dedicated another identical corselet at the sanctuary of Athena in the Rhodian city of Lindos.[136] According to Pliny, the corselet of Amasis survived for over five hundred years in that sanctuary, and its intricate cotton-and-gold embroidery appears to have attracted a fair few visitors, much to the detriment of the object:

> Those, no doubt, will be astonished at this, who are not aware that there is preserved in the Temple of Minerva, at Lindus, in the Isle of Rhodes, the cuirass of a former king of Egypt, Amasis by name, each thread employed in the texture of which is composed of three hundred and sixty-five other threads. Mucianus, who was three times consul, informs us that he saw this curiosity very recently, though there was but little then remaining of it, in consequence of the injury it had experienced at the hands of various persons who had tried to verify the fact.[137]

The use of cotton in a diplomatic gift suggests that it was a rare and costly material. Since there is no reliable evidence of cotton cultivation in sixth-century BCE Egypt, the fibers must have been procured from either Mesopotamia or eastern Arabia, which were the closest sources for cotton. Cotton is absent from the Egyptian archaeological record until the Greco-Roman period. The earliest archaeologically attested cotton fibers were found on the linen wrappings of a Ptolemaic mummy (second century BCE), but even these are thought to be intrusive.[138] Otherwise, cotton appears in the archaeological record of the Egyptian Nile Valley only from the second century CE on, by which time it was already a well-established cultivar further south in Nubia and in the oases of the Eastern and Western Deserts.

The evidence for cotton cultivation in the Levant is limited. The meaning of *karpas*, a *hapax legomenon* for cotton in the biblical book of Esther (VI.1), was forgotten by rabbinical times (third to fifth centuries CE), since it is glossed as a kind of cushion or mattress in the commentaries.[139] In rabbinical texts, cotton grown in the land of Israel is referred to as "vine wool" (*ṣemer gepen*), since its palmate leaves

136. Hdt. III.47. Quilted armor corselets made of cotton (Skt. *sūtrakaṅkaṭa*) were known in India, but in Amasis's corselet cotton is used only in the embroidery. For Indian parallels, see the *Kauṭilīya Arthaśāstra* II.18.16, II.23.10.

137. Pliny *HN* XIX.12.

138. Cockburn 1986.

139. TB, *Megillah* 12a.

were thought to resemble those of the grape vine.[140] The first reference to cotton cultivation in the southern Levant is encountered in the Mishnah (*Kil'ayim* 7.2), indicating that it was taken into local cultivation by the second century CE. The earliest archaeologically attested cotton textiles in Israel and the Transjordan, found at Masada, Jerusalem (Jason's Tomb), Qumran, and Khirbet Qazone, date to the first century CE and are thought to represent imported materials.[141] Further east, in the Syrian desert, cotton is amply represented in the rich tombs of Palmyra dating to the first and second centuries CE.[142] Modest evidence for cotton dating to the early centuries CE was also found in Dura-Europos.[143] It is often readily assumed that these finds are imports from India, since the Palmyrenes played a major role as middlemen in the trade with India through the Persian Gulf.[144] It is equally likely that the cotton textiles of Palmyra, Dura-Europos, and the wider Levant were sourced from nearby production centers in Mesopotamia and eastern Arabia.

Greece and Rome

The earliest appearance of cotton in mainland Greece is coeval with the Achaemenid Empire. This is not surprising as Persian material culture was adopted in piecemeal fashion by Greek elite consumers.[145] Cotton fibers have been identified in textile fragments recovered from an Alcmaeonid family tomb (35 HTR73) dating to the late fifth century BCE in the Kerameikos cemetery of Athens.[146] Not far from

140. Mishnah, *Kil'ayim*. 7.2; TJ, *Kil'ayim* 2:3; Tosefta, *Shabbat* 10.2.

141. A reel of cotton, stratigraphically dated to between the twelfth and tenth centuries BCE, was reported from the site of Tell es-Saidiyeh, in the central Jordan valley (Shamir 2001, 126; Tubbs 1988, 41). The string was recovered during the re-excavation of the site in the late 1980s under the sponsorship of the British Museum. The find is unlikely to be intrusive, since it derives from a layer (Level XII) that was destroyed by fire and found in conjunction with charred debris (Tubbs 1988, 41). However, there is no information in the archaeological report as to how this string was identified as cotton or whether it was subject to scientific analysis at all. For the cotton fragment from Masada, see Sheffer and Granger-Taylor 1994. On the cotton hairnet from Jason's Tomb in Jerusalem, see Rahmani 1967, 93–94. The Nabataean cemetery at Khirbet Qazone has yielded a child's tunic in cotton and a baby's linen tunic with cotton sewing-thread (Granger-Taylor 2000, 150, 155). The Qumran caves have yielded cotton textile fragments which, if not intrusive, are thought to date between the first century BCE and 68 CE (Müller et al. 2004). For later third-century cotton finds from Nessana and En-Boqeq, see Bellinger and Pfister 1962; Sheffer and Tidhar 1991.

142. Forty-two fragments of cotton textiles were found in the tombs of Kitot (40 CE), Iamblik (83 CE), Elahbel (103 CE), and Grave 69. See Schmidt-Colinet et al. 2000, 8–9, 12; Stauffer 2000, 249; Wild, Wild, and Clapham 2008, 145.

143. Pfister and Bellinger 1945.

144. Seland 2014.

145. Miller 2004.

146. Environmental scanning electron microscopy established that both textiles (Y1, Y2) have flat and convoluted cellulose microfibrils, a morphological trait which could only be consistent with cotton. See Margariti, Protopapas, and Orphanou 2010, 525.

Athens, the village of Trakhones at the foot of the Hymettus range in Attica has also yielded a cotton textile fragment from a grave dating to the fifth century BCE.[147] We have already made note of the carbonized cotton textile recovered from the tomb of Philip II of Macedon (Tomb II) in Vergina.[148] The association of cotton textiles with Hellenic elites like the Alcmaeonid family of Athens and the Macedonian royal house indicates that they were luxury commodities, presumably acquired through long-distance exchange and diplomatic networks. There is no evidence that cotton cultivation was attempted in the eastern Mediterranean in the mid-first millennium BCE.

The rarity of cotton in the Mediterranean is best exemplified by the lack of a consistent term for cotton in classical Greek sources. Although Herodotus conceives of cotton periphrastically as "tree wool," much like the Assyrian *iṣṣū nāš šīpāti,* he uses various terms to express this concept: "wool from trees" (III.47), "fruit which bears wool" (III. 106), and "garments made from trees" (VII.65). Nonetheless, Herodotus's comment that cotton surpassed sheep's wool "in beauty and goodness" (III.106) suggests firsthand knowledge of the fibers and textile.[149] Theophrastus, or rather his source, Androsthenes of Thasos, speaks of cotton in India, Arabia, and Bahrain as "wool-bearing trees" in the Herodotean vein.[150] Theophrastus compares cotton leaves to those of the grape vine and the mulberry tree, its fruit to the spring apple, and the whole plant to the dog rose.[151] The description is, on the whole, accurate, but the eclectic comparanda invoked by Theophrastus make it clear that the cotton plant was a strange sight to the Mediterranean observer of the fourth century BCE.

By the Hellenistic period, Greek had adopted eastern loanwords for cotton. But they were not used regularly, and some scholars, like Aristobulus (fourth century BCE) and Eratosthenes (third century BCE), continued to describe cotton as wool from trees.[152] *Kárpasos,* loaned from the Sanskrit *kārpāsa,* is known in Greek from the third century BCE on. Manfred Brust suggests that the word may have been adopted for cotton when the Greeks came into direct contact with Indians during Alexander's campaign.[153] But the terms *kárpasos* and its Latin equivalent *carbasus* were not used consistently to designate cotton. They were regularly conflated with linen cloth. Even the *Periplus Maris Erythraei,* a first-century CE merchant-mariner's guide to western Indian Ocean trade, describes *kárpasos* (cotton) as the material out

147. Zisis 1955.
148. Moraitou 2007.
149. Hdt. III.106.
150. Theophr. *Hist pl.* IV.7.7.
151. Theophr. *Hist. pl.* IV.4.8, 7.7.
152. Aristobulus and Eratosthenes, ap. Strabo XV.1.20–21.
153. Brust 2005, 313.

of which Indian "linen cloth" (*othónion*) is made, for lack of other words.[154] While the Periplus is clearly referring to raw cotton as *kárpasos,* in most other cases *kárpasos* and the Latin *carbasus* were simply used as a generic term for a fine textile.[155]

Another loanword for cotton attested in Hellenistic sources is *pámbax* or *pambakís,* a borrowing from the *Old Persian *pampaka/i/u* (Middle Persian *pambag,* New Persian *pambah*).[156] The word is rarely used in Greek, and its earliest known appearance is in the work of the late Hellenistic epigrammatist Myrinus (second to first centuries BCE).[157] The *Suda,* a tenth-century Byzantine lexicon-encyclopedia, holds an entry for this word. The *Suda*'s sources on *pámbax* include Eudemus of Argos, a poorly known Greek lexicographer of the second century CE, and Myrinus himself, since it quotes his epigram.[158] The context in which *pambakís* appears in Myrinus's work evokes the rarefied world of a rich androgynous courtesan who possessed cotton dresses.[159] It appears therefore that cotton was still a fiber of prestige in the Mediterranean during the late Hellenistic period. Nonetheless, a single cotton seed recovered from the Hellenistic levels of Aşvan in eastern Turkey, if not intrusive, may indicate localized cultivation of cotton in parts of the eastern Mediterranean.[160]

Bambakoeidḗs ("like cotton"), an adjectival form derived from *pámbax,* is also used in the context of Dioscorides's account of the *akánthion,* either the cotton thistle (*Onopordum acanthium*) and/or the Illyrian thistle (*Onopordum illyricum*), whose leaves are covered by a white cottony down.[161] From this analogy, it follows that the cotton plant was not a complete rarity in the eastern Mediterranean of the first century CE. It is also likely that the Greek word *bámbakos,* attested in the *Antatticista,* a second-century CE Greek lexicon, as the name of a drug among the Cilicians, is another variation on *pámbax,* referring not to the fibers but to another part of the cotton plant used in medicine.[162]

Even though cotton was cultivated in the Middle East, the textile was strongly associated with India in Greek ethnographic discourses. Herodotus is the first to remark that the Indians wove garments from cotton.[163] The fine white cottons

154. *PME* 41. While *othónion* usually denotes a linen cloth, the *Periplus* consistently uses it to describe an ordinary-grade cotton textile; higher-grade cotton textiles were referred to as *sindṓn* (Wild and Wild 2014a, 214–15, 2014b, 101).

155. Brust 2005, 312–13; Wagler 1897.

156. Brust 2005, 155; Hemmerdinger 1970, 64.

157. Albiani 2006.

158. For fragments of Eudemus's *On Rhetorical Language,* see Niese 1922; *Suda,* s.v. *Bámbax.*

159. *Greek Anthology* VI. 254.

160. Nesbitt et al. 2017, 117.

161. Dioscorides, *Mat. Med.* III.16; cf. Oribasius, *Collectiones medicae* 11.21; Serv. Aen. I. 649.

162. *Antatticista,* ed. Bekker 1814, 85.

163. Hdt. III.106.

(muslins) of Indian manufacture must have been distinctive enough for the Greeks to recognize that the Indian soldiers led by Pharnazathres in the army of Xerxes were dressed in cotton.[164] Early Hellenistic authors, including those like Nearchus who had firsthand experience in India during Alexander's campaign, also took notice of Indians wearing white cotton garments.[165]

Apart from the loanwords *kárpasos* and *pambakís,* Koine Greek also used the terms *bussós, othónion,* and perhaps *túlē* for cotton.[166] Like *kárpasos,* the semantic range of these terms extended well beyond cotton to include linen and other textiles. The Semitic loanword *bussós* (Akk. *būṣu,* Heb. *būṣ,* Aram-Syr. *būṣā*) originally referred to very fine linen textiles, but in Greek it was applied to cotton and other fine-textured textiles, like those manufactured from the fibers of the *Pinna nobilis* mollusk.[167] Pollux, observing cotton cultivation in Egypt in the late second century CE, refers to cotton as *bussós,* as does his contemporary Philostratus.[168] Given the extended semantic field of these terms, it would be rash to draw any conclusions regarding the availability of cotton in the ancient Mediterranean based on later Greco-Roman references. Pausanias's references to *bussós* growing in Elis are ambiguous and probably do not refer to cotton.[169] Only in the local documents (papyri, ostraca, wooden boards) from the cotton-producing centers of Roman Egypt is the compound word *erióxulon* or *ereóxulon* (lit. "tree wool") consistently used to designate cotton.

Leaving aside the problematic references to *carbasus* in early Latin authors (e.g., Caecilius Statius in Pausimachus, c. 190 BCE), Pliny is the first Latin author to offer substantial comments on cotton.[170] Pliny uses the terms *xylon* (Greek *xulon* "wood") and *gossypinus* or *gossypium,* a word of unknown etymology, to describe the cotton plant.[171] He also uses the term *xylinum* for the fabric.[172] Otherwise, he is content to describe it as wool (*lana*) from plants.[173] Pliny never saw a cotton plant,

164. Hdt. VII.65.

165. Arr. *Ind.* XVI.1; Curtius VIII.9.15, 21, 24; Strabo XV.1.20, XV.1.71.

166. Brust 2005, 313.

167. Burke 2012; Oppenheim 1967; Wild, Wild, and Clapham 2008, 143.

168. Pollux *Onomasticon,* 7.75–76; Philostr. *Vita Apoll.* II.20.

169. Pausanias V.5.2, VI.26.6, VII.21.14.

170. Caecilius Statius, ap. Nonius M.548, ed. Lindsay 881.

171. Pliny *HN* XIX.14, XII.38–39. Semitic philologists have suggested a link between *gossypium* and the Arabic *kursūf* for cotton. *Kursūf* certainly derives from the Sanskrit *kārpāsa* via Persian (Ducène 2019, 4). *Gossypium* likewise probably derives from the Sanskrit *kārpāsa,* perhaps via a Greek intermediary (hypothetical form: **korsípion*). See Fraenkel 1886, 145; Löw 1924, 236.

172. Pliny *HN* XIX.14.

173. The Latin geographer Pomponius Mela, a contemporary of Pliny the Elder, also describes cotton in India as wool from trees (III.62). The oldest extant Latin authority to cite cotton as "wool-bearing" trees is Varro (first century BCE), who cites Onesicritus, a companion of Alexander, as authority (ap. Serv. Aen. I.649). His younger contemporary Vergil also refers to cotton as "wool" in a Nubian context (*G.* II.120).

and his descriptions of the plant and the fabric are hopelessly confused. In one instance he claims that the Nubians and Indians made thread out of fleshy fruits, and elsewhere that the "wool" was derived from the leaves.[174] Pliny's descriptions of what appears to be cotton in Central Asia are conflated with silk.[175] He is not alone in this regard, since Strabo also claims that *sērikà* or silken cloth was made from the bark of some plants, while Vergil speaks of the Seres combing silky fleece from leaves.[176] But Pliny does recognize that he is discussing the same plant which grew in Egypt, Nubia, Arabia, and India.[177]

Cotton textile finds in the western provinces of the Roman Empire mostly date to the third and fourth centuries CE, suggesting that it was a rare textile in earlier periods.[178] Clearly cotton was not cultivated in the western Mediterranean in antiquity, since Latin authors remark of the plant in unfamiliar terms and the imported textile was regularly confused with silk and other finely woven wares. Hence Posidonius's remark that the (New) Carthaginians wove beautiful garments from the bark of the thorns of a tree must refer to some kind of barkcloth rather than cotton.[179]

THE APPEAL AND LIMITS OF COTTON CULTIVATION

As cotton absorbs dyes better than bast fibers like linen, it is not surprising that it appealed to the Middle Eastern and Mediterranean textile aesthetic, which valued multicolored fabrics.[180] Also, white undyed cottons were purer and more luminous than wool or linen in coloration, making them more attractive in ritual and funerary contexts.[181] Herodotus avows that cotton is superior to wool in beauty and goodness, and Nearchus also claims that it appears brighter than any other textile.[182] The early use of cotton for the manufacture of sacred and royal garments in

174. Pliny *HN* XIX.15, *HN* XVI.88.

175. Pliny *HN* VI.54–55.

176. Strabo XV.1.20; Vergil *G.* II.121.

177. India: *HN* VII.25, XII.25. Bahrain: *HN* XII.38–39. Egypt and Nubia: *HN* XIII.90, XIX.14.

178. The earliest cotton textile remains in the western Mediterranean are the fragments recovered from Pompeii (first century BCE to 79 CE; Borgongino 2006, 72–73). Otherwise, cotton textile fragments have typically been reported from late Roman contexts in sites like Munigua, Aquincum (Budapest), Rome, Pisa, Chew Stoke, and Damblain. See Raschke 1978, 651, 907–09; Wild, Wild, and Clapham 2008, 145.

179. Posidonius, ap. Strabo III.5.10.

180. Barber 1991, 33; McCorriston 1997, 523. Wool, however, absorbs dyes even better than cotton (Dalley 1991, 120).

181. Zawadzki 2006, 28.

182. Hdt. III.106; Nearchus, ap. Arr. *Ind.* XVI.1. Nearchus or perhaps Arrian, in this instance, also suggests that cotton may appear brighter to the observer since it is worn by dark-skinned Indians.

Mesopotamia was perhaps seen as enhancing the divine radiance (*melammu*) and terrifying splendor (*namurratu*) of both god and king.[183]

The ancient Mesopotamian aesthetic preference for white and luminous cotton garments in ritual contexts was inherited by the Islamic world.[184] According to the *Muwaṭṭa'* of Mālik ibn Anas, the earliest collection of *hadith*, the Prophet Muhammad was buried with three white cotton cloaks from Saḥūl (*saḥūliyya*), a town in Yemen renowned for these textiles.[185] Cotton was held in equally high regard in Zoroastrian tradition, even over silk, which was tainted by its association with the obnoxious worm.[186] The *Dādestān ī Mēnōg ī xrad* (Judgements of the Spirit of Wisdom), a didactic Zoroastrian text transmitted orally for centuries and then written down (in Middle Persian) sometime in the late Sasanian period, avows for the ritual purity of cotton over that of silk:[187] "Of the garments that people wear, polychrome silk is good for the body and cotton for the soul, because polychrome silk arises from a noxious creature, and the nourishment of cotton is from water and its growth from earth."[188]

Laundering cotton is also much easier than laundering wool, particularly as the latter displays a marked tendency to shrink. Cotton clothing also easily adapts itself to a variety of climes, unlike wool. But despite the promising qualities of the new fiber, cotton remained a marginal element in the overall fiber output of the ancient Middle East. Sheep's wool (Akkadian *šipātu*) and to a lesser extent goat's wool (Akkadian *šipāt enzi* or *šārtu*) remained the most important materials for textile manufacture.[189] Even linen (Akkadian *kitû*), despite having been known for millennia, was far less common a material than wool.[190] Hartmut Waetzoldt estimates that linen accounted for only 10% of the textile production in the voluminous corpus of economic texts from the Ur III period (2112–2004 BCE).[191] The status of linen in the Middle East of the early first millennium BCE was no different. Louise Quillien's study of first-millennium BCE Babylonian textual sources concerning linen suggests that it remained of secondary importance at least until the Achaemenid period, when private entrepreneurs actively promoted flax cultivation.[192]

183. On Mesopotamian "aesthetics of luminosity" discussed with Indian parallels, see Winter 1994.

184. Halevi 2007, 87, 95.

185. Halevi 2007, 85.

186. Andrés-Toledo 2013, 28.

187. Tafażżolī 1993.

188. *Dādestān ī Mēnōg ī xrad* 16.64–66 (Andrés-Toledo 2013, 28).

189. On the use of wool in the ancient Middle East and the eastern Mediterranean, see Breniquet and Michel 2014.

190. On the history of linen cultivation in the Middle East and Europe, see Zohary, Hopf, and Weiss 2012, 101–06.

191. Waetzoldt 1980–83, 585.

192. Quillien 2014, 273, 292.

The chief risk in cultivating fiber crops like cotton or flax is the reduction of arable land and the diversion of labor and water resources from subsistence crops. Fiber crops demand a greater labor investment than the rearing of caprids for wool production.[193] Caprid husbandry typically exploits agriculturally marginal lands like steppe zones, desert meadows, hilly flanks, and highland pastures in summer. The herd's consumption of post-harvest stubble and provision of manure make caprids a welcome seasonal presence in cultivated fields. Caprids also provided meat, tallow, milk, horn, and hides. Consequently, the cost- and labor-effective pastoralist strategy prevailed over fiber crops in the Middle East of the first millennium BCE. Nonetheless, cotton was easily incorporated into the Middle Eastern agricultural repertoire, as the labor regimes and irrigation infrastructure required for cotton cultivation were already in place. Flax provided a model for the cultivation of cotton. Mesopotamia and adjacent regions, which had a long and illustrious history of textile production, applied the prevailing knowledge of fiber processing, textile production, and ornamentation to the new fiber.

Yet the agricultural practice associated with cotton was not identical to flax. Cotton has a long growing season (about two hundred days) and is a water-intensive crop in the early growing phase, requiring at least 500 millimeters of rainfall.[194] In India this is provided by the seasonal monsoon, but in Mesopotamia and the eastern Mediterranean cotton cultivation would be entirely dependent on irrigation in the summer months. During the last two months, the plant also needs drier conditions to allow the fibrous fruit to grow free of decay.[195] The medieval Iberian agronomist Ibn Baṣṣāl of Toledo (d. 1105) underscores the meticulous care required for cotton cultivation in a Mediterranean context:

> Watering should be stopped until it grows to the length of a finger or about a hands' breadth. Then it should be tended, pruned, straightened, and moved again and again, then watered, then singled and weeded, then watered, continuing this practice until the beginning of August. It should be watered every fifteen days. At the beginning of August watering should stop, for fear lest the plant go soft . . . giving no produce.[196]

The dense irrigation networks made Mesopotamia an ideal environment for the growing of cotton, since the water supply could be easily regulated. Cotton is a relatively versatile crop in spite of its demands on water and labor resources. Tree cotton tolerates a range of environmental stresses, including saline soils, periodic droughts, and slight temperature fluctuations, ideally requiring a minimum of

193. McCorriston 1997, 523–24.

194. Fuller 2008, 4,6; van der Veen 2011, 89; Wild and Clapham 2007, 16; Wild, Wild, and Clapham 2008, 144.

195. Burkill 1997; Langer and Hill 1982, 262; Robbins 1931, 497.

196. Ibn Baṣṣāl, *Dīwān al-filāḥa*, trans. Lewis 1976, 146–47.

15 °C.[197] While annualized photoperiod-neutral forms of cotton which are able to thrive in temperate climatic zones are a first-millennium CE development, Sennacherib's description of cotton cultivation in Nineveh suggests the presence of a hardier strain of the perennial cotton crop in the early first millennium BCE.[198] Cotton is also known to grow in marginal environments like the arid plains of Khorasan and Golestan in Iran, and the saline littoral zone of Gujarat and Baluchistan.[199] It is on account of the latter adaptation that cotton is also called *samudrāntā* ("reaching to the sea") in Sanskrit. While summer cultivation of cotton in the Middle East meets its climatic requirements, agricultural pests may have been a significant concern for ancient cotton cultivators in this region. The high cellulose content of cotton fibers makes them particularly attractive to virulent local pests like the Egyptian stemborer (*Earias insulana*) and locusts, in particular the desert locust (*Schistocerca gregaria*) and the Moroccan locust (*Dociostaurus maroccanus*).[200]

It is the processing and not the growing stage of cotton which is far more painstaking, since it involves several time-consuming processes, beginning with picking, dehusking, ginning (removal of seeds), carding or bowing to align the fibers, spinning, and finally weaving. Fiber crops are by nature cash crops, as they yield more produce than is necessary for the consumption of individual households. The complicated processing of the fibers also means that cotton cultivation must involve a large organized community of cultivators and weavers. It comes as no surprise, therefore, that cotton is first attested in the context of an institutional textile industry, supported by the palace in the case of Assyria and the temple in the case of Babylonia.

The fledgling local production of cotton in the ancient Middle East was ultimately never large or prestigious enough to compete with the cotton textiles of India, which were highly valued in the trading system of the ancient Indian Ocean world. The *Periplus Maris Erythraei* of the first century CE bears testimony to the continued import of Indian cotton textiles.[201] Perhaps the quality of Middle Eastern cottons fared poorly against their Indian counterparts in antiquity. In India, the creation of extraordinarily fine cotton fabrics began with a painstaking mode of spinning the thread. Using a very light spindle under humid conditions, "the hand-spinners of India were able to stretch a single pound of cotton into well over

197. Bouchaud, Yvanez, and Wild 2019, 3, 6; Bouchaud, Tengberg, and Prà 2011, 414; Brite and Marston 2013, 41.

198. Bouchaud, Yvanez, and Wild 2019, 5–6; Brite and Marston 2013, 41–42, 44; Watson 1983, 162 n2. Primitive cottons are photoperiod sensitive: flowering is triggered by seasonal changes in day length. Cotton requires short and mild winter days to trigger flowering (Brite and Marston 2013, 41).

199. Brite and Marston 2013, 41.

200. Samuel 2001, 381. On locusts in the ancient Middle East, see Radner 2004.

201. E.g., *PME* 6, 14, 31–2, 41, 48, 51, 61.

200 miles of thread, a feat not possible on the best of modern machinery."[202] Felicity and John Wild, who examined Indian cotton textiles at the Red Sea site of Berenike, note that "some pieces were exceptionally fine, and, after two thousand years in the ground, have the texture and appearance of old hand-made paper."[203] The tenth-century CE Arab traveler Sulaiman reported that the muslins of eastern India were so fine and delicate that a dress made of this fabric could pass through a signet ring.[204] One variety of diaphanous cotton produced in Bengal was named "evening dew" (*śabnam*), as it was reputed to become invisible when laid over dewed grass.[205] Such diaphanous cottons were also known in antiquity. The *Āyāraṅga-sutta*, a Jain canonical text (fourth to third century BCE), refers, for instance, to a "pair of robes so light that the smallest breath would carry them away."[206] It was perhaps the ethereal quality of the finest Indian cottons that captivated the earliest Middle Eastern and Mediterranean consumers of this new fiber and prompted them to attempt its cultivation in spite of environmental and economic impediments.

The ubiquity of cotton in our time makes it difficult to recognize the novelty or uniqueness of this fiber. The ninth-century CE Arabic poet al-Buḥturī's ruminations on the unfading radiance of the ruined Sasanian palace at Madā'in (Ctesiphon) invoke, among other things, luminous white billowing robes of cotton, an image perhaps familiar to the earliest connoisseurs of cotton textiles in the Middle East:

> It was not disgraced by the robbery of carpets
> of silk brocade, or plunder of curtains of raw silk;
> Lofty its battlements soar,
> Raised on the summits of Radwa and Quds;
> Clothed in white, so that
> you see of them but cotton robes.[207]

202. Barber 1991, 43.
203. Wild and Wild 2014b, 102.
204. Elliot 1867, 5.
205. Chakraborti and Bari 1991, 57.
206. *Āyāraṅga-sutta*, trans. Jacobi 1884, 196.
207. Serrano 1997, 83.

3

The Golden Grain: Asiatic Rice

Asiatic rice (*Oryza sativa*), the proverbial golden grain, is a native of tropical and subtropical Asia (figure 10). Alongside wheat and maize, rice accounts for almost half of global calorie intake.[1] The significance of rice to Asian diets is best reflected in the synonymity of words for rice and food across South, Southeast, and East Asia. Domesticated rice, derived from the wild progenitor *Oryza rufipogon*, has four subspecies distinguished along phenotypic and genotypic lines: *indica, japonica, aus,* and *aromatic.* Of these, the *indica, aus,* and *aromatic* varieties find their origin in the lowlands between the middle and lower Ganges plains and the lower Brahmaputra basin, while the broad, thick-grained *japonica* originates in southern China.[2] The precise chronology and trajectory of rice domestication remains mired in scholarly disagreement, but a collective reading of the genomic, linguistic, and archaeological data favors three protracted, independent domestication events for *japonica* in the Yangtze valley (c. 4500–4000 BCE) and *indica* and *aus* in northeastern South Asia (c. 2500 BCE), followed by a history of introgression.[3] The genetic history of the crop is further complicated by post-domestication

1. Reeves, Thomas, and Ramsay 2016, 3. According to the data collected by Prescott-Allen and Prescott-Allen (1990, 368), rice ranks second only to wheat in the global production and consumption of cereal crops.

2. Londo et al. 2006.

3. Civáň and Brown 2017, 2018; Civáň et al. 2015; Fuller 2002, 297–300, 316, 2003b, 194–96, 2006a, 39–44, 2007, 396–99; Fuller, Castillo, and Murphy 2017, 714–15; Fuller et al. 2010; Nesbitt, Simpson, and Svanberg 2010, 320–22; van Driem 2017. Some researchers have interpreted the genetic and archaeological data relating to early rice cultivars as suggesting that the *indica* subspecies of rice was not an independent domesticate but the result of hybridization between domesticated *japonica* rice from East

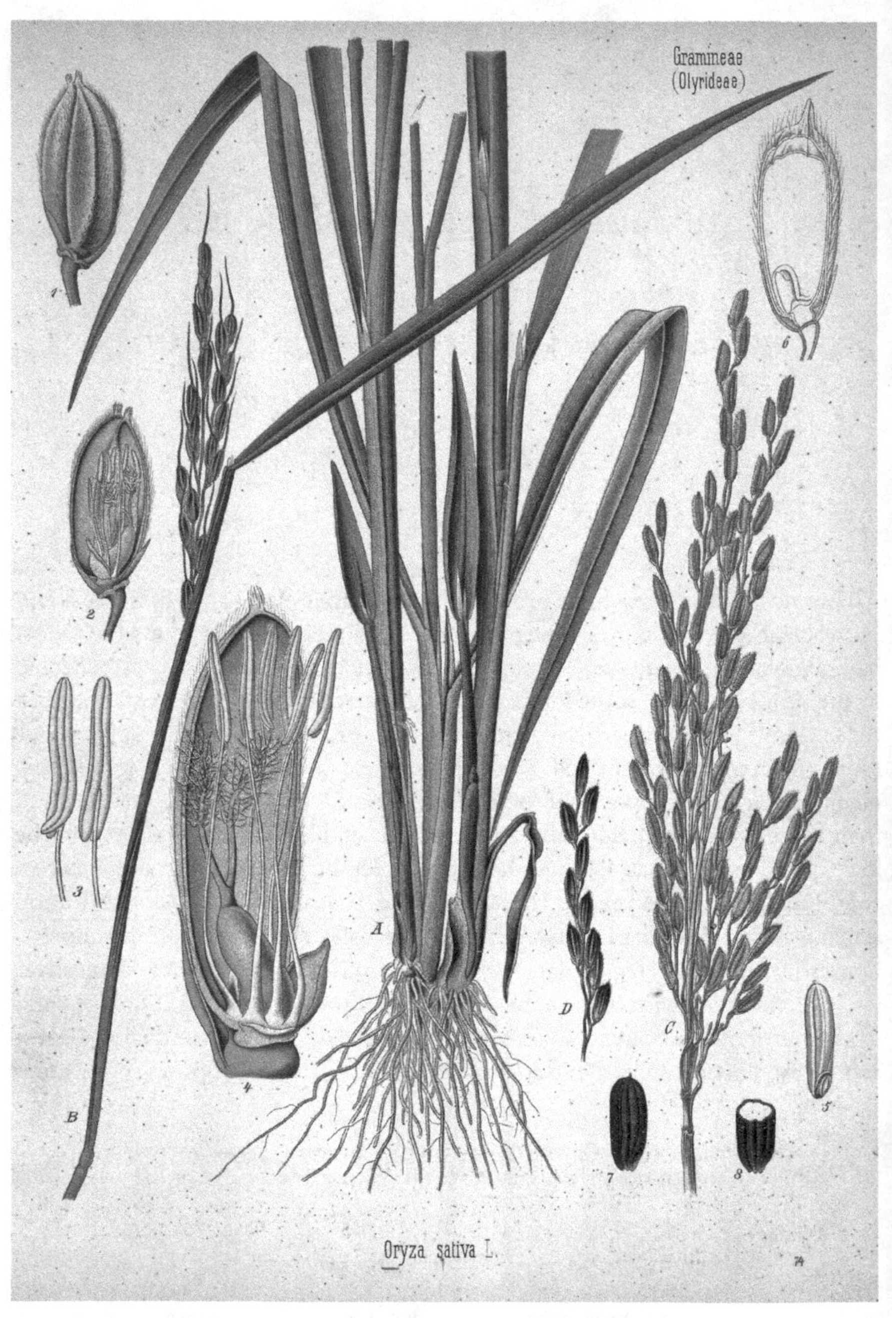

FIGURE 10. Asian rice (*Oryza sativa*), illustration from *Köhler's Medizinal-Pflanzen* (Gera, 1898). Peter H. Raven Library, Missouri Botanical Garden (public domain).

hybridization between the domesticated varieties and wild rice. In this regard, the aromatic variety, represented by the likes of *basmati* in South Asia and *sadri* in Iran, may have developed through an admixture of *japonica* with wild rice cultivars ancestral to the *aus* type.[4]

Like other grain crops, domesticated rice (*Oryza sativa*) is typically distinguished from its wild form by a reduction in awn size, seed shattering functions and seed dormancy, inflorescence enlargement, and changes in seed pigmentation.[5] In South Asia, the earliest directly dated evidence for the anthropogenic management of wild rice derives from a mid-seventh-millennium BCE context (c. 6400 BCE) at the site of Lahuradewa, in the Gangetic valley (Uttar Pradesh).[6] Rice is subsequently attested at other Neolithic sites in the wider Gangetic belt (e.g., Tokwa, Kunjhun, Jhusi, Chopani-Mando), although it is not clear when it can be considered domesticated, since non-shattering traits and grain size increase do not appear to be significant domestication markers in early Indian rice cultivars.[7] But the appearance of non-shattering rice spikelet bases at early-second-millennium BCE sites like Masudpur and Bahola in Haryana and Mahagara in Uttar Pradesh (c. 1650 BCE) provides a *terminus ante quem* for the protracted domestication process.[8]

Rice cultivation spread to northwestern India from its Gangetic epicenter by the mid-third millennium BCE, since rice is encountered in mid-to-late-third-millennium Harappan sites like the eponymous Harappa, Balu, Banawali, Binjor, Kunal, and Masudpur, where direct dating has yielded dates between 2430 and 2140 BCE for the earliest rice samples.[9] It subsequently materializes in the archaeological records of Pirak, on the north Kachi Plain, in the early second millennium BCE.[10] The arrival of rice at Pirak, a settlement on the Harappan periphery along the eastern Iranian plateau, heralds its slow westward movement along the Iranian

Asia and the wild predecessor of *indica* rice ("proto-indica") in the Gangetic Valley (Choi and Purugganan 2018; Choi et al. 2017; Fuller 2011; Kingwell-Banham, Petrie, and Fuller 2015, 274; Silva et al. 2018). Note also the proponents of a monophyletic model, wherein *indica* is considered an offshoot of *japonica* rice (Molina et al. 2011, 8351–56; for a critique of Molina et al., see Fuller 2011, 82–83).

4. Civáň et al. 2019.

5. Ishikawa, Castillo, and Fuller 2020.

6. Fuller 2006a, 41–43; Kingwell-Banham, Petrie, and Fuller 2015, 273–74; Pokharia 2011; Tewari et al. 2008.

7. Murphy and Fuller 2016, 350; Pokharia, Pal, and Srivastava 2009; Ray, Chakraborty, and Ghosh 2020, 3–4.

8. Bates, Petrie, and Singh 2017; Ray, Chakraborty, and Ghosh 2020, 3–4.

9. Bates, Petrie, and Singh 2017, 195; Petrie et al. 2016; Saraswat and Pokharia 2002–03, 109–10, 124–25; Sharma et al. 2020, 1363–64.

10. The rice finds at Pirak have been suggested to be of the *japonica* variety on the basis of grain size. See Bates, Petrie, and Singh 2017; Costantini 1981; Fuller 2006a, 36; Fuller and Madella 2001, 336–37, 354–55; Nesbitt, Simpson, and Svanberg 2010, 325; Petrie et al. 2016, 1496–1501; Sato 2005.

plateau by overland and perhaps even coastal routes into western Iran and Mesopotamia.

In late antiquity and the early Islamic period, rice figures as an important crop in the Middle East and the eastern Mediterranean, growing in Mesopotamia and Susiana, in the lowlands south of the Caspian Sea (Gīlān, Daylam, Tabaristān), in Jordan and Palestine, notably around the well-watered districts of Bet She'an and Banias in the Golan, in the Nile Delta and Egyptian oases like the Fayyūm, and also in the low-lying plains of Anatolia, like those of the Seyhan River delta in Cilicia.[11] Rav Ashi, the fourth-century CE editor of the Babylonian Talmud, notes that rice was a staple crop of Susiana, in southwestern Iran (modern Khuzestan).[12] The geographer Yāqūt al-Ḥamawī, commenting on the same region in the thirteenth century, maintains that the 50,000 ovens baking rice bread there warmed the entire country.[13] Ibn Ḥawqal, a tenth-century geographer, affirms, no doubt with hyperbole, that some people, especially in the riparian regions south of the Caspian sea, were so used to eating rice bread that they suffered terrible colics, and even died, from consuming wheat.[14]

The culinary history of the Middle East also abounds with recipes involving rice, from the elaborate rice-jelly and Greek (*rūmī*) rice pudding served at the court of Balāš (484–488 CE) to the Sasanian spiced-rice dish served with poultry and lamb, about which the Abbasid poet Abu al-'Abbās al-Adīb waxes lyrical: "Sasan in his days invented it and Kisra Anu Shirwan loved it."[15] The culinary arts reached a new peak among the urban leisured class in the Abbasid period (750–1258 CE), as evinced by the flowering of gastronomic poetry and cookbooks, some ascribed to no less than the caliphs themselves.[16] The tenth-century poet al-Ḥāfiẓ al-Dimashqī's eulogy on *aruzziyya*, an Iranian-inspired rice porridge simmered in milk with cassia, galangal, sugar, and clarified butter, demonstrates that rice preparation had evolved into a complex culinary art form, much as it remains in present-day Iranian cuisine:

11. Canard 1959; Feliks 1963b; Watson 1983, 17, 156; Decker 2009, 195–96; Nesbitt, Simpson, and Svanberg 2010, 313–14.

12. TB, *Pesaḥim* 50b–51a. Hozae is the term for Susiana in the Babylonian Talmud.

13. Yāqūt al-Ḥamawī, *Mu'jam al-buldān* I, 413.

14. Ibn Ḥawqal, *Kitāb ṣūrat al-arḍ* 272 (381). On rice consumption in the riparian region south of the Caspian sea, see also *Ḥudūd al-'ālam* 134, 137; al-Iṣṭakhrī 212; al-Muqaddasī 354.

15. Al-Tha'ālibī, *Ta'rīkh ghurar al-siyar*, 585 (rice at the court of Balāš); Abu al-'Abbās al-Adīb, ap. Ibn Sayyār al-Warrāq, *Kitāb al-Ṭabīkh* 50, trans. Nasrallah 2007, 259–60. Sāsān is the eponymous ancestor of the Sasanian kings, usually considered to be the grandfather of Ardašīr I (r. 224–242), the founder of the Sasanian dynasty. Kisra Anu Shirwan is none other than the celebrated Ḵosrow I Anušīrwān (r. 531–79), whose patronage of learning, Gibbon (1788, 238) thought, even convinced exiled Greek philosophers from Athens that a "disciple of Plato was seated on the Persian throne."

16. Nasrallah 2007, 15–22 (introduction in text edition for Ibn Sayyār al-Warrāq's *Kitāb al-Ṭabīkh*); Perry 1994.

What a wonderful *aruzziyya*, cooked to perfection,
Like a full moon in the middle of the sky.
Purer than the doubly condensed snow that the winds and dew tinted.
As white as a large pure pearl spread in a bowl.
It dazzles the eyes with its sheen.
Behold moonlight even before the evening is seen.
The sugar on its sides, like lustrous light projected from the skies.[17]

Rice remains a common food across the Middle East today, especially in southern Iraq, Iran, the Arab states of the Persian Gulf, and deltaic Egypt. The fragrant rice of southern Iraq (*timman ʿanbar*), served with meat or fish stew, is a hallmark of contemporary Iraqi cuisine, as are the elaborate *polows* and *chelows* of Iran—rice and meat dishes delicately flavored and colored with aromatics, dried fruits, and nuts.[18] In other parts of the Middle East where rice was not a staple before the twentieth century, it was typically considered a prestige food reserved for the affluent. The lower classes consumed rice irregularly, customarily on festive occasions. In eighteenth-century Aleppo rice cost twice as much as wheat.[19] The status of rice as an enviable luxury is echoed in an old Arab folk saying: "What do the people of paradise eat?—rice in butter."[20]

Despite the prominence of this crop in later times, the earliest history of rice in the Middle East and the Mediterranean has elicited scant scholarly attention, and its appearance in antiquity has been largely discussed with the aid of Greco-Roman or Hebraic texts.[21] Yet the materials for the study of rice cultivation in the ancient Middle East are already to be found in Akkadian and Elamite, the languages of Mesopotamia and southwestern Iran. A collation of these texts, ranging in genre from the lexicological to the epistolary, has hitherto been unrealized, not least on account of philological impediments. The archaeobotanical finds for rice, while meager for much of the first millennium BCE, will be synthesized in this chapter with textual sources to understand the spatial and chronological distribution of rice cultivation as well as to assess the appeal rice had for local producers and the factors which limited its widespread cultivation. I will also consider the ancient food-processing methods associated with rice and its peculiar role in local medical and magical traditions.

17. Al-Ḥāfiẓ, ap. Ibn Sayyār al-Warrāq, *Kitāb al-Ṭabīkh* 51, trans. Nasrallah 2007, 262.
18. On *polows* see especially the chapters on rice in the *Māddat al-ḥayāt*, a Safavid cookbook authored by Nūr-Allāh, the chef of Shah ʿAbbās I (r. 1588–1629), in Fragner 1984, 342–60.
19. Marcus 1989, 103.
20. Zubaida 1994, 93.
21. Feliks 1963b; Hehn 1887, 368–76; Konen 1999.

RICE BETWEEN ASSYRIA AND THE AEGEAN

In the spring of 318/7 BCE, amid the clamor of the Second War of the Diadochi, the troops of Eumenes of Kardia, fleeing the armies of Seleukos and Pithon, the Greco-Macedonian satraps of Babylonia and Media, crossed the Tigris into Susiana, where they found themselves "completely without grain" and subsisted instead on rice, sesame, and dates, which were said to grow aplenty there.[22] In the second century BCE, Zhang Qian, the Han ambassador to Central Asia, observed that rice grew in Parthia (Anxi) and Mesopotamia (Tiaozhi).[23] Strabo (c. 63 BCE to 23 CE), probably citing Aristobulus (fourth century BCE), notes that rice grew in Bactria, Babylonia, Susiana, and Lower Syria.[24] Rice may have been familiar to the Greek world by the fifth century BCE, since a fragment of Sophocles's *Triptolemos* appears to mention bread made of rice flour (*oríndēs ártos*).[25]

The casual references of Greek and Chinese commentators to the cultivation of rice in Mesopotamia and Susiana in the last centuries of the first millennium BCE suggest a longer history of rice cultivation in these regions. But attempts to trace rice in the textual records of the ancient Middle East and the eastern Mediterranean prior to the earliest secure attestation in Greek authors of the fourth century BCE (Theophrastus and Hieronymus of Cardia) are fraught with philological problems. It is imperative therefore to deal at length with the complex vocabularies relating to rice in the written records of the first millennium BCE.

THE TERMS FOR RICE IN AKKADIAN AND ELAMITE

Reginald Campbell Thompson (1876–1941), a pioneer in the study of Mesopotamian natural sciences, was the first to convincingly identify the cereal named *kurângu* in Neo-Assyrian lexical texts with rice on the basis of Iranian cognates.[26] Thompson's 1949 *Dictionary of Assyrian Botany,* which remains the only comprehensive treatment of Sumero-Akkadian plant names, has been criticized for relying too much on cognates and ethnographic parallels as well as ignoring the high mutability of ancient plant names.[27] Thompson thought the philological connection between *kurângu* and its Iranian cognate to be obvious, and the identification of *kurângu* with rice has been accepted by some Semitic philologists.[28] Yet the *Chi-*

22. Hieronymus of Cardia, ap. Diod. Sic. XIX.13.6.

23. Sima Qian, *Shiji,* Dayuan 123.

24. Strabo XV.1.18.

25. Ath. III. 110e.

26. Thompson, 1939. Variant spellings including the Middle Assyrian *kuriangu,* the Neo-Assyrian *kurâggu,* and the Neo-Babylonian *kuriaggu.* Note also the spelling ŠE *ia-an-gu* in the lexical series Murgud, known from a Hellenistic copy from Uruk (Recension B: SpTU III 116 iv 23).

27. E.g., Dafni and Böck 2019.

28. Borger 1971, 310; Rabin 1966; Thompson 1949, 106.

cago Assyrian Dictionary, the lexical standard for Assyriology, among other doubting voices, only identifies it as "a cereal."[29]

As we shall observe, Thompson's identification of the Akkadian *kurângu* with rice stands up to both philological and contextual scrutiny. Thompson's now-outdated positivist approach to the identification of ancient plants is perhaps not entirely without merit. Thompson was only aware of the appearance of this grain crop in the Neo-Assyrian lexical series called Uruanna = *maštakal*, a work largely of pharmaceutical interest.[30] There since have appeared other cuneiform sources for *kurângu*, including an earlier late Middle Assyrian record (c. 1100 BCE). The term *kurângu* is presently known from eight Akkadian texts, which roughly date between 1100 and 200 BCE.

To begin with, *kurângu* is not a term of great antiquity by Mesopotamian standards. No such cereal is listed in any of the copious cuneiform archives of the late third and early second millennia BCE. It is even absent from the cereal section (Tablet XXIV) of the Urra = *ḫubullu*, a comprehensive Sumerian-Akkadian lexical series of the early second millennium BCE. The earliest known appearance of *kurângu* (spelled *kuriangu*) is in an epistolary document dating to c. 1100 BCE from Tell Barri (ancient Kaḫat), along the Jaghjagh River in northeastern Syria.[31] Interestingly, the reference to *kurângu* here is rather casual and does not suggest that the crop was unfamiliar. *Kurângu* must therefore have a longer genealogy in the region—how long, one cannot say, since some crop introductions could take place within a cultivator's lifetime. Pliny in the first century CE reports that a kind of high-yield millet, possibly a variety of sorghum, had "been introduced from India into Italy within the last ten years."[32] The first Mesopotamian lexical text to make note of *kurângu* is the Murgud, a lexical commentary compiled in the early first millennium BCE to update the older Urra = *ḫubullu*.[33] The relatively late appearance of *kurângu* and its absence from early Mesopotamian texts indicates that this cereal was not one of the founder grain crops of the Middle East, and its introduction could not have been earlier than the late second millennium BCE.

29. CAD, s.v. *kurângu*. The arguments against the identification with rice, voiced notably by Salvini (1998, 188) and Jursa (1999/2000, 294), have largely been constructed on the basis that the eleventh century BCE is too early a date for an Indo-Iranian word to be attested in Mesopotamia. But there is no good reason to assume that *kurângu* must be of Indo-Iranian extraction. A number of other language families found in the Iranian plateau and South Asia could have been the source of *kurângu*.

30. BM 108860 = CT 37, 32; VAT 9000 = KADP 11. On the Uruanna = *maštakal* series, see Böck 2011, 692–93; Wilson 2005.

31. Salvini 1998: K9, T1.

32. Pliny *HN* XVIII.55; Dalby 2003, 306.

33. SpTU III 116 iv 23 (Recension B): še.ba.ri.gim = MIN *par-sik-ti* = kur[!](text: ŠE)-*ia-an-gu*. Most copies of Murgud derive from the seventh-century Assyrian royal library in Nineveh, but the passage referring to rice is found only in a Hellenistic copy from Uruk in Babylonia. For the dating of Murgud, see Vedeler 2002.

Unfortunately, the Mesopotamian lexical texts do not provide much data on the morphology of *kurângu.* The logographic reading ŠE.LI.A, rendered as a synonym of *kurângu* in the lexical series Uruanna = *maštakal*, written with the signs ŠE meaning grain and LI.A denoting grass (Akkadian *dīšu*, from *dešu* "to sprout, flourish"), offers little by way of a crop-specific description, although it confirms that we are dealing with a grain crop.[34] As the inflorescence of rice, like sorghum and oats, is panicled (loosely branched), while that of wheat and barley is spiked and compressed along a central rachis (stem), ancient observers could not have missed the striking visual difference between mature rice crops and the traditional cereal crops of the Middle East.[35] This morphological variation may have led to the ambiguous taxonomic status of rice in the view of some Mesopotamian scholars. A ritual text from Neo-Babylonian Sippar (ca. 626–539 BCE) prefixes the determinative Ú, referring generically to plants, rather than ŠE (grain) to *kurângu.* But the Mesopotamian cognitive approach to plant classification was governed by exigencies other than morphological resemblance, since even leguminous crops like chickpeas (*ḫalluru*) and peas (*kakkû*) were occasionally categorized with cereals (ŠE- group) in lexical texts like Uruanna = *maštakal.*[36]

Where lexical cognates are concerned, the identification of *kurângu* with rice is much more promising. The Akkadian *kurângu* has no cognates in Semitic languages, whose forms for rice, like the Hebrew *ʿōrez*, only appear in the Greco-Roman period and are related to the Greek *óruza.* Turning to the Iranian languages, Thompson observed the striking congruence of *kurângu* with *guranǰ* or *gurinǰ*, the New Persian word for rice found in Ferdowsī's *Šāh-nāma* (1010 CE), the great epic of Iranian-speaking peoples.[37] Importantly, Ferdowsī was in the habit of retaining older Persian words rather than turning to Arabic, the hegemonic language of the Islamic Middle East.[38] Further support for the identification of *kurângu* with rice can be found in lexical forms related to the Persian *guranǰ* or *gurinǰ* in the various Turkic languages of Central Asia. Turkic lexical cognates like the Uigur *kürüč, kurüč, krünč,* and *krüč,* the Kirghiz *küriš,* the Karakalpak *gurinǰ*, and the Tobol Tatar *kürüch* make the case for Akkadian *kurângu* denoting rice more convincing.[39] Finally, an older attestation in the form of the restored Middle Persian **gwrync* is known from a fragmentary Middle Persian–Sogdian glossary.[40]

34. Urra = *ḫubullu* 17, 3 ú-li-a = *dīšu;* Thompson 1949, 106.

35. Uruanna = *maštakal* (II 485); Thompson 1939, 181.

36. Thompson 1949, 95.

37. Thompson, 1939, 180, 1949, 106. The *Šāh-nāma* was composed between 975 and 1010 CE.

38. Perry 2005.

39. Bailey 1976, 306 (s.v. *rrīysū*); Hauenschild 2006, 42 (s.v. *kurinǰ*); Nazarova 2005, 82; Nesbitt, Simpson, and Svanberg 2010, 334; Rachmati 1932 (s.v. *kurinǰ*); cf. Japanese *urushine, uruchi.*

40. Henning 1977, 45 (Fragment M: 2a).

The survival of a term related to *kurângu* in Middle and New Persian, as well as the Turkic languages influenced by Persian, indicates that an ancient language spoken in the Iranian plateau must have been the source for both the Akkadian *kurângu* and the later Turko-Persian descendants. The specific linguistic affiliations of *kurângu* are unclear, but it is most likely a lexical borrowing from further east. The formative suffixes *-gu*, *-ṅgu*, and *-kku* are strongly redolent of a South Asian, especially Dravidian, linguistic context, although none of the extant homophonous nouns in early South Asian languages are semantically related to grain crops.[41] There are, however, several terms for millet species in South-Central Dravidian languages constructed with the root *ragu* (proto-Dravidian **iraki*), which may at the outset have functioned as a generic term for gramineous species.[42] A similar and perhaps related series of words with an infixed *-ṅ-* in the suffix is also used to describe millet species in Dravidian and Indo-Aryan languages.[43] The philologist Michael Witzel speculates that these lexical forms could have had precursors in the extinct languages of the Indus civilization.[44] The presence of South Asian linguistic isolates like Nahali and Kusunda in Nepal, Vedda in Sri Lanka, Shom Pen in the Nicobar Islands, and Burushaski in the Gilgit-Hunza Valley (northern Pakistan) indicates that many other language families could have been the source of *kurângu*.[45] Cumulatively, the temporal sequence of *kurângu* attestations, its clear cognates with Turko-Persian terms for rice, and a potential South Asian etymology connecting it with grain crops strongly support the identification of *kurângu* with rice.

While *kurângu* demands much effort in the linkage with rice, this is not the case with Elamite, the ancient language of southwestern Iran. The Neo-Elamite word *miriziš*, found in the fifth-century BCE Persepolis Fortification tablets, incontrovertibly derives from the well-known Indo-Iranian word for rice, **vrīziš*.[46] In Elamite, this word undergoes a minor phonetic change which is found elsewhere in Iranian loans into Elamite. For instance, the Old Persian name Vištāspa, the name of king Darius I's father, is rendered as Mišdašba in Elamite. The earliest sources

41. Caldwell 1856, 153, e.g. Ta. *kuraṅku*, Ka. *koraṅgi*, Tu. *kuraṅga* (monkey), Skt MBh *kuraṅga* (antelope), Skt *kuruṅga* (name of a chieftain in Ṛgveda).

42. E.g., Ka. *kadu baragu* (*B. ramosa*); Ka. *hāraku*, Te. *āruka*, *āruga* (*P. scrobiculatum*); Ka., Te., Tu. *rāgi*, Ta. *irāki* (*E. coracana*); Ta. *varaku* (*P. scrobiculatum*); Ta. *kēlvaraku* (*E. coracana*), etc. See DEDR 379, 812, 5260; Fuller 2003b, 201–02, 2006b, 193; cf. Ka. *āku* referring to young rice not yet transplanted or young sprouts of corn, and Te. *āku* referring to seedlings of paddy for transplantation (DEDR 335).

43. E.g. Ta. *iṟuṅku* (*S. vulgare*); Ka. *kaṅku* (ear of corn); Skt *priyaṅgu* (*S. italica*); Skt *kaṅgu* (*S. italica*), etc.

44. Witzel 1999b, 35–36. On the likely affiliation of the language recorded by the Indus script, see Muthukumaran 2021.

45. On the linguistic isolates and unclassified languages of South Asia, see Blench 2008.

46. PF 544:1.

for the **vrīziš* form for rice are in Vedic Sanskrit, where it appears as *vrīhi*. This term is first noticed in the *Atharvaveda Saṃhitā* (e.g., X.9.26, XI.1.18) and in the *Taittirīya Saṃhitā* of the *Black Yajurveda* (I.8.9.3). Apart from Sanskrit and the Prakrits, **vrīziš* forms for rice are also known in a range of other Indo-Iranian languages, including Nūristanī (*wrīc*), Kāmviri (*wrúji*), Khotanese (*rrīzu*), Sogdian (*rīza*), and Pašto (*vriži*). The form *rumiziš* appearing in another Persepolis Fortification tablet (PFNN 587: 1) represents a metathesis of *miriziš*.[47] The Elamite specialist Wouter Henkelman has also proposed that the form ^{GIŠ}*ru*^{MEŠ} may stand as an abbreviation for *rumiziš*.[48] Although the Elamite evidence for rice is scanty, it draws attention to the cultivation of the crop at important Iranian localities, like Liduma (PF 544) and Kurra (PFNN 587), on the royal route between Persepolis and Susa in the early fifth century BCE.[49]

EARLY MEDITERRANEAN RECORDS OF RICE

In contrast to Akkadian, the Hebrew and Aramaic terms for rice (Heb. *ʿōrez*, Aram. *ʿurzā*, *ʿarūzā*) are not in doubt. There appears, however, to be a significant time gap between the Akkadian and Elamite attestations for rice and their appearance in Hebrew and Aramaic records. Rabbinic texts indicate that rice was widely cultivated in Roman Palestine, especially after the lifetime of Hillel the Elder, a contemporary of the Roman emperor Augustus.[50] More than one variety of rice was known in Roman Syria-Palestine, since rabbinic sources observe that red rice grew in the plain of Antioch on the Orontes.[51] The history of rice cultivation in Syria-Palestine is evidently older, since Strabo already refers to the cultivation of rice in Lower Syria, and a benediction for rice dishes attributed to Simeon the Just, either the first or the second high priest of Jerusalem to bear that name in early Hellenistic times, indicates that rice was a notable crop in Syria-Palestine by the late third or second century BCE.[52]

As for Greek records, while the question of whether *óruza* (rice) is of Indo-Iranian or Dravidian etymology remains unsettled, the term itself presents no semantic conundrums and is undoubtedly related to, and most likely derived from,

47. On *vrīziš*, see PF, Glossary, s.v. *miriziš* (entry draws on Ilya Gershevitch's proposals); Hinz 1975, 270; Mayrhofer 1971, 58. On *miriziš*, see Tavernier 2007, 457.

48. Attested in an unpublished Persepolis Fortification tablet in Tehran (PFNN 7253). See Henkelman 2008, 526, 2010, 56.

49. Kurra (Kórra in Ptol. *Geog.* VI.4.6) and Liduma have been localized in the Fahliyān region of western Fars. Liduma has been identified with the site of Jenjān (Arfa'i 1999, 43; Potts 2008, 284, 295).

50. Mishnah, *Tevul Yom* 1:1 (Hillel the Elder), *Dema'i* 2:1, *Shevi'it* 2:7, *Ḥallah* 1:4, 3:7, 4:3; TJ, *Terumot* 1:2; Feliks 1963a, 151, 165; Konen 1999, 29–32; Safrai 1994, 117–18.

51. E.g., Tosefta, *Dema'i* 2.1.

52. Strabo XV.1.18; TJ, *Berakhot* 6:1.

West Semitic forms for rice like the Hebrew *ʿōrez*, Aramaic *ʿurzā* or *ʿarūzā*, Syriac *rūzā* or *ʿūrūzā*, Arabic *ʿaruzz*, *ruzz*, *ʿuruz*, or *ʿurz*, and Amharic *rūz*.[53] Much controversy, however, engulfs *oríndēs*, the earliest known Greek reference to rice.[54] The term *oríndēs* appropriately occurs in a lost Sophoclean oeuvre, composed in 468 BCE, on the mythical Eleusinian prince Triptolemos, who was instructed by the goddess Demeter in the arts of agriculture.[55] The botanical identity of *oríndēs* already appears to be disputed in ancient texts. Three Greek texts of the Antonine period (c. 138–193 CE), namely the gastronomic encyclopedia of Athenaeus and the lexica of Athenaeus's contemporaries, Pollux and Phrynichus, have left their impressions of the Sophoclean *oríndēs*. Phrynichus, following common opinion, thought *oríndēs* to be rice, but Athenaeus is uncertain and suggests that it was either bread made of rice or an Ethiopian grain resembling sesame.[56] Meanwhile, Pollux, the Naucratian grammarian, does not identify it with rice at all but only remarks that it was an Ethiopian bread made from grain resembling sesame.[57] The fifth-century CE Alexandrian lexicographer Hesychius remains undecided and glosses it as bread among the Ethiopians and a grain resembling sesame or, according to others, rice.[58]

The source of the conflicting statements about the identity of *oríndēs* may be connected to the recognition, in Greek ethnography, of a branch of "Eastern Ethiopians" localized in or near South Asia and the regular confusion between Ethiopia and India.[59] Herodotus himself alleges that both Ethiopians and an Indian tribe named the Kallantiae used the same grain.[60] The muddled identity of *oríndēs* could also stem from the superficial morphological similarities between rice and some millet species.[61] The native *Setaria* millet of Nubia (*Setaria sphacelata*) and a small-grained variety of sorghum, both archaeologically attested in the Napatan and Meroitic periods (c. 750 BCE–350 CE), fit well in size and color with the Greek descriptions of an Ethiopian grain resembling sesame and could have been

53. Boisacq 1938, 712; Caldwell 1856, 66; Frisk 1954–72, s.v. *óruza;* Karttunen 1989, 54; Krishnamurti 2003, 5; Mayrhofer 1956–80, s.v. *vrīhi*; Rabin 1966, 5; Witzel 1999a, 31.

54. Dalby (1996, 251–22, 2003, 60), for instance, does not accept this identification.

55. Pliny *HN* XVIII.65.

56. Phrynichus, *Praep. soph.* 93 B; Ath III.110e.

57. Pollux, *Onomasticon* VI.73.

58. Hesychius, s.v. *oríndēn.*

59. On the eastern Ethiopians, see Hdt. III.94, VII.70. This may be an attempt to interpret the Homeric dictum "*Aithíopas, toì dikhthà dedaíatai*" (Hom. *Od.* I.23). For the regular confusion between Ethiopia and India in classical sources, see e.g. Aeschylus, *Supp.* 284–86; Karttunen 1989, 86 n159; Pisani 1940.

60. Hdt. III.97.2. This could refer to some kind of grain used in both Africa and India, like pearl millet (*Pennisetum glaucum*), sorghum (*Sorghum bicolor*), or finger millet (*Eleusine coracana*). See Fuller (2006b, 190) on African sorghum and millets in South Asia.

61. Fuller 2006b, 191 (in some cases leading to semantic shifts in terms for millets into rice).

confused with rice by unfamiliar observers.[62] Jules Bloch proposes that the term *oríndēs* was extended to rice by Greco-Roman lexicographers on account of phonetic similarities with *óruza* but at the outset probably denoted a different grain.[63]

Oríndēs is said to refer to bread made of rice and not the grain itself, so some divergence from *óruza* is not anomalous. Phonetically, *oríndēs* suggests an Iranian root rather than a northeast African origin, and this supports an identification with rice rather than an Ethiopian (i.e., Nubian) millet species.[64] More than a century ago, the historian Victor Hehn drew attention to the fact that the nasalized *oríndēs* corresponds with the Armenian *brinz* and Persian *birinǰ* forms for rice, also borrowed into Mesopotamian Arabic as *pirinj*.[65] In any case, it is clear that *oríndēs* was not in regular use in Koine Greek. Chrysippus of Tyana (second to first centuries BCE) refers to flat rice bread as *oruzítēs plakoũs* rather than using the Sophoclean *oríndēs*, which could simply be a poetic neologism.[66]

Herodotus, a contemporary of Sophocles, may have known of the consumption of wild rice (Sanskrit *nivāra*) in South Asia from hearsay reports. He refers to itinerant Indian tribes consuming a grain the size of a millet which they gathered with the husk before boiling, but otherwise he appears to be altogether unfamiliar with the domesticated rice crop.[67] The earliest unambiguous references to *óruza* proper manifest in the works of Theophrastus and early Hellenistic authors (fourth to third centuries BCE), who accurately describe its cultivation and knew it to be a staple in India.[68] Aelian, drawing on an unknown Hellenistic source, perhaps Megasthenes, remarks of war elephants fed on rice wine.[69] Its cultivation in the Mediterranean is suggested by the reference to a flat bread or cake prepared from rice (*oruzítēs plakoũs*) in Chrysippus of Tyana's (second to first centuries BCE) *Artopoiïkós*, a treatise on breadmaking.[70] Horace refers to rice gruel as food for the invalid (*tisanarium oryzae*), while Dioscorides notes that it grew in marshes and

62. On sorghum and millets in Napatan and Meroitic Nubia, see Fuller 2004, 72–73, 2013, 165–69.

63. Bloch 1925, 45.

64. Schmitt 2002a; Tucker 2007, 778.

65. Hehn 1887, 370.

66. Chrysippus of Tyana, ap. Ath. XIV.647d. On neologisms in classical dramatists, see Collard 2005, 355–60; Smereka 1936; Stevens 1976.

67. Hdt. III.100. This description could apply to any one of the native Indian millets (e.g. *Panicum sumatrense, Brachiaria ramosa,* or *Setaria verticillata*), especially since the modes of processing millet are similar to those for rice. On South Asian millets, see Fuller 2006b.

68. Theophr. *Hist. pl.* IV.4.10; Aristobulus and Megillus, ap. Strabo XV.1.18; Eratosthenes, ap. Strabo XV.1.13; Megasthenes, ap. Ath. 153e, Strabo XV.1.53, 60; Diod. Sic. II.36.3.

69. Aelian *NA* XIII.8.

70. Chrysippus of Tyana, ap. Ath. XIV.647d. Chrysippus's treatise is lost and only survives in quotations.

wetlands and was considered moderately nutritious as well as useful for gastrointestinal ailments.[71]

Although rice never became a common article of consumption everywhere in the Roman Mediterranean, a sizeable corpus of texts attests to its use in food and medicine.[72] Rice cultivation was probably limited to the eastern Mediterranean for much of antiquity, since western Mediterranean authors like Turranius Gracilis, who wrote one or more geographical and agricultural works on Spain and Africa in the first century BCE, claimed that rice and *olyra,* a variety of emmer wheat, were the same species, and even the well-versed Pliny offers a confused description of the plant.[73]

Whether rice was known in Hellenistic Egypt, where it was eminently suited to grow in the wetlands of the deltaic region and oases like the Fayyūm, remains unclear. A long account of payments from Herakleopolis of the second century BCE may refer to a rice-seller (*oruziopṓlēs*) by name of Hiérōn, but the papyrus is too fragmentary for a decisive reading.[74] Pliny, quoting an Egyptian medical recipe, notes that a certain medicament was devised from crocodile innards mixed with the droppings of starlings fed only on rice, suggesting that rice may have been cultivated in Egypt sometime earlier than the first century CE.[75] Alternatively, the starlings may simply have fed on imported grain. Rice never seems to have become an important crop in pre-Islamic Egypt, since it is rarely named in the fairly voluminous papyrological documentation of the Greco-Roman period.[76] There also appears to be no native Egyptian term surviving for rice, since the later Coptic word for rice (*pi-arros*) derives from West Semitic forms. But the Nile Delta, where rice was better suited to grow, has not produced many papyri.[77]

THE ARCHAEOLOGY OF RIZICULTURE

Overall, the Elamite sources, owing to their use of an unambiguous Indo-Iranian loanword, provide a relatively secure *terminus ante quem* of the mid-first millennium BCE for rice cultivation in the Middle East. The case for rice in earlier Akkadian

71. Dioscorides, *Mat. Med.* II.95; Horace *Sat.* II.3.155.

72. Food: Apicius II.2.8, II 2.9; Vinidarius *Exc.* 7, 9; HA, Elagabalus 21.3; *Ed. Diocl.* I 23; André 1981, 54; Konen 1999. 37–41. Medicine: Galen *Aliment. fac.* 6.525 (K), 6.687, *Simpl. med.* 12.92; Celsus *Med.* II 18.10, 22.11; Aretaeus *De curat. acut. morb.* I 10.6.5, II 2.17.7; Archigenes of Apamea 23.5, 23.12; Aet. Amid. *Med.* I 305.1, II 97.9.

73. Pliny *HN* XVIII.75; cf. XVIII.82.

74. P. Tebt. III.2.890: 2.35.

75. Pliny *HN* XXVIII.110.

76. E.g., P. Hawara 245 = SB I 5224, Z. 36. 41 (first and second centuries CE); P. Tebt. II 612 (d) (first century CE); P. Freib. IV 67 (second and third centuries CE); see also Cappers 2006, 105.

77. Konen 1999, 34–35.

records is philologically sound but, at present, unlikely to satisfy all scholars—especially because the archaeological data for rice in the ancient Middle East and the Mediterranean are meager, if not disputed, for periods before the first century CE. A single charred grain of rice was reported from the Iranian site of Hasanlu (Gilzanu) in a pit dated by the excavators to 750–590 BCE.[78] But the archaeobotanist Marijke van der Veen suggests that it could be einkorn (*Triticum monococcum*), since subsequent archaeobotanical work at the site yielded no trace of rice at the first-millennium BCE levels.[79] Another solitary grain of rice dating to the twelfth century BCE was identified at Mycenaean Tiryns.[80] If taken as positive evidence, this find would complement the late Middle Assyrian reference to *kurângu* from Tell Barri, in northeastern Syria. Robert Sallares proposes that the grain was an exotic import rather than a locally cultivated species.[81] Tiryns, the port of Mycenae, may have received rice from a nearby source like Syria-Mesopotamia rather than distant India in this period. Serious history, however, cannot be written on the basis of a solitary grain of rice, and further finds are needed to confirm the presence of rice in Mycenaean Greece.[82]

Following a long drought in data, the first century CE is exceptionally well endowed with rice finds from various sites, including the Egyptian Red Sea ports of Myos Hormos (Quseir al-Qadim) and Berenike (Medinet el-Haras), Parthian Susa, and even sites as far afield as Roman Novaesium (Neuss am Rhein) and Mogontiacum (Mainz) in Germany, and Tenedo (Zurzach) in Switzerland (map 8).[83] A septic pit from an early-second-century context (120–130 CE) in Mursa (modern Osijek, Croatia) has also produced five grains of rice.[84] And an amphora from Herculaneum dated before the Vesuvian eruption of 79 CE may

78. Tosi 1975.

79. Van der Veen 2011, 77 no. 6. Einkorn grains recovered in the early 1970s by Japanese researchers at Sang-i Chakmak, a Neolithic site in northern Iran, were also misidentified at the outset as rice owing to superficial morphological similarities between einkorn and rice (Dorian Fuller, personal communication).

80. Kroll 1982, 469 (illustrated).

81. Sallares 1991, 23.

82. Note should also be made of an eighteenth-century claim for the presence of rice in ancient Egypt. The French antiquarian, Anne Claude de Caylus (1692–1765), and one M. de Bose, his colleague in the Académie des Inscriptions et Belles-Lettres, independently identified pieces of rice straw used as a binder on the gilded plaster covering of an undated statue of Osiris (de Caylus 1752, 14; Sonnini 1799, 253). Täckholm and Täckholm (1941, 412) believe the identification to be improbable, but at least two contemporary scholarly works (Darby, Ghalioungui, and Grivetti 1977, 493; Daressy 1922) appear favorable to the testimony of the French scholars, though the whereabouts of the statue are presently unknown.

83. Myos Hormos and Berenike: Cappers 1998, 305–06, 2006, 191; van der Veen 2011, 46–7. Susa: Miller 1981. Novaesium: Knörzer 1966, 433–43, 1970, 13, 28. Mogontiacum: Zach 2002, 104–05. Tenedo: Furger 1995, 171; Nesbitt, Simpson, and Svanberg 2010, 329.

84. Reed and Leleković 2019.

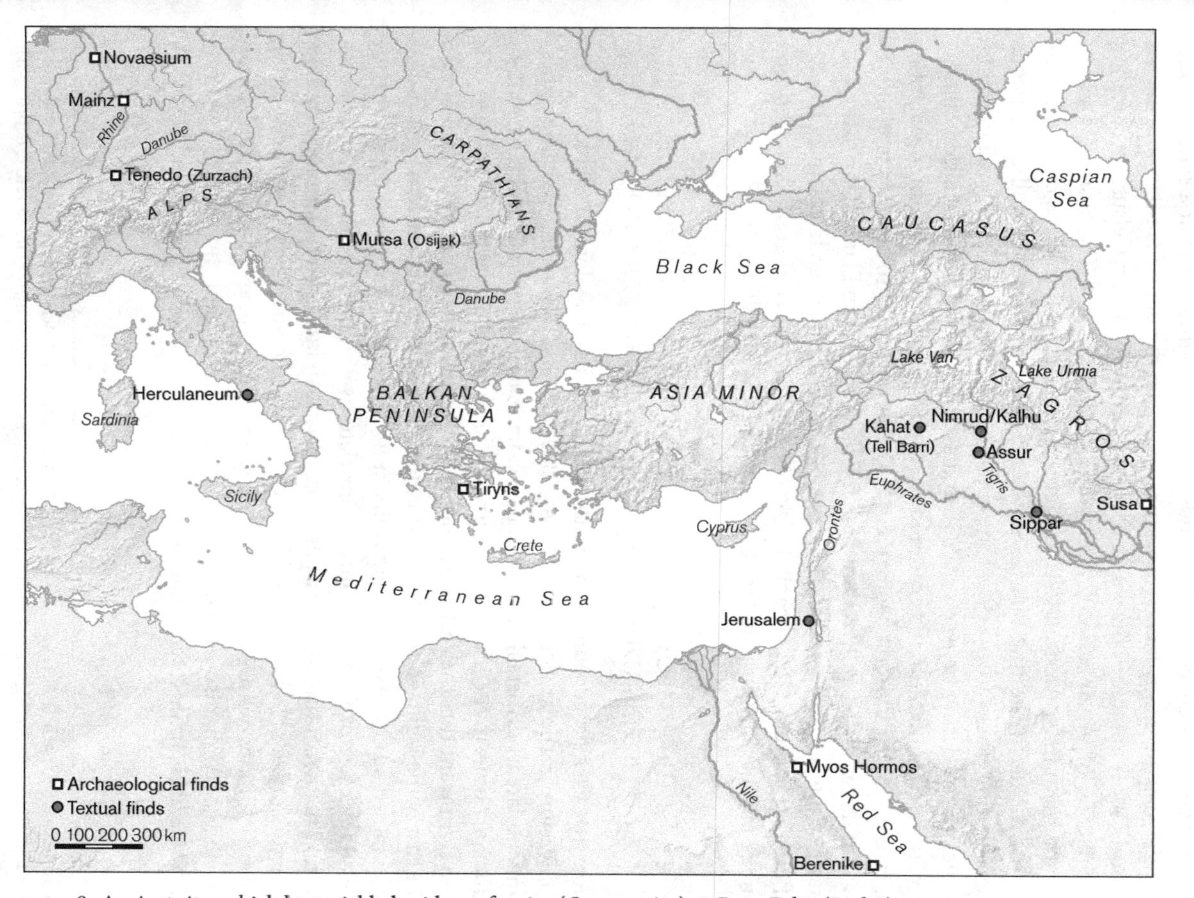

MAP 8. Ancient sites which have yielded evidence for rice (*Oryza sativa*). © Peter Palm (Berlin).

have been filled with rice, since it is inscribed with the label *orissa,* a vulgar variant of the Latin *oryza* for rice.[85]

The surge in archaeobotanical finds of rice belonging to the first century CE is paralleled by the many literary references to rice in the Julio-Claudian and Flavian periods, most notably in the works of Horace, Pliny the Elder, Celsus, Dioscorides, Archigenes of Apamea, and Aretaeus of Cappadocia.[86] The Roman military encampment in Novaesium (Neuss, Germany) alone produced 196 charred grains of rice dating to the first quarter of the first century CE. Rice-hull impressions identified on bricks from several sites in the South Dez plain of Susiana in southwestern Iran, dating between 25 BCE and 250 CE, also indicate localized cultivation of rice.[87] And Susa, the capital of the province, yielded 373 carbonized grains of rice dating to the first century CE.[88] The latter finds are hardly surprising in light of the mid-first-millennium BCE references to rice in Elamite, the native language of Susiana. Nonetheless, the high visibility of rice in archaeological records of the first century CE probably reflects deliberate efforts in grain diversification on the part of cultivators in the Roman and Parthian realms.

CHRONOLOGY AND ROUTES: WHO BROUGHT RICE TO THE MIDDLE EAST AND THE MEDITERRANEAN?

Now that we have accounted for all the textual and archaeological sources for rice cultivation in the ancient Middle East and the Mediterranean, we can begin to contemplate its dispersal history on a broader scale. The philological dissection of *kurângu* points to the Iranian plateau as the likely place of origin for this term. It is also significant that the earliest known reference to *kurângu* derives from Kaḫat, a site east of the Jaghjagh tributary in the Ḫābūr triangle of northeastern Syria. Kaḫat maintained strong Assyrianizing traits in its material culture throughout the so-called dark age of Aramaean migrations (late twelfth to eleventh centuries), suggesting that it continued to be a part of the Assyrian kingdom and profited from its easterly contacts.[89] The remarkable discovery of gold and silver vessels displaying strong affinities with Middle Assyrian iconography near Fullol, in Badakhšan (northeast Afghanistan), close to the lapis lazuli and tin mines, confirms Assyria's lively trade interactions with polities in the Iranian plateau and beyond.[90]

85. *Corpus Inscriptionum Latinarum* IV 10756.

86. Horace, *Sat.* II 3.155; Celsus *Med.* II 18.10, 20, 23, 24; II 22.11, 24; II 7.2; II 23; III 7.2; IV.14.3; Dioscorides, *Mat. Med.* II 80.6.6, 96.1.2; II 75; II 95; Aretaeus, *De curat. acut. morb.* I 10.6 = 25 A.234 (K); II 17.7 = 24 A.255 (K); Archigenes of Apamea, *Fragm.* 23.5, 23.12; *Fragm.* Inedita 69.16 (ed. Calabró 1961).

87. Nesbitt, Simpson, and Svanberg 2010, 326, 329.

88. The rice grains from Susa were recovered alongside remnants of storage jars (Miller 1981).

89. D'Agostino 2009, 17, 22, 32–35.

90. Olijdam 2000.

Archaeological and literary sources from the Middle East of the late second millennium BCE reveal an insatiable demand for lapis lazuli, which was imported from the mountains of Badakhšan.[91] This material, as we have already observed in chapter 1, is the most diagnostic and archaeologically visible element of east–west contacts along the Iranian plateau.

It seems likely, therefore, that the earliest rice was brought into Mesopotamia by traders moving along Iranian plateau routes. The grain may have initially been transported for personal consumption, and the surplus could have been bartered at the end destination. At some point, it may have become a trade commodity in its own right. This mode of transmission finds a parallel much later, in the early first century CE, when the monsoon trade brought Indian rice to Roman Egypt. The archaeobotanical finds of rice at the Roman Red Sea port of Myos Hormos (Quseir al-Qadim), which include husk fragments recovered alongside items of Indian provenance like Indian pottery and Tamil-Brahmi ostraca (Trenches 8 and 8A), indicate that rice was consumed on-site by Indian traders.[92] Similarly, the small quantities of rice recovered alongside Indian pulses like mung beans (*Vigna radiata*) from a first-century CE dump in Berenike (Medinat el-Haras), another Roman Red Sea port, suggest that the consumers were members of a South Asian trading diaspora rather than local inhabitants.[93] The movement of rice with traders also finds support in a tale relating to seafaring merchants in the *Nāyādhammakahão*, a Jain didactic text dating to the late centuries BCE.[94] The story relates that seafaring merchants from Campā, in eastern India, took with them essentials like rice, flour, oil, ghee, curd, freshwater, utensils, medicines, hay, wood, weapons, and clothing for a long-distance voyage (Naya. *Mallī* VIII.49).

While the Akkadian and later Turko-Persian terms for rice reflect transmission along overland routes, the straightforward Indic loanwords for rice in Elamite, West Semitic languages, and Greek suggest for another transmission event, perhaps along maritime routes, in the early Achaemenid period (late sixth to early fifth century BCE). The transmission of rice, as with other crops discussed here, was likely the result of multiple dispersals through space and time. This complexity is reflected in the divergent lexical trajectories of rice in the early Middle East and the Mediterranean.

91. Olijdam 1997.

92. Van der Veen 2011, 46–47.

93. Cappers 2006, 191; Wendrich et al. 2003, 64. For a mid-first-millennium BCE reference to a South Asian recipe involving rice cooked in milk with mung beans and sesame (*kuryāttilamudgamiśraṃ sthālīpākaṃ*), see the *Jaiminīyagṛhyasūtram* I.7 (trans. Caland 1922).

94. Upadhye (1983, 14) and Winternitz (1983, 418) suggest that parts of the Jain canon could date back to the period between Mahāvira and the Council of Pāṭaliputra (fifth to third centuries BCE).

THE SUBSISTENCE VALUE OF RICE

The non-lexical references to rice (*kurângu*) from early-eleventh-century BCE Kaḫat and eighth-century BCE Kalḫu in central Assyria occur in agricultural contexts and clearly indicate that we are dealing with a locally grown subsistence crop and not an exotic import meant for elite consumption. The dedication of rice (*kurângu*), alongside sesame, to the Ebbabar temple of Sippar by Iqīšāya, a tithe collector, in 575 BCE is also a sure sign that rice (*kurângu*) was grown in the Sippar region of southern Mesopotamia by the sixth century BCE.[95] Nevertheless, the scarce references to rice (*kurângu*) in Akkadian textual sources must be symptomatic of its relatively marginal role in the Mesopotamian agriculture and diet of the first millennium BCE. The grain *par excellence* in Mesopotamia was barley (*Hordeum vulgare,* especially the six-rowed variety), and following on its heels were an assortment of wheats (emmer, einkorn, timopheevoid), millets (broomcorn, foxtail), sesame, dates, flax, chickpeas, onions, garlic, and cress.[96]

The producer's incentives to embrace a new grain crop could include agricultural diversification, risk mitigation, and social and prestige reasons, especially if the crop was held in esteem for its taste, nutrition, and potential medical value. The statement of the Assyrian king Sennacherib (704–681 BCE) that he divided the irrigated meadowland around Nineveh into plots and gave them to the citizens to grow their own orchards suggests that individual producers in antiquity had greater leverage in crop choices than is usually supposed.[97] Two Neo-Assyrian letters of the second half of the eighth century BCE, both excavated at the capital of Kalḫu, throw some light on the desirability of rice (*kurângu*) as a subsistence crop to the individual producers and the state in the early first millennium BCE. The first letter is addressed to the governor of Kalḫu, either Bēl-dān (active 744–734) or Šarru-dūrī (active c. 734–728), by a subordinate official and refers to citizens fleeing a settlement named Ṣidqī, where they are reported to have "abandoned the rice (*kurângu*) they were sowing":[98]

> [. . .] they fled [. . .]. I myself chased (those) of the village (and) brought (them) back. [. . .] chased (those) of the village Ṣidqī, (but) they escaped. None of them agree to stay (and) to cultivate their seed grain; they have abandoned the rice they were sowing and have [. . .]. Of the cooks (and) of the [. . .], my lord will see the yield of the threshing-floor, how much they bring in.[99]

95. Jursa 1998, no. 13 (BM 63797); Jursa 1999/2000, 294 n25.

96. Fales 2010, 76–78; Jursa 2010, 362; Thompson 1949, 99–100; van Zeist 2008.

97. RINAP 3/1 15 viii.8–19; 16 viii 12–23.

98. On dating of the text, see Postgate 1973, 10–11. The location of Ṣidqī is not known, but it must be in the province of Kalḫu, since it appears in the archives of that governor.

99. Postgate 1973, no. 207 (ND 425).

As the letter is highly fragmentary, the factors impelling the runaway cultivators of Ṣidqī are thoroughly unclear. It may perhaps be a case of locust infestation, an agricultural woe well attested in other Mesopotamian epistolary and omen texts.[100] It is also not entirely certain whether rice was a crop of the cultivator's own choice or if its cultivation here was encouraged by the official writing the letter.

The second Neo-Assyrian letter, also dating to the reign of Tiglath-pileser III (r. 744–727), suggests more clearly that the state was indeed keen on incorporating rice (*kurângu*) into its agricultural repertoire and system of rations.[101] In this letter, a royal official tells the king that all the barley and rice has been harvested on land ostensibly belonging to the state. The grain was most probably used to provision the building construction referred to in subsequent fragments of the same letter.

> [As to the order] that the king, my lord, gave me: "Break the fallow ground and cut the hay!", the fallow ground has been broken and two cubits high hay has been cut. All the barley and rice has been harvested. Really, I [. . .] one homer of sesame oil to [. . .]. Ṣabu-damqu [. . .] the fort [. . .] of the Arsazaeans [. . .] each [x] cubits to the four directions. 6 depots of broad bricks for 6 towers, 10 brick (depots of) broad bricks for 4 towers (and) 10 (depots of) bricks before the city wall for . . . each 20 cubits between them. (Of) the nearest tower before the gate, 6 cubits of each wall set has been laid down.[102]

It is likely that the letter found at the Assyrian provincial center of Kaḫat, in the Ḫābūr triangle, dating to c. 1100 BCE, in which a certain official Erīb-īli writes to his subordinate in Kaḫat asking whether there was enough rice (*kuriangu ibašši laššu*) and for someone to irrigate (*lišqi*) the fields, also represents a state-sponsored enterprise:

> Speak to Kalbu, thus Erīb-īli: I am well. Is there rice or not? Why have you not written news to me? Let someone go to Qalliya and ask him for water and let him irrigate (the fields). Bring an *abaruḫu*-tool to Ṭab-ṣiya.[103]

The incorporation of rice in a system of rations for provisioning the bureaucracy, the labor force, and their dependents is also evidenced in the Persepolis Fortification archives of the Achaemenid period.[104] Yet the rarity of rice in official

100. Radner 2004.

101. *Kurângu* is written with the logogram ŠE.LIL.MEŠ; cf. Practical Vocabulary of Aššur for the equation with *kurângu:* Landsberger and Gurney 1957 (VAT 14264, 14260; Aššur 13956 (Istanbul); SU 51/131).

102. SAA 19, 20 obv., lines 3-18 (ND 2675).

103. Salvini 1988, K9. T1. The *abaruḫu* is some kind of metal object, most likely an agricultural implement (CAD, s.v. *abaruḫḫu*).

104. E.g., PF 544; PFNN 587.

documentation indicates that rice was a crop with little or no tradition of taxation across much of the ancient Middle East, so it left a light paper trail. A comparable situation may be observed in late medieval Spain and Italy (c. thirteenth century CE), where rice was cultivated, primarily by the poor, in low-lying marshes too damp for wheat and at the outset bore little or no tradition of taxation.[105]

THE LIMITS OF RICE PRODUCTION

Despite the apparent adoption of rice cultivation by the Assyrian state as early as the eleventh century BCE, if the identification of *kurângu* with rice is recognized (as I advocate), rice never emerged as a rival to the traditional grain plants of the Middle East and the Mediterranean in antiquity. This was certainly due in part to environmental constraints. Although rice is cultivated at a wide range of latitudes, it is best grown in areas of low-lying fertile soils like naturally or artificially inundated river valleys, marshlands, and desert oases, with recommended growing temperatures of 20 to 38 °C.[106] Ancient and medieval sources indicate that rice was a summer crop in the Middle East and the Mediterranean, typically planted in spring and harvested between August and November.[107] Ibn al-ʿAwwām, writing in twelfth-century Spain, suggests that rice may be sown twice a year, but admits that winter sowing yields far less than summer.[108] Furthermore, as the sprouted seedlings of rice grow best in standing water, rice cultivation offers potential breeding sites for mosquito malaria vectors, in particular *Anopheles superpictus* and *Anopheles maculipennis,* which are known to prefer rice fields in the contemporary Middle East.[109] But it is not clear whether the ancients made any causal connection between rice cultivation and malarial fever.

Like wheat, rice is high in yield, relatively disease resistant, and of high calorific value, making it an attractive cultivar for producers. But the intensive labor and water requirements of rice circumscribed its potential ecological niche in the ancient Middle East and eastern Mediterranean. In antiquity, rice cultivation was

105. For rice in medieval Spain, see Lagardère 1996, 71–87; Levi-Provencal 1932, 165–66; Montanari 1994, 101–02, 131. For Italy, see Lecce 1958; Messedaglia 1938, 2–15, 50–64; Motta 1905; cf. rice-bread (*khubz al-aruzz*) as poor man's food in medieval Basra and Wāsiṭ: Ibn Qutaybah, *ʿUyūn al-akhbār* I, 221 (Cairo edition); Ibn Baṭṭūṭa II.5; Al-Jāḥiẓ, *Kitāb al-bukhalāʾ*, 100, 108 (Cairo edition).

106. Van der Veen 2011, 77.

107. Feliks 1963b; Samuel 2001, 388–90. The Mishnah (*Shevi'it* 2.7) notes that rice, like millet and sesame, is a summer crop, planted before Rosh Hashanah (September/October). The Jerusalem Talmud (TJ, *Shevi'it* 2:2) more specifically remarks that rice was grown in irrigated fields three months before Rosh Hashanah. Strabo (XV.1.18) notes that rice was harvested at the setting of the Pleiades (October/November).

108. Ibn al-ʿAwwām, *Kitāb al-filāḥa* II.1, 59.

109. Hanafi-Bojd et al. 2018; Samuel 2001, 390.

largely limited to regions of higher temperatures and better irrigation infrastructure, like the marshlands of southern Mesopotamia, Susiana, and the Ḫābūr triangle of northeastern Syria. Beyond the naturally inundated marshlands in the extreme south of Mesopotamia, where the distributaries of the Euphrates and Tigris form a delta that leads into the Persian Gulf, rice cultivation in the flat semiarid alluvial plain of southern Mesopotamia would have been entirely dependent on irrigation. The searing summers and high evapotranspiration rates in southern Mesopotamia, up to 3,400 mm/y in the Baghdad region, are unfavorable for large-scale rice production, especially since the crop is typically grown in damp soils.[110] Talmudic and early Islamic sources indicate that rice was predominantly grown in lower Mesopotamian districts with easy access to abundant water supplies, like the Kaskar region, where the Shaṭṭ al-Ḥayy, a major canal, links the Tigris and Euphrates; or Maysan (Greek Mesene), in the eastern part of the Great Swamp (Arabic *al-Baṭīḥa*) along the lower reaches of the Tigris.[111]

Even in the rain-fed moist steppe zone of northern Mesopotamia, where dry farming of cereal crops was practiced, summer cultivation of rice would have required much more water than was provided by winter and spring precipitation. The estimated water requirement of paddy in modern Iran is 9,000 m³ per hectare, whereas barley, the staple crop of the ancient Middle East, only needs between 4,000 and 7,500 m³ per hectare.[112] Irrigation was, of course, also adopted in northern Mesopotamia to circumvent water shortages and to bring more land into cultivation. The Assyrian king Aššurnasirpal II (883–859 BCE) brags in his famed "banquet stele" of the elaborate irrigation scheme devised for Kalḫu, his new capital, which involved diverting the waters of the Greater Zab into fields lying contiguous to the city:

> I dug out a canal from the Upper Zab, cutting through a mountain at its peak, (and) called it *Patti-ḫegalli* ("canal of plenty"). I irrigated the meadows of the Tigris (and) planted orchards with all kinds of fruit trees in its environs.[113]

It is not surprising therefore to find textual references to rice cultivation in the central Assyrian provinces in the eighth century or at early-eleventh-century Kaḫat, in the Assyrian-held eastern Ḫābūr region, where the adoption of rice was probably stimulated by a combination of higher precipitation rates and large-scale state-sponsored irrigation schemes. The epistolary text from eleventh-century Kaḫat, in any case, explicitly refers to irrigation.

110. See Sanlaville (2002, 95) on evapotranspiration rates in southern Iraq.

111. Simpson 2015, 21.

112. Rice: Nesbitt, Simpson, and Svanberg 2010, 313. Barley: Balland 1989, 802–05.

113. RIMA II A.0.101.30: 36–38. On Aššurnasirpal's canal in Kalḫu, see Oates 1968, 46–48.

THE CULINARY USES AND SOCIAL FUNCTIONS OF RICE

While we have established that rice was cultivated in the Middle East and the eastern Mediterranean much earlier than is generally acknowledged, we now need to discuss its consumption, uses, and cultural associations in antiquity. In the ancient Middle East, apart from bread (and beer, in the case of barley), cereal crops were used to concoct a variety of foods for daily consumption and ritual offering, ranging from cereal porridges or mash (Akkadian *aṣūdu* or *pappāsu*) to garish confections like the *siqqurratu*, a tiered cake resembling a Mesopotamian temple tower.[114]

It appears that South Asian food-processing techniques and culinary traditions did not migrate along the trade routes which first brought rice to the Middle East and the Mediterranean, since Greek and Hebrew sources indicate that rice was primarily consumed in the form of bread, porridge, and cake, much like barley or wheat varieties, or even adapted to older recipes.[115] A scholium to Aristophanes's *Hippeĩs*, notes, for instance, that rice was employed to make *thrĩon*, a traditional dish of food wrapped in fig leaves, variously filled with eggs, milk, lard, flour, honey, cheese, and so on.[116] Ancient opinions on the nutritious value of rice appear to be ambivalent. The physician Galen (c. 129–c. 216/17 CE) was not enthusiastic: "Everyone uses this grain for restraining the stomach, producing a boiled version like groats. But it is less digestible and less nourishing than groats, as it is also inferior to groats in its pleasantness as food."[117]

Oribasius, a fourth-century CE Greek physician, shared Galen's view that rice was not easily digested.[118] The statements of both Roman physicians could reflect a more widespread Mediterranean belief concerning the dietary inferiority of rice. Accompanied by conservative foodways, this belief might account for the relatively small role rice had as a subsistence grain in the ancient Middle East and the Mediterranean.

Quite apart from subsistence, rice (*kurângu*) was also thought to have magico-medical properties, although only one Neo-Babylonian ritual text from Sippar

114. On porridges, see ND 5461:2; SAA 12, 68:30. On confectionaries, see TFS 87 r.2; Gaspa 2012, 45–91, 295–96.

115. On rice bread in antiquity, see Galen, *Aliment. fac.* I 16 = 6.524 K.; Oribasius, *Synopsis ad Eustathium* IV 13.6.3; Aet. Amid. *Med.* VIII 31.18; TB, *Pesaḥim* 35a, *Pesaḥim* 50b-51a, *Berakhot* 37a; Mishnah, *Ḥallah* 3.7–10; TJ, *Ḥallah* 3.1, 3.5. On rice cakes, see Aet. Amid. *Med.* IX 20.55d; Tosefta, *Berakhot* 4.7; TJ, *Berakhot* 6.1; Ath. XIV.647d.

116. *Schol.* Ar. *Eq.* 954b.1.

117. Galen, *On the Properties of Foodstuffs* I.17 (trans. Powell 2003). Cf. Ibn al-ʿAwwām, *Kitāb al-filāḥa* II.1, 62. Yanbūšād, a Kasdānī (Chaldean?) author of late antique Iraq (c. 300–600 CE), attributes to rice the humoral quality of dryness and alleges that the consumption of rice interfered with one's cognitive faculties (ap. Ibn Waḥshīyah, *Al-filāḥah al-nabaṭīyah*, 482; El Faïz 1995, 155).

118. Oribasius, *Collectiones medicae* I.16.

(ca. 626–539 BCE) avows for this role.[119] Rice appears in this text alongside various amulet stones and *materia medica* used against malevolent disease-inducing entities. The appearance of rice alongside such constituents in an apotropaic prescription is far from anomalous, since grains and lentils like *arsuppu, šeguššu* (a variety of barley), emmer, bread wheat, and chickpeas (which were processed into flour) were often used in anti-witchcraft and other magico-medical rituals.[120] The association of rice with magical ritual even appears to have persisted into Late Antique Mesopotamia.[121] Glimpses of late pagan agricultural life in the rural areas of northern Mesopotamia are preserved in Ibn Waḥshīyah's *Al-filāḥah al-nabaṭīyah*, or "Nabataean Agriculture" (tenth century CE), an Arabic reworking of a Syriac original compiled circa 600 CE by one Qūthāmā, who, according to the text, relied on earlier Mesopotamian authors.[122] In this text, rice bears an interesting association with tricksters and "people of illusions and conjurers":

> It (rice) is also used by tricksters who take a handful (of rice) and throw it in a bowl where there are snakes. The snakes will then rise up on their tails and dance in the bowl. This is done by magicians and the people of illusions and conjurers. [123]

Anûkh (c. 300–600 CE), one of the Mesopotamian authors cited by Qūthāmā, places rice under the influence of Mars and Saturn, planetary bodies which in Mesopotamia were long believed to exercise a nefarious influence.[124] The supernatural astral affiliation of the crop, albeit known from a later text, is also consistent with its use in magical traditions. The peculiar cultural associations gained by rice in the Middle East ultimately indicate that the crop successfully adapted itself to the local cultural milieu.

DIVERGENT TRAJECTORIES

While in South Asia, rice became the object of lyrical musing and religious ceremony in the first millennium BCE and beyond, it does not appear to have had any such great sway on the cultures of the Middle East and the Mediterranean before the Islamic period. It is nevertheless certain that rice has a longer history in the

119. BM 93084 = CT 14, pl. 16: obv. 3.

120. Abusch and Schwemer 2011, 264, 285, 287.

121. It may be significant that rice is extensively mentioned by late Roman medical authors hailing from the eastern provinces (e.g., Aetius of Amida, Paulus of Aegina, and Oribasius of Pergamon). This suggests greater familiarity with the crop in the Roman East. For references, see Konen 1999, 40.

122. Hämeen-Anttila 2006, 3–52.

123. Ibn Waḥshīyah, *Al-filāḥah al-nabaṭīyah* 487; Hämeen-Anttila 2006, 189.

124. El Faïz 1995, 154–55. Note, for instance, the astrological report of Balasi from seventh-century Nineveh (SAA 8, 82), in which he suggests that the conjunction of Mars and Saturn portended evil and necessitated an apotropaic *namburbû* ritual.

Middle East and the Mediterranean than previously assumed and was a notable constituent of an increasingly diversified agricultural regime. The neglect of rice as a component of ancient Middle Eastern and Mediterranean agriculture partly stems from the obscure terminology relating to rice in local languages, an uncertainty that has been clarified but perhaps not completely resolved in this chapter. The divergent names for rice in Middle Eastern and Mediterranean languages hint at varied and complex dispersal histories following either maritime or overland routes from the Indus Valley and further east. The lack of a straightforward Indic loanword for rice in the earliest texts from the Middle East reveals a highly stochastic pattern of circulation and suggests a slow westward transmission of rice via intermediaries in the Iranian plateau rather than direct transmission from India.

The evidence for rice in ancient Middle Eastern and Mediterranean sources also indicates that an interplay of ecological, economic, and cultural forces affected this crop's diffusion. Rice thrives in wet soils and warm climates; hence cultivation is limited by latitude, altitude, and topography. In the right environment, rice makes for an important summer crop complementing the winter grain staples of the Middle East. Culturally, ancient Middle Eastern peoples appreciated rice, but less so in Mediterranean contexts. As a result, it was apparently not economical to transport rice across long distances in the Mediterranean region the way it was for wine, olive oil, and wheat. In the wider Indian Ocean region, however, the commodity potential of rice from India was not dampened, since the *Periplus Maris Erythraei* cites it as an article of trade in the first century CE.[125] The selective export of Indian rice only to ports at the entrance of the Red Sea (northern Somalia and Socotra) could suggest that local rice production was sufficient in contiguous regions, making the Indian exports redundant.[126] Alternatively, rice was simply not known to many peoples inhabiting the Indian Ocean rim. The novelty of rice to some denizens of the Indian Ocean world even as late as the seventh century CE is revealed by Ibn al-Faqīh al-Hamaḏānī's anecdotal account of 'Utbah ibn Ghazwān's troops mistaking a basket of non-hulled rice for a poisonous substance left by the enemy, suggesting that rice remained unfamiliar to some Peninsular Arabs even on the eve of the Islamic conquest of Mesopotamia.[127]

125. *PME* 14, 41, 31. Production of rice in Gedrosia: *PME* 37.

126. *PME* 14, 31.

127. Al-Hamaḏānī, *Mukhtaṣar Kitāb al-buldān,* s.v. al-Baṣra, (trans. Massé 1973), 227. A late-fourteenth-century Yemeni agricultural treatise by al-Malik al-Afḍal Al-ʿAbbās suggests, however, that rice was known in pre-Islamic Arabia (Meyerhof 1943–44, 53).

4

Persian "Apples": Citruses

INTRODUCTION TO THE GENUS *CITRUS*

The globally familiar citrus family includes many commercially important comestible species, like the mandarin (*Citrus reticulata*), sweet orange (*C. sinensis*), Seville or bitter orange (*C. aurantium*), pomelo or shaddock (*C. maxima*), satsuma (*C. unshiu*), citron (*C. medica*), lime (*C. aurantiifolia*), lemon (*C.* × *limon*), grapefruit (*C. paradisi*), and makrut or kaffir lime (*C. hystrix*). All citrus varieties are of tropical and/or subtropical Asia-Pacific origin, with a natural range from northwest India through to southwestern China, Southeast Asia, and northern Australia.[1]

Citrus taxonomy is convoluted, owing to interspecies hybridization and high mutation frequency. As citrus species are able to reproduce apomictically (i.e., asexually) or through vegetative propagation, the interspecific hybrids are able to maintain their distinct characteristics.[2] Most commercially significant citrus species do not occur in the wild and are anthropogenic hybrids or selections from wild species.[3] Recent phylogenetic evidence, as well as earlier studies of citrus morphology and biochemical features, indicate that all commercially important citruses can be traced back to a handful of wild amphimictic or interfertile progenitor species, including citron (*C. medica*), mandarin (*C. reticulata*), pomelo (*C. maxima*), biasong (*C. micrantha*), Ichang papeda (*C. ichangensis*), Nagami

1. Fuller et al. 2011, 549; Gmitter and Hu 1990; Mai and Girard 2014, 171; Pagnoux et al. 2013, 422; Weisskopf and Fuller 2013b, 1481; Wu et al. 2018; Zohary, Hopf, and Weiss 2012, 146.
2. Curk et al. 2016; Mabberly 1997, 167; Pagnoux et al. 2013, 423; Zohary, Hopf, and Weiss 2012, 146.
3. Bayer et al. 2009, 679; Mabberly 2004, 485–91.

kumquat (*Fortunella margarita*), mangshanyegan (*C. mangshanensis*), and perhaps extinct varieties of wild citrus.[4]

Only two varieties of citrus were known in the ancient Middle East and Mediterranean: citrons and lemons. The citron (figure 11) is indisputably present in these regions by the mid-first millennium BCE, while the lemon (figure 12) arrives slightly later, in the last quarter of the first millennium BCE. Other citrus species, like Seville or bitter orange (*C. aurantium*) and lime (*C. aurantifolia*), dispersed westward in the early medieval period with the Arab expansion, while sweet oranges (*C. sinensis*) were only familiar in the Mediterranean from the Renaissance on.[5] Although the citron (*C. medica*), one of the "true" citrus species, was the first citrus species to arrive in the Middle East and the Mediterranean, it is less familiar today on account of its displacement by other sour citrus cultivars like lemon and lime. The citron is a thorny, short-lived (fifteen to twenty years) shrub or small tree growing to a height of up to three meters.[6] The fragrant yellow fruit of the citron is distinguished by minimal flesh, a thick aromatic rind, and above all its knobby skin. The natural habitat of the citron extends from the eastern Himalayan foothills to monsoonal northeast India, with a disjunctive population in the Western Ghats of South India.[7] Meanwhile, the lemon is of hybrid origin, with the citron being its direct male parent and up to three or four other ancestral citrus taxa (*C. aurantium, C. micrantha, C. reticulata*) contributing to its genome.[8] This hybridization took place in South Asia, although the chronology and specific circumstances leading to the creation of lemon cultivars are hazy.[9]

CITRUSES IN EARLY SOUTH ASIA

Citrus cultivation has a long genealogy in South Asia. The citrus charcoal from the Harappan site of Banawali, dating to the last quarter of the third millennium BCE, is the earliest find of cultivated citrus in the Indian subcontinent.[10] Citrus seeds were also recovered from an early-second-millennium BCE context at the Harappan site of Sanghol, in the Punjab.[11] Madho Vats proposes that an elliptic-ovate, leaf-shaped

4. Barkley et al. 2006; Barrett and Rhodes 1976; Bayer et al. 2009, 669, 679; Carbonell-Caballero et al. 2015; Curk et al. 2016; Froelicher et al. 2011, 50, 58; Pagnoux et al. 2013, 423; Pang, Hu, and Deng 2007; Scora 1975, 371–75; Wu et al. 2018.

5. Ramón-Laca 2003, 507–08, 510; Ruas et al. 2015 364–65; Watson 1983, 42–50.

6. Bayer et al. 2009, 675, 680; Curk et al. 2016; Gmitter and Hu 1990, 273–74; Yang et al. 2015.

7. Asouti and Fuller 2008, 114; Singh 2017, 109, 119.

8. Bayer et al. 2009, 675, 679; Curk et al. 2016; Gulsen and Roose 2001; Ramadugu et al. 2013, 12.

9. Nasrallah 2007, 634 (glossary entry in the text edition for Ibn Sayyār al-Warrāq's *Kitāb al-Ṭabīkh*); Watson 1983, 46.

10. Asouti and Fuller 2008, 126.

11. Asouti and Fuller 2008, 114–15; Fuller and Madella 2001, 341; Saraswat 1997, 2014, 208–09 (illustrated); Weisskopf and Fuller 2013b, 1482.

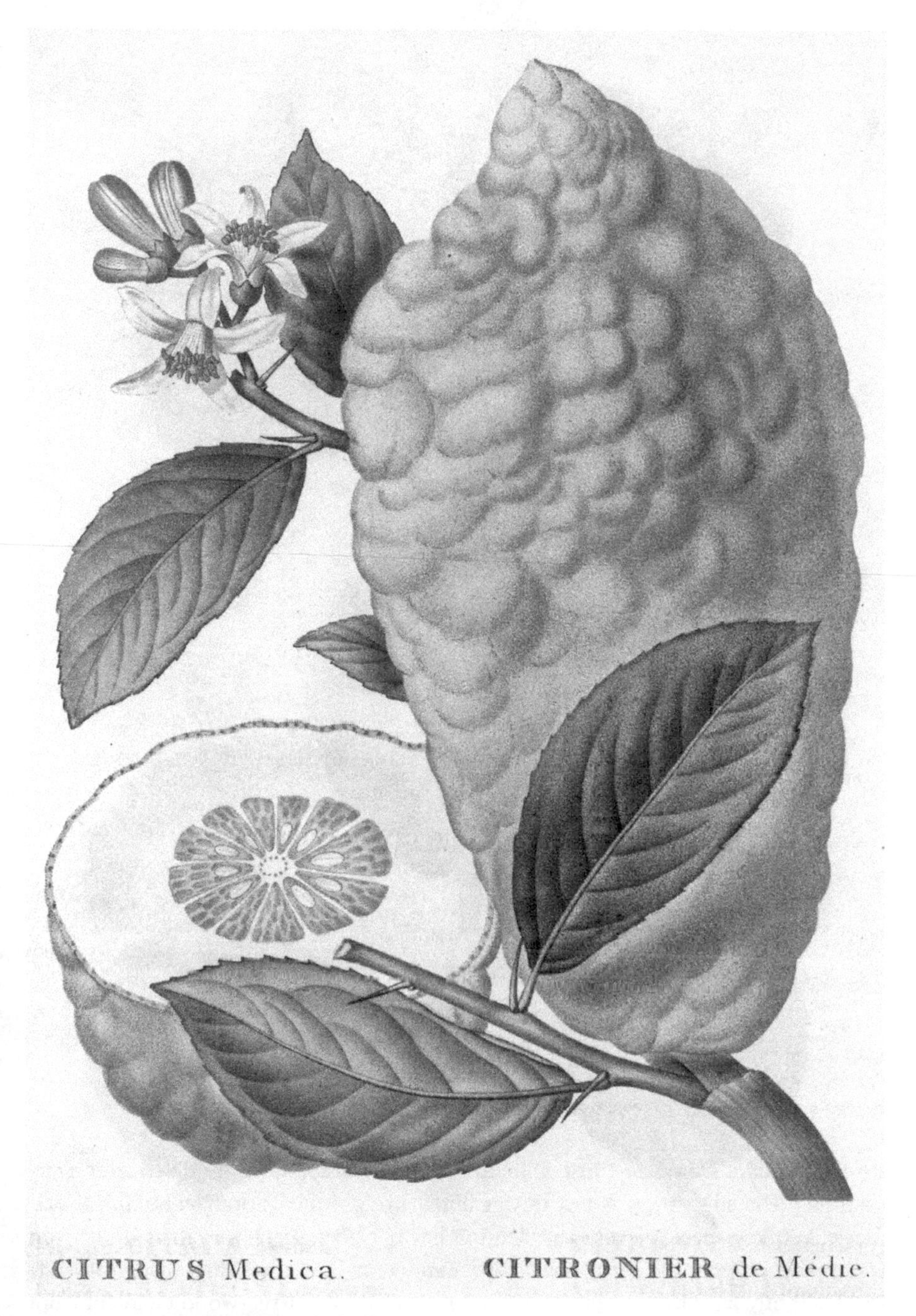

FIGURE 11. Citron (*Citrus medica*), illustration by Pierre-Joseph Redouté in Duhamel du Monceau, *Traité des arbres et arbustes que l'on cultive en France en pleine terre* (Paris, 1801–19). New York Public Library (public domain).

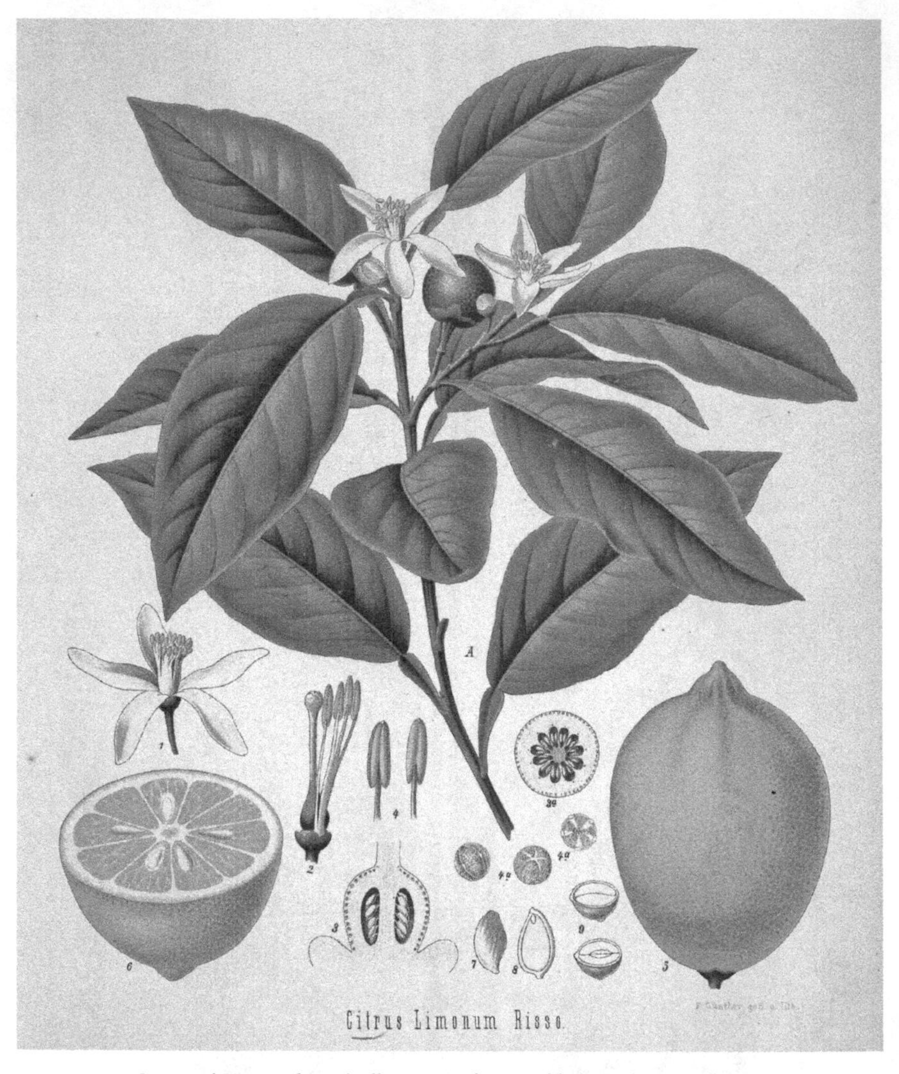

FIGURE 12. Lemon (*Citrus × limon*), illustration from *Köhler's Medizinal-Pflanzen* (Gera, 1887). Peter H. Raven Library, Missouri Botanical Garden (public domain).

steatite pendant from late-third-millennium BCE Harappa was modeled after a citrus leaf.[12] The suggestion is not unwarranted in light of the archaeobotanical evidence. While a species-level identification has not been made for citrus finds from Harappan contexts, the citron is a likely candidate in light of its early dispersal westward. Slightly later finds of citrus in South Asia include citrus-wood charcoal from

12. Vats 1940, 467.

Sanganakallu in South India (c. 1400–1300 BCE) and citron peel fragments from Gopalpur in Odisha, eastern India (c. 1300–1100 BCE).[13] The scanty but geographically dispersed finds of early citruses, probably citrons, indicate that citriculture had a pan-Indian distribution by the middle of the second millennium BCE.

The scent of the citron was so familiar to South Asian audiences that it could be cited as a benchmark for incense. In Kauṭilya's *Arthaśāstra,* a treatise on political economy dating in part to the late centuries BCE, incense from the region of Suvarṇakuḍya is likened to the odor of citrons (II.11.64). Citrons were regarded as ideal greeting gifts in ancient South Asia in a manner not unlike the Chinese custom of gifting oranges during the Chinese New Year.[14] In the Sanskrit playwright Kālidāsa's *Mālavikāgnimitram,* a historical play set in the second century BCE, a female attendant procures citrons (*bījapūraka*) from the royal garden as greeting gifts for Queen Dhāriṇī; while in the *Apadāna,* a compendium of Buddhist biographies, a monk is nicknamed a "giver of the citron" (*mātuluṅgaphaladāyaka*).[15] The importance of citrons in South Asian cultures is also suggested by the multitude of names for this plant in both classical and vernacular languages. In Sanskrit, its many names include *mātuluṅgā, bījapūra, cholaṅga, vetasa, sukesara, gila, jantumārin, jambīra, dvijaketu,* and *rucaka.* A few of these are descriptive labels (*bījapūra,* "seed-filled"; *jantumārin,* "worm-killer"), while others are borrowings from Austroasiatic and Dravidian languages (*cholaṅga, mātuluṅgā*).[16]

THE ARCHAEOLOGICAL AND VISUAL DATA FOR CITRONS AND LEMONS

If the history of citruses in the ancient Middle East and the Mediterranean were reconstructed on the basis of the textual data alone, it would appear that only one species, the citron, was known in either region by the mid-first millennium BCE. The archaeological data, while patchy for earlier periods and biased in favor of the Mediterranean region, suggest an earlier date. More significantly, the archaeological evidence allows us to distinguish the lemon from the citron and date its arrival in the Mediterranean basin to the last quarter of the first millennium BCE. The earliest archaeological finds of citrus in the Mediterranean consist of a handful of charred seeds, a few with remnants of fruit flesh, unearthed in a stratum dated to c. 1200 BCE at Hala Sultan Tekke, a city on the southeast coast of Cyprus with pan-

13. Sanganakallu: Fuller et al. 2011, 549. Gopalpur: Kingwell-Banham 2015, 180; Kingwell-Banham, Petrie, and Fuller 2015, 279; Kingwell-Banham et al. 2018, 7, 10.

14. Welch 1997, 14–15.

15. *Apadāna* II.446; Kālidāsa, *Mālavikāgnimitram,* Act 3. See Sharma (1979, 49–50) on other references to citrons in Sanskrit sources.

16. Ghosh 2000, 156; Southworth 2005, 215, 221.

Mediterranean trading links (map 9).[17] While the remains have not been subject to direct dating, Ayelet Gilboa and Dvory Namdar remark that the presence of several seeds in a secure archaeological context leaves little room for doubt on the dating.[18] But even if the early dating is accepted, these seeds do not necessarily represent local cultivation, as the fruits could have been sourced from regions further east, perhaps Mesopotamia or Iran.

The earliest and surest sign of citrus cultivation in the Mediterranean zone is provided by pollen samples from mid-first-millennium BCE sediment deposits in Kyme (Cumae), Carthage, and Jerusalem. Citrus species depend on entomophilous pollination (i.e., pollination by insects), so airborne pollen is limited. Consequently, even minute traces of citrus pollen in the archaeological record are significant, as they suggest the presence of a sizeable number of trees in the immediate vicinity in order for the pollen to be detectable in the sediment samples. The pollen data from an ancient lagoon at Kyme/Cumae on the Bay of Naples provide evidence for citron cultivation, possibly from as early as the foundation of the site in the seventh century BCE, if not by the mid-first millennium BCE.[19] A total of seventy citrus pollen grains were observed regularly throughout the entire sediment sequence dated between the seventh century BCE and the fifteenth century CE, attesting to the great antiquity and continuity of citrus cultivation in southern Italy.[20]

Elsewhere in the Mediterranean, citrus pollen dating to the mid-fourth century BCE was found in a silted channel at the harbor of the renowned Phoenician city of Carthage on the North African coast.[21] Monte Sirai, another Phoenician colony-site in southern Sardinia, has also yielded evidence for citrons from a chronological context (sixth century BCE) closer to the finds from Cumae. Chemical analysis of organic residues from a wine jug found among funerary offerings at the necropolis of Monte Sirai (Grave 158) yielded traces of polymethoxyflavones, a group of flavonoids (organic chemical compounds) found exclusively in the *Citrus* genus.[22] They are found especially in the essential oil of citrus peels, suggesting the use of citrus peels in the flavoring of wine in antiquity.[23]

17. Hjelmqvist 1979, 113–14.

18. Gilboa and Namdar 2015, 273. Dafna Langgut (2017), however, raises doubts about the dating and archaeological context of these finds. She also reports that the seeds can no longer be located.

19. Mai and Girard 2014, 173–74; Pagnoux et al. 2013, 425.

20. Mai and Girard 2014, 174.

21. Mai and Girard 2014, 173; van Zeist et al. 2001, 32.

22. Pagnoux et al. 2013, 425. On polymethoxyflavones, see Li et al. 2009.

23. Some ancient sources do note the use of citrons as an additive to wine, albeit in the context of using it as a purgative for poisons (Oppius, ap. Macrobius *Saturnalia* III.19.4; Pliny *HN* XXIII.105; Theophr. *Hist. pl.* IV.4.2; Vergil *G.* II. 126–35).

MAP 9. Ancient sites which have yielded evidence for citruses (*Citrus* spp.). © Peter Palm (Berlin).

Closer to the Phoenician homeland in the Levant, citron pollen grains dated stratigraphically and by optically stimulated luminescence to the fifth or fourth century BCE were identified in the layers of plaster in a garden attached to the Persian satrapal residence in Ramat Raḥel, Jerusalem.[24] As the citron pollen appears at relatively high frequencies at the site (up to 32% of the palynological assemblage of Layer II), there is no doubt that citron trees grew there.[25] Apart from micro-remains in the form of chemical compounds and pollen, a tomb in Tamassos in central Cyprus, dating to the Archaic period (c. 750–480 BCE), has yielded an entire desiccated citron fruit in an assemblage of unburnt funerary offerings of fruits.[26] The citron, presently in the collection of the Fitzwilliam Museum of the University of Cambridge, was found during Ohneflasch-Richter's 1889 excavations in Tamassos.[27] As the exact tomb from which the fruit derives is presently unknown, no precise dating can be accorded to the find.[28]

The westward spread of citrons has frequently been credited to Persian elites transplanting the trees to the eastern Mediterranean, and this interpretation is reinforced by the earliest Greek terms for the citron: Median or Persian apple.[29] The evidence for citrons from Cumae, Monte Sirai, and perhaps even Tamassos predates, or is contemporary with, the earliest phase of Persian expansion into the Mediterranean, suggesting that the introduction of citrons into the Mediterranean region was not necessarily the handiwork of the Persians or even the result of imperial agency. The temporal sequence of archaeological finds of citrus in the Mediterranean suggests that the citron may have been familiar in the eastern Mediterranean by the second quarter of the first millennium BCE. Once established on the Levantine coast, the citron spread across the Mediterranean through the colonial and trading networks of the Phoenicians and Greeks. This mode of dispersal is consistent with the presence of citrons at Cumae, Carthage, and Monte Sirai, key colony sites of the Greeks and Phoenicians in the western Mediterranean.

Macro and micro-botanical finds of citrus become much more copious toward the end of the first millennium BCE and the early centuries CE. The lemon, previously thought to be a medieval Arab introduction to the Mediterranean, makes its

24. Langgut 2014, 5–6, 8–9; Langgut et al. 2013, 122–24. See Lipschits, Gadot, and Langgut 2012 for an overview of the site of Ramat Raḥel.

25. Langgut 2014, 5–6, 8–9.

26. Buchholz 1988, 123; Schoch 1993, 95; cf. Euripides Fr. 912.

27. Inventory no. GR 275.1892.

28. Although the fruit under discussion from Tamassos has been illustrated in Buchholz's publication and certainly appears to superficially correspond to the morphology of a citron, a re-examination of the find is in order given the early date of discovery. A carbonized citrus-like fruit discovered at sixth-century BCE Ischia in the Gulf of Naples was later reidentified as the fruit of *Sorbus domestica* (true service tree), a native of southern Europe (Coubray, Zech-Matterne, and Mazurier 2010).

29. Langgut 2014, 5–6, 8–9; Langgut et al. 2013, 122–24.

earliest appearance in the archaeological and iconographic records of the Mediterranean between the late centuries BCE and the early centuries CE. The later archaeological finds of citrus are almost entirely concentrated in Campania (in southern Italy) and Egypt. This is no doubt a result of more intensive archaeological work in these regions and inherent preservational biases, namely the dry climatic conditions of desert regions in Egypt and the Vesuvian eruption of 79 CE, which preserved an extraordinary quantity of organic matter at sites like Pompeii, Herculaneum, and Oplontis.[30]

Egypt has yielded several complete and well-preserved desiccated citron fruits. Renate Germer draws attention to an entire citron fruit and several fragments found in an undated and poorly documented context at Thebes.[31] While Germer has arbitrarily dated the find to the Ptolemaic period (third to second century BCE), an earlier dating is not implausible in light of the citron's appearance in mid-first-millennium BCE contexts elsewhere in the Mediterranean. Several complete, albeit unprovenanced, citron fruits of Greco-Roman date are also to be found in the collection of the Dokki Agricultural Museum in Cairo.[32] The only near-complete citron fruit with a well-documented archaeological context in Egypt appears to be the desiccated fruit recovered from the site of Mons Claudianus in the Eastern Desert (first century CE).[33] Citron seeds have also been identified at the Eastern Desert sites of Mons Claudianus (second century CE) and Mons Porphyrites (second century CE), and further afield at the Red Sea port of Myos Hormos (first to second century CE).[34] Later finds of citron in Roman Egypt include rind and seed fragments from Kellis in the Dakhla Oasis (third to fourth century CE) and leaves from the funerary garlands of mummies in the necropolis of Antinoë (c. third century CE).[35]

In southern Italy, pollen samples recovered from sediments in Lake Avernus, west of Naples, indicate that citrus species were grown around the lake by the late centuries BCE.[36] Citrus pollen dated to the first century CE was also recovered from the silted harbor of Naples (ancient Neapolis).[37] The city of Pompeii in the Vesuvian region has yielded a wealth of evidence for citrus species in the form of macro-remains, pollen, and iconographic evidence dating from the third century BCE to the destruction of the site by Vesuvius in 79 CE. The earliest citrus finds at Pompeii,

30. Borgongino 2006.

31. Germer 1985, 106.

32. Darby, Ghalioungui, and Grivetti 1977, 704–06.

33. Van der Veen 2001, 193.

34. Mons Claudianus: van der Veen 2001, 180–81. Mons Porphyrites, van der Veen and Tabinor 2007, 94–95. Myos Hormos: van der Veen 2011, 86.

35. Kellis: Thanheiser 2002, 307, 309; van der Veen 2011, 86. Antinoe: Germer 1985, 106.

36. Grüger and Thulin 1998, 38; Grüger et al. 2002, 251.

37. Ermolli et al. 2014, 9, 12.

six citron seeds found in a well under the Temple of Venus, date to the pre-Roman Samnite period of Pompeii's history (third to second century BCE).[38] Citrus seeds dating to a second-century BCE context were also found in the House of the Wedding of Hercules and the House of the Vestals.[39] Palynological analysis at the House of the Wedding of Hercules confirms local cultivation of citrus trees.[40] Later finds in Pompeii include citrus pollen grains from a grave in Porta Nocera and seeds at the Temple of Fortuna Augusta (late first century BCE).[41] Wilhelmina Jashemski has also noted that the trees planted in holed pots along garden walls in Pompeii (e.g., House of Polybius, Garden of Hercules, and the House of the Ship Europa) may have been citruses, given Theophrastus's remark that citrons were sown in pots with holes.[42] The early-third-century CE Roman agricultural author Florentinus also notes that citrons were planted along the wall for ease of roofing them in winter.[43]

Beyond Pompeii, two charred citrus fruits of the first century CE, not identified at the species level, were found at Herculaneum during eighteenth-century excavations.[44] Another unpublished citrus fruit of the first century CE, now housed in the Antonino Salinas Museum in Palermo, was retrieved from a Roman villa (Villa Sora) in Torre del Greco, on the Bay of Naples, during excavations in 1797.[45] Further afield in Rome, fifteen waterlogged citrus seeds and a rind fragment, AMS-dated to the Augustan period (27 BCE to 14 CE), have been identified in a votive deposit under the Carcer Tullianum, an ancient cistern and jail at the northern end of the Forum Romanum.[46]

Archaeobotanical studies and iconographic sources from southern Italy provide the first unambiguous evidence for the presence of lemons rather than citrons anywhere in the Mediterranean or the Middle East. The morphology of the exine or outer layer of the citrus pollen grains recovered from the House of the Wedding of Hercules at Pompeii suggests an identification with lemons (*Citrus × limon*) rather than citrons (*C. medica*).[47] Similarly, the citrus seeds from the Augustan votive deposit at the Carcer Tullianum in Rome, examined with a scanning electron microscope, display an irregular rugose seed surface cell pattern characteris-

38. Pagnoux et al. 2013, 426, 434.

39. House of the Wedding of Hercules: Ciaraldi 2007, 112–13; Coubray, Zech-Matterne, and Mazurier 2010, 278. House of the Vestals: Pagnoux et al. 2013, 425.

40. Lippi 2000; Mai and Girard 2014, 174; Pagnoux et al. 2013, 425.

41. Porta Nocera: Mai and Girard 2014, 174. Temple of Fortuna Augusta: Coubray, Zech-Matterne, and Mazurier 2010, 278; Mai and Girard 2014, 174; Pagnoux et al. 2013, 426.

42. Theophr. *Hist pl.* IV.4.3; Jashemski 1979, 79, 240, 285.

43. Florentinus, ap. *Geoponica* X.7.

44. Van der Veen 2011, 86. These fruits are now at the Archaeological Museum in Madrid.

45. Borgongino 2006, 154.

46. Celant and Fiorentino 2017; Pagnoux et al. 2013, 426.

47. Lippi 2000, 208–10, 2012, 108–09.

tic of lemons (figure 13).[48] A citron seed's surface, on the other hand, exhibits regular longitudinal striations.[49] Also, wood recovered from an amphora in the sculpture garden of the Villa of Poppaea at Oplontis (constructed c. 50 BCE) has been identified as belonging to the lemon, and the exine morphology of the citrus pollen found in the garden here corresponds to the lemon as well.[50]

The presence of lemons in southern Italy is also affirmed by contemporary iconographic sources which depict both lemons and citrons. The House of the Orchard in Pompeii contains two panels with paintings of fruiting lemon trees (figure 14), while another Pompeiian panel, now preserved at the Naples Museum, depicts Eros bearing a floral and fruit garland containing lemons.[51] A Roman mosaic housed at the Palazzo Massimo in Rome, probably originating from a first-century CE villa near Tusculum, contrasts the smooth lemon with the knobby citron, both nestled in a basket of fruits.[52]

While the lemon is present in the Mediterranean and, by extension, the Middle East by the last quarter of the first millennium BCE, ancient texts do not distinguish between lemons and citrons. In the earliest period of its introduction, the lemon was probably perceived as a variety of citron, hence leaving no lexical traces.[53] The Kew botanist Clifford Townsend notes that the terminology for citrons and lemons is confused even in modern Iraq.[54] In Basra, lemons are locally known in Arabic as *atruǧ*, a term that is clearly related to the Hebrew and Aramaic terms for citron (*etrog*).[55] In this respect, medieval Arabic agricultural authors like Ibn al-Awwām who describe several types of citrons probably included lemons with thick mesocarps as a kind of citron.[56]

References to an edible variety of citron in ancient Greek sources may refer to the lemon rather than the citron, whose edible pulp is minimal.[57] While Theophrastus pronounces the citron inedible in the fourth century BCE, authors from the early Roman Imperial period describe an edible variety of citron, which could refer to the lemon.[58] Yet there are a few unambiguous references to the consumption of citrons in

48. Celant and Fiorentino 2017; Pagnoux et al. 2013, 434.

49. Pagnoux et al. 2013, 431.

50. Wood: Jashemski et al. 2002, 102; Lippi 2012, 109; Pagnoux et al. 2013, 425. Pollen: Ermolli and Messager 2013; Ermolli, Messager, and Lumaga 2017; Lumaga et al. 2020.

51. Jashemski et al. 2002, 101; Pagnoux et al. 2013, 425. See Andrews (1961, 41, 44) on other Roman depictions of citrus fruits.

52. Palazzo Massimo (Museo Nazionale Romano), inv. no. 58596; Andrews 1961, 44; Jashemski et al. 2002, 101.

53. Andrews 1961, 44.

54. Townsend 1980, 467.

55. Townsend 1980, 469.

56. Brigand and Nahon 2016.

57. Pagnoux et al. 2013, 436.

58. Ath. III.85c ; Plutarch *Quaest. Conv.* VIII.9; Theophr. *Hist. pl.* IV.4.2.

FIGURE 13. Seeds of *Citrus × limon* recovered from a votive deposit in the Carcer Tullianum, Rome, c. 27 BCE to 14 CE. Courtesy of Dr. Alessandra Celant.

non-Greek sources, which caution against a facile interpretation of all edible citrons as lemons. The Mishnah, for example, remarks that children ate their citrons at the end of the Sukkot ceremony.[59] The earliest known Middle Eastern–Mediterranean sources which explicitly distinguish citrons from lemons date no earlier than the seventh century CE.[60] Ibn Waḥshīyah's *Nabataean Agriculture* (tenth century CE), an Arabic translation of a Syriac text by one Qūthāmā (c. 600 CE), quotes Arabicized Syriac words for the lemon (*ḥasbanā, khashīthā*), suggesting that the distinction between lemons and citrons was made at some point in the pre-Islamic period.[61]

CITRUS TERMINOLOGY IN GREEK AND LATIN

Citrons: These fruits are either from the branches of Corcyra's garden or they were the Massylian dragon's.

MARTIAL, *EPIGRAMS*, XIII.37

In describing citrons as Persian or Median apples, the earliest Greek terms for the citron emphasize its eastern origin.[62] Theophrastus, who offers an accurate account of the

59. Mishnah, *Sukkah* 4.7.

60. Glidden 1937, 382; Ramón-Laca 2003, 504, 508–09.

61. Nasrallah 2007, 634 (glossary entry in the text edition for Ibn Sayyār al-Warrāq's *Kitāb al-Ṭabīkh*); Ramón-Laca 2003, 508.

62. Note Pliny's variant *malus Assyria*, or the "Assyrian apple" (*HN* XII.15).

FIGURE 14. Painting of a fruiting lemon tree (*Citrus* × *limon*), Casa del Frutteto (House of the Orchard), Pompeii, Italy, c. 25 BCE to 50 CE. Courtesy of the Archaeological Park of Pompeii, Ministry of Culture, Italy.

citron tree's morphology, is the earliest known author to describe the plant as the Persian or Median apple.[63] At least one Hellenistic author recognized the Indian origins of the tree, as the fifth-century CE lexicographer Hesychius, who drew on now-lost Hellenistic sources, glosses the citron (*kítrion*) as the Indian apple (*tò Indikón mẽlon*).[64] Despite long-standing familiarity with citrons and localized cultivation in the

63. Theophr. *Hist. pl.* IV.4.2: *tò mẽlon tò Mēdikòn ȅ tò Persikòn kaloúmenon,* I.11.4, I.13.4.
64. Hesychius, s.v. *kítrion.*

Mediterranean from the mid-first millennium BCE at the latest, the memory of the citron's eastern origin remained strong in the Mediterranean zone down to the early centuries CE. The second-century CE gastronomic author Athenaeus notes, for example, that the citron came to the Greeks from the "upper country"—that is, the interior of Asia.[65] Citruses were perhaps encountered with greater frequency in regions east of the Mediterranean, thus reinforcing its status as an eastern fruit *par excellence*.

From the fourth century BCE on, a few Greco-Roman authors identified citrons with the mythical "golden apples" (*pankhrúsea mē̃la*) guarded by a fearsome serpent in the garden of the Hesperid nymphs.[66] The retrieval of these apples was counted as either the penultimate or the last of Herakles's famous labors.[67] In his discussion of citrons, Athenaeus invokes *The Boeotian Woman*, a fragmentary play by the fourth-century BCE Attic comic playwright Antiphanes, as a witness to citrons.[68] While the identity of the fruit in question is not made explicit in the passage cited by Athenaeus, the reference to the "apple" seeds having come to Athens from the Persian king supports their identification with the citron.[69] Both Antiphanes and his contemporary imitator Eriphus, who is also quoted by Athenaeus, liken the Persian-derived apples to the Hesperid "golden apples."[70] The scholar-king Juba II (c. 48 BCE to 23 CE) also believed citrons to be the golden apples of the Hesperides.[71] The horticultural author Pamphilus of Alexandria (first century CE) describes Hesperid apples as fragrant and inedible, mirroring Theophrastus's remarks on citrons.[72] Pamphilus was not describing the mythical tree but actual fruits which the Spartans used as divine offerings. The early mythic associations gained by the citron are suggestive of the high value accorded to the tree, whose cultivation in the Mediterranean was probably restricted to the estates of elites in the earliest periods.

The growing of citruses must, however, have become more widespread by the first century CE, since Dioscorides claims that everyone knows the citron (*Mat. Med.* I.115.5: *pā̃si gnṓrima*). It is roughly in the same period when a more specific terminology for citruses evolves in both Greek and Latin, perhaps as a result of increased familiarity with citrons. From about the first century BCE on, terms relating to coniferous tree species were applied to the citron in both Greek and Latin. Diophanes of Nikaia (first century BCE) and Dioscorides (first century CE) speak of the citron as the *kedrómēla* or "cedar-apple," probably on account of the

65. Ath. III.84a.
66. Andrews 1961, 38; Hesiod *Theogony* 216, 335.
67. Panyassis, *Herakleia*, *PEG* I, fr. 11; Pherekydes of Athens, *BNJ* 3, fr. 17.
68. Ath. III.84a-b.
69. Ath. III.84b.
70. Ath. III.84a-c.
71. Juba, ap. Ath. III.83b-c; cf. Martial XIII.37.
72. Andrews 1961, 38; Ath. III.82e.

superficial resemblance of green, unripe knobby citron fruits to unripe cedar cones.[73] Isidore of Seville (d. 636 CE) notes that the "Greeks call it *kedrómēlon,* and Latin speakers *citrea* because its fruit and leaves bring to mind the smell of cedar (*cedrus*)."[74] The application to citrons of terms relating to coniferous trees was a source of great confusion for later commentators attempting to identify citrons in older texts, which only cited them as Persian, Median, or golden apples.[75] Athenaeus discusses this lexical confusion at length, ending with the suggestion that the presence of spines around the leaves of both juniper trees (*kedríon*, another coniferous species) and citrons (*kítrion*) may have led to the shared terminology.[76]

In Latin, the citron fruit was described as the *citrium* or *malum citrium* (variants: *citreum, citrum*), and the tree itself *citrus,* from which derives the modern botanical Latin name for the genus.[77] In addition to the term *kedrómēla,* later Greek authors and Greek papyrological records from Roman Egypt extensively used the Latinate forms *kitréa, kítron,* and *kítrion* to describe citrons.[78] The Latin *citrus* for citrons appears to derive from *citrum,* the name of the coniferous sandarac tree (*Tetraclinis articulata*), a native of Iberia and the southern Mediterranean which was highly valued for its fragrant wood and resin. The extension of terms relating to fragrant coniferous species to describe citrons in both Greek and Latin emphasizes the ornamental and odoriferous value of citrons rather than its gustatory quality to Mediterranean cultivators.

THE CITRON IN ISRAEL

> *Just as the citron is fair and praised among the wild trees, and all the world acknowledges it, so the Lord of the World was fair and praised among the angels.*
>
> TARGUM SONG OF SONGS II.3 (COMPILED SEVENTH TO EIGHTH CENTURY CE)[79]

The citron (*etrog*) is one of the four species (*arba'at ha-minim*), alongside myrtle twigs (*hadassim*), date-palm fronds (*lulavim*), and willow twigs (aravot), associated with rituals performed during the Jewish Feast of the Tabernacles (Sukkot), an autumnal harvest and thanksgiving festival which has as its mythological underpinning the Israelites' dwelling in tabernacles (tents) during their forty years

73. Diophanes, ap. *Geoponica* X.76; Dioscorides *Mat. Med.* I.115.5; Grant 1997, 193.

74. *Etymologiae* XVII.vii.8.

75. See, for example, the discussion in Macrobius's *Saturnalia* 3.19.3–5, where Homer is erroneously cited as providing evidence for citrons.

76. Ath. III.83–84d.

77. Andrews 1961, 42–43.

78. Bouchaud et al. 2017, 15–18; Pagnoux et al. 2013, 423.

79. Trans. Alexander 2003, 98.

of wandering in the wilderness.[80] The book of Leviticus, which enumerates the four species to be used to rejoice before the Israelite God, only speaks of the "fruit of the goodly tree" (*peri eẓ hadar*, Lev. 23:40), which rabbinic exegetes universally identify as the citron (*etrog*).[81] The book of Nehemiah's discussion of Sukkot is altogether unaware of this tradition and notes instead that foliage from olives, date palms, and thick trees was used to construct the ceremonial tabernacle (*sukkah*; Neh. 8:15).[82] Nehemiah also claims that traditional rituals associated with Sukkot had been disrupted for a long time (8:17), undoubtedly as a result of the Jewish exile in Babylonia. The divergence between Leviticus and Nehemiah on Sukkot traditions and the vague reference to the "goodly fruit" in Leviticus indicate that the use of the citron in Sukkot festivities was a post-exilic innovation.[83]

Citrons were associated with Sukkot by the Hellenistic period at the latest, since Josephus and rabbinical sources note that Alexander Yannai (Jannaeus), the Hasmonean ruler of Judaea and high priest of the Temple in Jerusalem (r. 103–76 BCE), was pelted with citrons by worshippers for disregarding the libation ritual.[84] Josephus explicitly notes that the citron was one of the four species used in Sukkot rituals, as does the Targum Onkelos, the Aramaic translation of, and commentary on, the Torah dating to the second century CE.[85] Rabbinical sources indicate that the citron was a common cultivar in Roman Palestine and that the fruit was cheap, except during Sukkot, when fine specimens could fetch high prices.[86]

The archaeological and linguistic data suggest that citrons may have already assumed a ritual role in Sukkot festivities at an earlier date. The early archaeological finds of citrons from across the Mediterranean establish that the introduction of citrons in the Levant could not have been later than the mid-first millennium BCE. The earliest finds of citrus pollen in the Levant are those belonging to a fifth-century BCE Persian satrapal garden at Ramat Raḥel, in Jerusalem, where citrons were grown alongside other local and foreign species like Lebanese cedar, Persian walnut, olive, fig, grape, willow, poplar, myrtle, and birch.[87] Even if the Persians were not necessarily responsible for the introduction of citrons to the Mediterranean, the use of a Sanskrit-derived Persian loanword (*tōrang*, from the Sanskrit *mātuluṅga*) to describe the citron in Hebrew (*etrog*) strongly suggests Persian agency in the widespread dissem-

80. Rabinowitz 2007.

81. Isaac 1959, 182; Löw 1924 vol. 3, 284–90; Rabinowitz 2007; Rokach and Shaked 2007.

82. Late fifth century BCE; Marcus 2007.

83. Ben-Sasson 2012, 16–17; Feliks 2007a; Rokach and Shaked 2007.

84. Josephus *Ant.* 13:372; TB, *Sukkah* 48b; Tosefta, *Sukkah* 3.5.

85. Josesphus *Ant.* 3.10; Targum Onkelos, *Leviticus* 23.40. On the date of the Targum Onkelos, see Sarna et al. 2007, 590–91.

86. E.g., TJ, *Sukkah* 3:10; Feliks 2007a.

87. Langgut 2014; Langgut et al. 2013, 122–26.

ination of the citron.[88] The use of the citron in Sukkot festivities perhaps already came into vogue sometime in the late Achaemenid Persian period (fourth century BCE).

Apart from its role in Sukkot festivities, the citron became a prominent marker of religious affiliations in both Hebraic and Levantine Christian art from around the first century CE on, appearing in a variety of media, including synagogue and church mosaic floors, tomb reliefs, coins, oil lamps, and ritual objects.[89] The appeal of the citron in religious ritual and imagery lies not only in its pleasant odor but also in its ability to flower and bear fruit throughout the year.[90] This feature rendered the citron an ideal symbol of fertility and endless spring in both Hebraic and Christian traditions.[91] Late antique and medieval Jewish commentators even identified the forbidden fruit of the garden of Eden with the citron.[92] The early-twelfth-century Jewish scholar Rabbeinu Tam had a copy of an Aramaic translation and commentary on the Song of Songs which equated the forbidden fruit of Eden with the *etrog*.[93] Medieval rabbinical tradition also claimed that the citron was held sacred on account of the fruit's morphological resemblance to the human heart.[94] As a sacred tree in Middle Eastern traditions, the citron had to be protected from perceived ritual pollution; hence the Syriac author Qūthāmā (c. 600 CE), preserved in Ibn Waḥshīyah's tenth-century CE Arabic translation, notes that menstruating women were forbidden to touch the citron tree.[95]

CITRUSES IN MESOPOTAMIA

When pomegranates hovered around me,
Lemons came to my rescue,
O that sweet one, I do not want him anymore!
Take me back home.

IRAQI (ARABIC) FOLK SONG[96]

While Mediterranean textual and archaeological records leave no doubt of the citron's presence in that region by the mid-first millennium BCE, the plant is oddly

88. A'lam 2011a; Ciancaglini 2008, 105; Daryaee 2006–07, 81. Variants of *tōrang* include *wādrang, wārang, bāzrang, bādrang,* and *bālang*; cf. Arabic *otranj* and *toranj*.

89. Ben-Sasson 2012; Isaac 1959, 182–83; Langgut 2017; van der Veen 2001, 194.

90. Ben-Sasson 2012, 16; Isaac 1959, 179. This feature is prominently noted by ancient Greco-Roman authors as well (Dioscorides *Mat. Med.* 1.115.5; Pliny *HN* XII.15–16, XVI.107; Servius, *Commentaries on Vergil,* II.127 (*Georgics*); Theophr. *Hist. pl.* IV.4.3, *Caus. pl.* I.11.1, I.18.5).

91. Ben-Sasson 2012, 19–20; Watson 1983, 167 n2.

92. Ben-Sasson 2012, 21.

93. Trans. Alexander 2003, 42 (7.9); cf. citrons in the gardens of Paradise in Niẓāmī, *Makhzan al-Asrār* 4.4.

94. E.g., Isaac 1959, 183; Leviticus Rabah 30.14; Rabinowitz 2007.

95. Nasrallah 2007, 641 (glossary entry in the text edition for Ibn Sayyār al-Warrāq's *Kitāb al-Ṭabīkh*).

96. Nasrallah 2013, 401.

unaccounted for in the visual, textual, and archaeological records of ancient Mesopotamia and Iran. But the present invisibility of the citron is almost certainly a result of our patchy knowledge of plant terminology in cuneiform texts rather than the unimportance of citruses in that region. For instance, many of the fruit trees listed in the "banquet stela" of the Assyrian king Aššurnasirpal II (883–859 BCE) remain unidentified, and it is plausible that the citron lurks behind one of these.[97]

R. C. Thompson equated the synonymous terms *ildakku* and *adāru* in Akkadian texts with the citron, but this is thoroughly unconvincing.[98] The *ildakku/adāru* tree was growing in Mesopotamia by the early second millennium BCE, and its wood was used in the manufacture of furniture.[99] The wood of the citron is hardly suitable for craft manufacture beyond a walking stick. The earliest Greek names for the citron, namely Median or Persian apple, provide a possible lead for Mesopotamian words relating to the citron. The translation of the Sumerian *hašhur* or the Akkadian *ḫašḫuru* as "apple" in a species-specific sense is misleading. These terms, just like Greek *mēlon,* Latin *malum,* Ugaritic *tpḥ,* Hebrew *tappūaḥ,* and Egyptian *dpḥ,* encompassed a wide range of rounded, fleshy arboreal fruits, including true apples, sorb apples, quinces, apricots, peaches, pomegranates, and citrons.[100] The Targum Song of Songs, an Aramaic translation of, and exegesis on, the Hebrew Song of Songs, explicitly cites the citron (*etroga*) as the equivalent of "apple" (*tappūaḥ*).[101]

97. RIMA II A.0.101.30, 36–52. For an identification of some of these fruits, see Postgate 1987, 128–32.

98. Thompson 1949, 312–14.

99. CAD, s.v. *adāru; ildakku.*

100. Andrews 1961, 38; Dalby 2003, 19, 275; Gelb 1982, 80; Hardy and Totelin 2016, 73; Isaac 1959, 183; Lambert 1987, 30–31. Cf. the Latin *pomum* and its Romance derivatives, e.g. French *pomme* (> Middle English *pome*), for a spherical fruit. True cultivated apples (*Malus pumila*), one of whose wild progenitors was of Central Asian origin (Cornille et al. 2012), were known in the Middle East by the late third millennium BCE. The earliest archaeobotanical sampling of apples in Mesopotamia consists of a garland of dried, diced apples from a grave (PG1054) in Early Dynastic Ur (Ellison et al. 1978). These are thought to be European wild apples or crab apples (*Malus sylvestris*) rather than domesticated apples (*Malus pumila*). Miller (2000, 154, 2013) proposes that the three-leaved, three-fruited pendant on the "diadem" of Puabi, also from the Royal Cemetery at Ur, represents apples. Pre-Sargonic texts from Tello attest to the use of "apple" wood in the construction of trunks and pegs and ribs for carts (Potts 1997, 108). This could well refer to the wood of a true apple species, since an apple (or pear) wood shaft for a bronze macehead is reported from Iron Age Hasanlu (Level IVb, c. 800 BCE; Dyson 1962, 646). Both wild and domesticated varieties of the apple were probably not a common sight in the warm Mesopotamian lowlands, since the apple grows best in temperate climes or at higher altitudes (Zohary, Hopf, and Weiss 2012, 137–38). While other fruits, like pomegranates, grapes, and figs, can be identified with certainty in Egyptian and Mesopotamian iconographic sources, apples are conspicuous by their rarity if not absence, suggesting that most of the "apples" appearing in ancient Middle Eastern textual sources were not true apples (Bleibtreu 1980, 187; Darby, Ghalioungui, and Grivetti. 1977, 697–99). Tengberg, Potts, and Francfort (2008, 925–26) even dispute the identification of apples on the Puabi diadem from Ur—as they note, "several other sub-globular fruits with adhering floral pieces exist in the Middle East."

101. Trans. Alexander 2003, 42, 97–98 (II.3).

The use of the term "apple" for a variety of plump arboreal fruits in the Mesopotamian textual tradition is suggested in the bilingual (Sumero-Akkadian) Urra = *ḫubullu* lexical series, whose manuscript tradition goes back to the Old Babylonian period (c. 1800 BCE). The lexical series lists various kinds of "apples," whose accompanying epithets suggest different fruit species rather than varieties of the same plant: white apple (*hašhur babbar*), mountain apple (*hašhur kurra*), conifer apple (*hašhur arganum*), *damšilum*-apple (*hašhur damšilum*), *kurdilum*-cucumber apple (*hašhur kurdilum*), and armanu-plant apple (*hašhur armanu*).[102] A few of these "apples" have been tentatively identified on the basis of lexical cognates in other Semitic languages.[103] The mountain apple (*hašhur kurra*) equated with the term *šapargillu* (variant: *supurgillu*) in the lexical series is identified, for instance, with the quince (Aramaic *səfargəlā*, Arabic *safarjal*).[104] The *armanu*-apple (*hašhur armanu*) is probably a pomegranate (Egyptian *alhammān*).[105] The precise botanical identity of most of the other "apples" is unclear. There is a remote possibility that the *hašhur arganum* or "conifer apple" cited in a Neo-Assyrian manuscript of the Urra = *ḫubullu* is the citron.[106] The Greek *kedrómēlon* (cedar/conifer-apple), first attested in Diophanes of Nikaia (first century BCE) for the citron, is perhaps a calque of a Semitic word.[107]

Regarding Mesopotamian visual sources, Emanuel Bonavia suggested that the knobby and oblong cone-like fruits held in a mode of blessing by winged protective spirits (*apkallu*) in Assyrian palace and temple reliefs (ninth to seventh centuries BCE) might be citrons (figure 15).[108] The fruit in question has typically been interpreted in the scholarly literature as a large conifer cone, the male flower of the date palm, or, less likely, an artichoke.[109] The presence of a bucket (Akk. *banduddû*) in the other hand of the protective spirit suggests that the Assyrian "cone" functioned as a sprinkler conveying blessings and good fortune, much like the rosewater sprinklers used across the Indo-Islamic world. The "cone" has been identified with the Akkadian term *mullilu* or "purifier," reinforcing the interpretation of its function as a sprinkler.[110] The role of the "cone" as a sprinkler makes an identification with the citron less likely but not impossible.

There is in fact a remarkable congruence between the Assyrian "cones" and depictions of citrons in medieval Indian art, where they are frequently found as

102. Landsberger et al. 1957–74; MSL XI, 110 (B); Thompson 1949, 302.

103. Powell 1987, 146–47.

104. CAD, s.v. *supurgillu*; Postgate 1987, 130–31; Thompson 1949, 307.

105. CAD, s.v. *armannu*; Hoch 1994, 25. Thompson (1949, 304–05) argues for an identification with the apricot.

106. Rm. 0367, obverse 8.

107. Diophanes of Nikaia, ap. *Geoponica* X.76.

108. Bonavia 1894, 66–68, 72.

109. Bleibtreu 1980, 187–88; Collins 2010, 184; Wiggermann 1992, 67; Ziffer 2019.

110. Collins 2010, 184; Wiggermann 1992, 67.

FIGURE 15. Detail of a winged protective spirit (*apkallu*) with a knobby "cone," Northwest Palace, Kalḫu (Nimrud), 883–59 BCE. Yale University Art Gallery, New Haven (public domain).

attributes of Hindu, Buddhist, and Jain deities and semi-divine beings.[111] The puckered surface of the Assyrian "cones" could be, like the Indian depictions of citrons, an attempt to evoke the small knobs and ridges on citrons. Citrus fruits, particularly lemons and limes, are used across South Asia to counteract the effects of the evil eye and malignant spirits. This can take the form of ritually waving a citrus around a person, tying citruses above the doorway, or halving them and laying them on either side of the threshold, the idea being that the use of sour or bitter objects averts misfortune.[112] If the Assyrian "cones" were indeed citrons, they may have been part of a specific ritual to protect the king's person rather than generic sprinklers of blessings.

The evidence for citrons in Mesopotamia, whether lexical or iconographic, is at present largely speculative. Archaeological evidence for citruses in Mesopotamia is not forthcoming either. Citron seeds were recovered from excavations at Nippur, in southern Iraq, in the last decade of the nineteenth century.[113] The present whereabouts of the seeds, the excavation context, and the dating remain poorly

111. McHugh 2012, 68–69; Shah 1987, 49–50, 116–17, 125–26, 219, 229, 245, 250, 264–65, 269, 280, 281, 284, 287–88, 289–90. Johannessen and Parker (1989), puzzled by these depictions of citrons, have spuriously identified them as New World maize.

112. Maloney 1974, 174.

113. Frimmel 1913, 187–88; Killermann 1916, 201; Townsend 1980, 468; Weisskopf and Fuller 2013b, 1482.

defined. Yet these finds are not entirely isolated, since citron seeds are known from a Late Bronze Age context in Cyprus. It is not improbable therefore that the Nippur finds could date back as early as the late second millennium BCE, or they may simply be a misidentification of taxa with morphologically similar seeds, most notably those of the Amygdaloideae subfamily.

THE CITRON IN IRAN

Kisrā of his golden citron, Parvīz and his golden quince
Were swiftly carried off by the wind, and became as one with the earth.

KHĀQĀNĪ (D. 1199), *ELEGY ON MADĀ'IN* (PERSIAN)[114]

There are no native Iranian records of the citron until the late Sasanian period, but foreign sources (Greek, Arab, Hebrew) on Persia dating from the Achaemenid period onward (post-fifth century BCE) indicate that the citron was long familiar as an odoriferant, flavorant, and comestible. It also acquired, as Khāqānī's elegy on the ruins of Ctesiphon indicates, strong associations with court ritual and royalty. The earliest Greek terms for the citron, Median or Persian apple, already testify to its importance as a cultivar in Iran. The plant was so esteemed that its seeds were seen as suitable gifts for envoys visiting the Achaemenid court.[115] Palynological analysis of a fifth-century BCE garden attached to an Achaemenid satrapal residence in Jerusalem indicates the localized cultivation of citrons.[116] The citron probably grew in ornamental gardens across the length and breadth of Persian imperial domains.

The earliest author who cites the use of citruses in Iranian cuisine is Pliny, who states that citron pips were added to Parthian dishes to flavor the breath.[117] While the citron does not have much flesh, it was occasionally consumed. An apocryphal story in the Babylonian Talmud relates that a Sasanian Persian king named Shapur offered slices of citron to his Jewish guests.[118] A prognosis adds that women who ate citrons produced fragrant children. In the same passage, the daughter of one king Shapur, whose mother consumed citrons during her pregnancy, is curiously described as a diffuser of fragrance in her father's court.[119] In the Middle Persian court romance *Ḵosrow, Son of Kavad and His Page* (*Xusraw ī Kawādān ud rēdag-ē*), which dates to the eighth or ninth century but draws on traditions from the late Sasanian period, the citron is listed among the best fruits for candying.[120]

114. Trans. Meisami, in Sperl and Shackle 1996, 167.
115. Antiphanes, ap. Ath. III.84a-b.
116. Langgut 2014, 5–6, 8–9; Langgut et al. 2013, 122–24.
117. Pliny *HN* XII.16, *HN* XI.278.
118. TB, *Avodah Zarah* 76b (probably the first or second king bearing the name Shapur).
119. TB, *Ketubot* 61a.
120. A'lam 2011a.

The citron also had ceremonial roles in the Persian Zoroastrian tradition, although the evidence for this only extends as far as the Sasanian period. During the feast of the autumnal equinox (*mihrajān*), the chief Zoroastrian priest, who was the first to enter the royal audience hall, bore offerings of citron (*utrujja*), quince, sugarcane, jujube, apple, grapes, and myrtle branches to the king.[121] In the Middle Persian *Bundahišn*, a Zoroastrian cosmological text compiled in the ninth century but embodying a tradition from Sasanian times, the citron is described as one of ten fruits said to be edible both inside and outside.[122] By the Islamic period, textual references to citrons are fairly pedestrian. The esteemed twelfth-century poet Niẓāmī refers to the citron no less than forty-one times in his work.[123] The citron is also one of the most commonly cited comestibles in Ferdowsī's *Šāh-nāma*.[124]

THE USES OF CITRUSES AND THEIR MODES OF GROWING

She fetched bright candles to dispel the night
And laid a little feast on which to dine,
Red pomegranates, citrons, quinces and wine.

FERDOWSĪ, *ŠĀH-NĀMA*, PROLOGUE TO THE STORY OF BIZHAN AND MANIZHEH (TRANS. DAVIS 2016)

Arboriculture is a long-term investment, since fruit trees become productive only after a few years and require constant attention, especially frost-intolerant tropical species like citruses growing in subtropical and temperate zones. The allure of citrus cultivation, however, outweighed the potential difficulties of transplantation. While citrus species gained a multiplicity of functions, there is no doubt that the novelty and agreeability of its scent, the result of high concentrations of the organic compound limonene in the fruit, was its chief attraction. Much like today, the ancients devised numerous ways to transfer citrus scents onto the body and garments. Theophrastus remarks that the citron was placed among clothes to keep them from being moth-eaten, a custom perhaps taken over from the Persians.[125] The chewing of citrus rind or the consumption of its juice, as Theophrastus observes, was the surest means for a fresh breath.[126] The *Kāmasūtra* (third century CE) likewise lists the citron peel as one of the necessary accoutrements of a young

121. al-Nuwayrī, *Nihāyat al-arab fī funūn al-adab* 1.188; Shaked 1986, 83.
122. A'lam 2011a.
123. Van Ruymbeke 2007, 84–88.
124. Ghamarzadeh and Ghasemov 2015, 30.
125. Theophr. *Hist. pl.* IV.4.2; cf. Pliny *HN* XII.15.
126. Ibid.

urban dweller's bedchamber.[127] Apart from the rind, the fragrant flowers and leaves of citruses were also used in the production of aromatic oils for perfumery.

The presence of citrus species in the context of funerary libations (Monte Sirai), votive deposits (Rome), religious ritual (Zoroastrian, Jewish), and myth (Greco-Roman Hesperid apples) argues for symbolic and spiritual roles which should not be overlooked in the spread and cultivation of citruses. The need for fresh citrons for use in the festival of Sukkot meant that diasporic Jewish communities throughout the Mediterranean and the Middle East cultivated citrons locally.[128] The sacred associations of citruses are already to be found in South Asian cultures. In South India, citruses are commonly strung into garlands for fearsome goddesses as a "cooling" fruit, impaled on weapons wielded by deities as sacrificial offerings, or have their pulp removed and are filled with ghee for use as a ritual lamp.[129]

Citruses were also used to flavor food and beverages, but they were rarely consumed on their own, at least in the Greco-Roman Mediterranean. Plutarch notes that the elderly were still not used to consuming citruses.[130] Pliny speaks of the citron having a "harsh taste."[131] Athenaeus in the second century CE claims that "as recently as our grandfathers' times no one ate it, but they stored it away like a great treasure in their chests along with their clothes."[132] The presence of organic residues derived from a citrus species in a wine jug from sixth-century BCE Monte Sirai in Sardinia indicates that citruses were used in the flavoring of wine. The late Roman collection of recipes attributed to Apicius (second to fourth centuries CE) provides one such recipe for flavoring wine with citrus leaves, with the resulting flavor said to imitate rose-wine.[133] In the modern Mediterranean, citrons are used to flavor liqueurs in Greece and Corsica (*cédratine*).[134]

Dried citrus peels may have also been used to make aromatic herbal teas in antiquity, much as they are throughout the modern Middle East.[135] The use of citruses in beverages is also evident in ancient India. The *Suśruta-Saṃhitā*, a Sanskrit medical treatise variously dated from the first to the fourth century CE, exhorts the reader to "prepare delightful confectionaries and beverages from hog-plums, mangoes, pomegranates and citrons."[136] It is highly probable that the first acquaintance

127. Vātsyāyana, *Kāmasūtra* I.4.4.
128. Isaac 1959.
129. Hiltebeitel 1991, 72, 75, 223, 444.
130. Plutarch *Quaest. Conv.* VIII.9.
131. Pliny *HN* XV.110.
132. Ath. III.84a.
133. Apicius, *De re coquinaria* I.4.
134. Van Wyck 2005, 142.
135. Nasrallah 2013, 553.
136. *Suśruta-Saṃhitā*, Uttaratantra, 25.2

with citruses in the Middle East and the Mediterranean was in the form of dried peels and leaves imported from South Asia. The ability to pickle and desiccate citruses made them a tradable commodity and calorific source beyond the summer months. In any case, the citron enjoys a longer shelf life without any kind of preservative treatment owing to its thick mesocarp (albedo).

Citrus flavors are useful in masking strong-flavored red meats like lamb or gazelle. Much of the information on the use of citruses in the flavoring of food, however, derives from medieval Arabo-Persian sources. Given the early presence of citrons in the Middle East, it is not unlikely that later recipes and food preparations have earlier precedents. In the tenth-century CE Abbasid cookbook (*Kitāb al-Ṭabīkh*) of Ibn Sayyār al-Warrāq, citron leaves, peels, and pulp find use in fish and meat dishes, and for the making of chutney (*maqra*), jams (*murabba*), and condensed juices (*rubb*) for medical and culinary use.[137] An entire chapter is also dedicated to dishes using a sour stew made of citron pulp (*ḥummāḍiyya*).[138] Lemon pulp is likewise recommended for flavoring various meat dishes.[139] Rabbinical sources from the mid-first millennium CE suggest that the rind of the citron was either consumed in its pickled form or boiled into a pulp.[140] Pickled citrons (Tamil *kaṭārai, nārttaṅkāi*) are incidentally still widely consumed in South India.[141]

Ancient Greco-Roman sources from the fourth century BCE on also mention the pharmacological potential of citruses. It was valued as an antidote to poison, a cure for shortness of breath, and a remedy for the nausea of pregnancy, indigestion, and various stomach ailments.[142] Beyond their ritual, festive, comestible, pharmacological, and odoriferous uses, citrus trees, like other fruit trees, also had secondary roles like providing shade, preventing erosion, and marking boundaries.

Citruses were, as Pliny notes, grown by sowing seed or through vegetative methods.[143] Citriculture may have even introduced the new agricultural technology of detached scion-grafting to the Middle East and the Mediterranean.[144] As citruses are highly susceptible to frost, their spatial distribution was limited to zones with

137. Ibn Sayyār al-Warrāq, *Kitāb al-Ṭabīkh* 23, 33, 36, 40-2, 49, 125-6, trans. Nasrallah 2007, 150, 178, 189, 205, 211–13, 216, 249, 485, 487, 489; cf. citron jam in 'Arīb ibn Sa'd's tenth-century *Calendar of Cordoba*, in Sato 2015, 31. On the use of citrons in medieval and modern Mediterranean cuisine, see Brigand and Nahon 2016. On similar uses in modern Iraq, see Townsend 1980, 468.

138. Ibn Sayyār al-Warrāq, *Kitāb al-Ṭabīkh* 58, trans. Nasrallah 2007, 278–81.

139. Ibn Sayyār al-Warrāq, *Kitāb al-Ṭabīkh* 46, 58, 78, 84-5, trans. Nasrallah 2007, 234, 280, 324, 347, 351.

140. E.g., TB, *Sukkah* 36a; Feliks 2007a.

141. Kothari 2007, 92; Naik 1963, 145.

142. Ath. 84d-85a; Dioscorides *Mat. Med.* 1.115.5; Galen *Aliment. fac.* 2.37; Nicander of Colophon, *Alexipharmaca* 533; Nicolaus of Damascus, *De plantis* 121; Oppius, ap. Macrobius, *Saturnalia* III.19.4; Pliny *HN* XII.15–16, XXIII.105; Theophr. *Hist. pl.* IV.4.2; Vergil *G.* II. 126–135. For medieval sources on the pharmacological properties of citrus species, see Arias and Ramón-Laca 2005.

143. *HN* XVII.64

144. Zohary, Hopf, and Weiss 2012, 115.

frost-free winters.[145] This, then, accounts for the distribution of citrus finds along the littoral zone of the Mediterranean, where mild winters allow open ground cultivation. But late antique horticultural texts from the Mediterranean also indicate that the citron's intolerance to frost could be circumvented by the indoor storage of potted trees during winter.[146] As citruses need moist, well-drained soil, irrigation was required in the absence of sufficient rainfall.[147] In modern Iraq, citruses are typically grown in irrigated date palm groves or, on a smaller scale, in domestic gardens.[148] This was probably no different in antiquity. A late-third-century CE papyrus from Oxyrhynchus in Egypt, detailing the lease of a fruit garden, notes that citrons were cultivated alongside dates, olives, peaches, figs, and melons.[149]

Whether in Seville or in Sicily, it is hard to imagine Mediterranean agricultural and garden landscapes devoid of citruses. Citrus species dominate fruit crop production today and account for over eight million hectares of cultivated land globally.[150] The present ubiquity of citrus fruits and their multifarious uses can ultimately be traced back to the dispersal of citrons from South Asia to the Middle East and the Mediterranean in the first millennium BCE. The lemon was carried westward along the same routes toward the end of that millennium. Were it not for the archaeological data, the lemon would have remained invisible to historians of agriculture, since it was not distinguished lexically from citrons until the middle of the first millennium CE. While citrons and lemons were not major calorific contributors in antiquity, they held important functions as ornamentals, flavorants, odoriferants, medicinal plants, and ritual accoutrements.

145. Townsend 1980, 466.
146. Florentinus, ap. *Geoponica* 10.7; Palladius *Agr.* 3.24.14, 4.10.14.
147. Theophrastus *Hist pl.* IV.4.3.
148. Nasrallah 2013, 401; Postgate 1987, 122; Townsend 1980, 466.
149. P. Oxy. 1631.
150. Shan 2016, 7.

5

Familiar but Foreign: Eastern Cucurbits

Here we examine how the humble cucumber, whose wild counterparts survive in the Himalayan foothills, started its journey to becoming a ubiquitous element of global gastronomy. The story of the cucumber is intimately connected, if not confused, with its cousin, the melon, another tropical Asian cultivar whose globular and sweet rather than elongated and non-sweet forms are more familiar today. In addition to the cucumber and melon, I will investigate the westward dispersal of other Indian cucurbits in antiquity, most notably the luffa.

SOUTH ASIA, A CENTER OF CUCURBIT DIVERSITY

Old World cucurbits are typically annual, herbaceous, frost-sensitive climbing or prostrate vines distributed throughout the tropical and subtropical zones of Asia and Africa. Many cultivated cucurbit species are native to northern India and adjacent regions, especially the Himalayan belt and the Indo-Gangetic Valley. These include, most prominently, the cucumber (*Cucumis sativus*), snake gourd (*Trichosanthes cucumerina*), bitter gourd (*Momordica charantia*), luffa or sponge gourd (*Luffa aegyptiaca*), angled or ridge luffa (*Luffa acutangula*), pointed gourd (*Trichosanthes dioica*), and ivy gourd (*Coccinia grandis*).[1] The cucumber (*Cucumis sativus*, figure 16), which is one of the most widely consumed fruit-vegetables today, traces its origins to northern India, where a closely related wild species (*Cucumis sativus* var. *hardwickii*) survives in the foothills of the western

1. Chomicki, Schaefer, and Renner 2020, 1241–50; Decker-Walters 1999; Fuller 2006a, 39–41, 2007, 419, 2009, 165–66; Harvey et al. 2006, 23–24; Schaefer, Heibl, and Renner 2009; Singh 2017, 85–96.

Himalayas.[2] Unlike its domesticated relative, this wild Himalayan cucumber features fruits that are small, spiny, and bitter, traits which were ironed out by artificial selection.

The melon (*Cucumis melo*, figure 17), whose history regularly intersects with that of the cucumber, also traces part of its ancestry to South Asia. Recent phylogenetic data indicate two or more domestication events for the melon, with at least one localized in India.[3] The substantial genetic diversity of melon germplasm in India, alongside the presence of wild melon species easily crossed with *C. melo*, supports the Indian genealogy proposed for the cultivated melon.[4] In this regard, the Himalayan wild melon species *Cucumis trigonus* has been identified as a likely wild progenitor of the domesticated melon.[5] Africa, which older scholarship believed to be the place of origin for cultivated melons owing to the presence of numerous wild *Cucumis* species and the identical chromosomal number of *C. melo* with those of African *Cucumis* species, is now considered a secondary center of melon domestication with a limited impact on melon diversification.[6]

The rich vocabulary relating to cucurbits in South Asian languages, with over three hundred terms in Sanskrit alone, testifies to their cultural importance in ancient South Asia.[7] The ubiquity of cucurbits in South Asian agricultural regimes was expressed in ritual, poetic, and everyday language, whether it was a prayer for liberation likened to the quick snapping of a cucumber from its vine (*Ṛgveda* VII.59.12: *urvārukamiva bandhanān*), a poem describing proverbially beautiful lips in the crimson shade of the ivy gourd (*Gītagovinda* III.14.3: *bimbādharamādhurī*), or even a verb likening the state of being useless to a putrid wax gourd (Sanskrit *pūtikuṣmāṇḍāyate*). Today, many native and imported New World cucurbits are commonly met with in South Asian markets and continue to find extensive culinary and pharmacopoeic uses.

EARLY MELONS

Two subspecies are recognized within cultivated melons today on the basis of phenotypic and genotypic variations: *melo* and *agrestis*.[8] The former is thought to be

2. Bisht et al. 2004; McCreight et al. 2013, 1081; Sebastian et al. 2010, 14269.

3. Chomicki, Schaefer, and Renner 2020, 1248; McCreight et al. 2013; Paris 2016; Renner, Schaefer, and Kocyan 2007; Sebastian et al. 2010; Serres-Giardi and Dogiment 2012; Zhao et al. 2019.

4. Gonzalo et al. 2019; McCreight et al. 2013, 1078; Sebastian et al. 2010; Serres-Giardi and Dogiment 2012; Tzitzikas et al. 2009, 1820. See also the archaeobotanical evidence (mostly seeds) for *Cucumis* species from Bronze Age Indian sites in Fuller and Madella 2001, 339.

5. Endl et al. 2018; Paris, Amar, and Lev 2012b, 24; Sebastian et al. 2010, 14272.

6. Sebastian et al. 2010; Serres-Giardi and Dogiment 2012.

7. Decker-Walters 1999.

8. Esquinas-Alcazar and Gulick 1983; McCreight et al. 2013, 1078; Paris 2016.

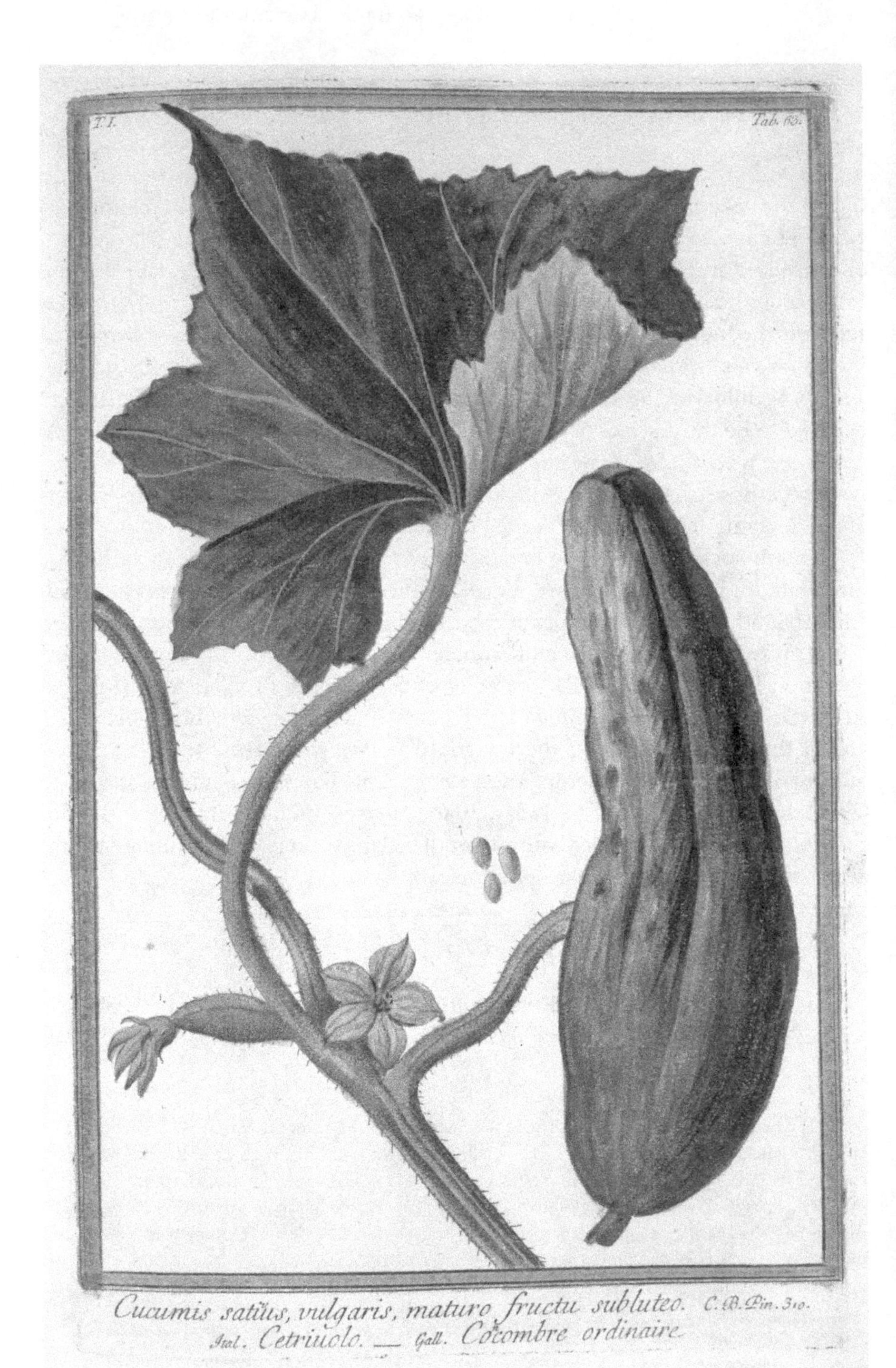

FIGURE 16. Cucumber (*Cucumis sativus*), colored etching by Magdalena Bouchard, 1772. New York Public Library (public domain).

FIGURE 17. Melon (*Cucumis melo*), illustration by Genevieve Regnault in Nicolas Regnault, *La botanique mise à la portée de tout le monde* (Paris, 1774). New York Public Library (public domain).

a late development in Africa based on Asian *agrestis* melons.[9] More recent DNA sequencing has, however, proposed a reordering of melon taxonomy to include an Asian *Cucumis melo* subsp. *melo* group and an African *C. melo* subsp. *meloides* group.[10] Within these subspecific classifications, scholarly taxonomies acknowledge up to sixteen botanical varieties of melon.[11] But the chronology and pathways for the westward dispersal of the Asian cultivated melon are unclear, not least because much of this process takes place in prehistoric horizons.

Cultivated melons are already found in the archaeological and textual sources of Egypt and Mesopotamia from the mid-fourth millennium BCE on.[12] The finds of melon seeds in Predynastic Hierakonpolis, Maadi, and Adaïma in Egypt (c. 3700–3300 BCE) offer the earliest archaeological evidence for melon cultivation in North Africa.[13] These early finds probably represent a separate domestication event based on locally available wild melons.[14] But the earliest melons in Mesopotamian records are more likely to be of South Asian origin, in view of the region's proximity to South Asia, the primary domestication center for the melon. The earliest archaeobotanical data for cucurbits in this region appear in the form of ten seeds belonging to an unidentified cucurbit, most likely melon, from Old Akkadian Tell Taya (c. 2300–2150 BCE), in northern Mesopotamia.[15] Meanwhile, the first Mesopotamian textual references to melons (*ukuš*) are found in the earliest Sumerian lexical texts, dating to the middle of the third millennium BCE.[16] Further afield, one *Cucumis melo* seed was recovered from an Early Bronze Age kitchen area excavated at Tell Hammam et-Turkman, in Syria (2500–2000 BCE), and a single grain of *Cucumis* sp. pollen dated to c. 2300 BCE was recovered from lake sediments in northwestern Crete.[17] It is not clear, however, whether these finds of melon were the result of dispersals via Mesopotamia or Egypt.

9. Pitrat 2013, 278; Serres-Giardi and Dogiment 2012.

10. Endl et al. 2018.

11. The *C. melo* subsp. *agrestis* group includes *chinensis, makuwa, momordica, acidulous,* and *conomon* (oriental pickling melon), while the *C. melo* subsp. *melo* group is composed of *adana, ameri, chandalak, flexuosus* (snake melon), *cantalupensis* (cantaloupe), *inodorus* (winter melon), *reticulatus* (muskmelon), tibish, *chate, chito,* and *duda'im* types. See Pitrat et al. 2000 on these classifications.

12. Paris 2016; Zohary, Hopf, and Weiss 2012, 154–55. See Sabato, Esteras, and Grillo 2019 for archaeological data on the melon in the Bronze Age Mediterranean.

13. Fahmy 2001; Newton 2007, 140–41; van Zeist and de Roller 1993; Zohary, Hopf, and Weiss 2012, 155. For later remains in Egypt, see Germer 1985, 129; Vartavan and Amorós 1997, 88–89. For textual sources, see Darby, Ghalioungui, and Grivetti 1977, 694–95.

14. See El Tahir and Yousif 2004 on wild melons in northeastern Africa.

15. Renfrew 1987, 162.

16. These texts belong to the Early Dynastic IIIa horizon in Mesopotamian chronology. See PSD, s.v. *ukuš* (ED Metals 79–80, where *ukuš* occurs in a list of metals in the context of a *gir ukuš* or melon knife).

17. Bottema and Sarpaki 2003, 745; van Zeist et al. 2003.

Non-sweet, vegetable-type melon varieties predominated in the ancient Middle East and the Mediterranean.[18] Melon fruits exhibit substantial polymorphism, ranging in shape from the globular to the elongated, with a variety of exocarp textures and patterns (netted, warted, wrinkled, smooth, striped, etc).[19] Several elongate, non-sweet varieties of melon, especially those of the *flexuosus*, *chate* (*adzhur*), and *conomen* groups, resemble cucumbers in morphology, color, and taste, leading to substantial confusion in the identification of the true cucumber in early Middle Eastern and Mediterranean texts.[20]

The large-fruited globular sweet "dessert" melons (e.g., casabas, cantaloupes, and muskmelons), which are the best-known melon types today, are a relatively late (ninth century CE) Central Asian phenomenon carried westward to the Mediterranean around the eleventh century CE.[21] Central Asia still has the greatest variety of sweet and aromatic melon cultivars.[22] But the Mesopotamian lexical series Urra = *ḫubullu,* whose manuscript tradition goes back to the eighteenth century BCE, is already familiar with sweet and sour varieties of the melon (Sumerian *ukuš,* Akkadian *qiššû*).[23] Perhaps Bronze Age cultivators were already privileging traits like sweetness in melon cultivars. But sweetness, like other taste sensations, is a culturally relative notion, so a comestible described as sweet in antiquity may simply have been non-sour or non-bitter. Sarah Hitch, remarking on the Homeric epics, notes that sweetness is applied to non-food-related activities like music, laughter, and sleep, suggesting that in antiquity the semantic field of sweetness embraced anything pleasurable, much as it does in many modern languages.[24] Furthermore, sweetness was not necessarily a marker of nutritive value, for Theophrastus remarks that not all sweet flavors are nutritious.[25]

THE CUCURBIT CONUNDRUM

The westerly dispersal of the cucumber has hitherto remained nebulous on account of the confused lexicology pertaining to cucurbits in ancient languages and

18. Janick, Paris, and Parrish 2007, 1447.

19. Avital and Paris 2014, 203; Paris, Amar, and Lev 2012a, 2012b, 24.

20. Avital and Paris 2014, 203; Paris, Amar, and Lev 2012b, 23–24; Pitrat 2008, 284.

21. Paris 2016; Paris, Amar, and Lev 2012b.

22. Paris, Amar, and Lev 2012b, 29.

23. The earliest manuscript of the bilingual (Sumerian-Akkadian) lexical series Urra = *ḫubullu* is from Old Babylonian Nippur. The references to sweet cucumbers are also known from the Middle Babylonian recension from Alalakh and Neo-Assyrian copies which give the fullest (and no doubt updated) version of the text. See MSL 11, 110 (A) (CBS 03918 + CBS 03928) VI r v. 10 ; MSL 11,110 (B) (CBS 11082 + N 7737) d ii 23; MSL 10, 36 W r iv 59; Neo-Assyrian Urra = *ḫubullu* XVII 369–70.

24. Hitch 2018, 28–29. See also Totelin 2018 on relative notions of sweetness in antiquity, especially in relation to honey.

25. Theophr. *Caus. pl.* VI.4.5.

modern scholarly misinterpretations of ancient cucurbit terminology. Semitic and Indo-European philologists alike have indiscriminately translated many terms relating to melons (*Cucumis melo*), like the Sumerian *ukuš,* Akkadian *qiššu,* Hebrew *qišūt,* Arabic *qithā,* Greek *síkuos,* and Latin *cucumis,* as "cucumber," leading to the false impression that the true cucumber (*Cucumis sativus*) was widespread and well known in antiquity.[26] Worse still are identifications of ancient Old World cucurbits with presently widespread New World species like squashes, pumpkins, and marrow (*Cucurbita maxima, Cucurbita pepo, Cucurbita moschata*).[27]

Ancient authors were themselves stymied by cucurbit terminology. Athenaeus, writing in the second century CE, was already confounded by the identities of cucurbits mentioned in late Classical (fourth century BCE) Attic sources:

> Speusippus in his *On Similar Things* refers to the melon (*pépōn*) as a *sikúa.* Diokles (of Karystos) uses the word *pépōn* but then never refers to a *sikúa.* Whereas Speusippus uses the word *sikúa* but never mentions a melon (*pépōn*).[28]

Elsewhere, Athenaeus remarks that the polymath Nicander of Colophon (second century BCE) called what his contemporaries identified as a *kolokúnthē,* a *sikúa.*[29] Ordinarily the *kolokúnthē* denoted the well-known bottle gourd (*Lagenaria siceraria*), a native of Africa and one of the earliest plants to be taken into cultivation (figure 18).[30] The *kolokúnthē* was also distinguished from other cucurbits by its large size. A fragment quoted by Athenaeus from the Athenian comic poet Hermippus (fifth century BCE) makes this feature rather explicit: "What a size of head he has—as big as a gourd (*kolokúnthē*)!"[31]

However, *sikúa* and its masculine equivalent *síkuos,* regularly translated as "cucumber" in modern scholarship, are trickier to translate, for their semantic range covers gourds, melons, and, slightly later, true cucumbers. Unlike Nicander and Athenaeus's contemporaries, Theophrastus in the fourth century BCE distinguishes between the *kolokúnthē* and the *sikúa.*[32] But this nomenclature was based on the morphological differences between cucurbit fruits and was not an interspecific distinction in the modern sense. Theophrastus's description of the *sikúa* as one that "takes the shape of the vessel in which it has been placed" indicates that his *sikúa* may simply have been a morphological variant of the bottle gourd, whose

26. E.g., Amigues 2007, 91; Andrews 1956; Beck 2011; Löw 1928, vol. 1, 530–35; Stol 1987; Thompson 1949, 82–86. See fuller references to erroneous translations in Janick, Paris, and Parrish 2007, 1453.

27. Dalby (2003, 162); LSJ, s.v. *kolokúntē.*

28. Ath. II.68e-f.

29. Ath. IX.372d.

30. Konen 1995, 46–47. See N'dri et al. 2016 on the genetic data pertaining to the dispersal of bottle gourds.

31. Ath. II.59c.

32. Theophr. *Hist. pl.* VII.2.9.

FIGURE 18. Morphologically diverse bottle gourds (*Lagenaria siceraria*), northern Mozambique. Photograph by Ton Rulkens via Wikimedia Commons (CC BY-SA 2.0).

polymorphism and sheer malleability in growing into the shape of a mold has even rendered it eminently suitable as sculptural material for contemporary art.[33] Much like Theophrastus, an earlier anonymous Hippocratic author (fifth to fourth centuries BCE) remarks that the *síkuos* grows into any shape it is forced to take.[34] Athenaeus records that the people of the Hellespont called long gourds *sikúa* and round ones *kolokúnthē*.[35]

The identity of *sikúa* and *síkuos* is further complicated by the polymorphic nature of melons. Non-sweet varieties of the melon, much like the bottle gourd, are frequently encountered in elongate forms (figures 19–21). Consequently, the extant

33. Theophr. *Hist. pl.* VII.3.5. The Greek papyrological evidence from Ptolemaic-Roman Egypt also indicates that both terms were used to describe the bottle gourd (Konen 1995, 47–48).

34. Hippocrates, *Generation* 9; cf. Pliny *HN* XIX.23.65.

35. Ath II.59a.

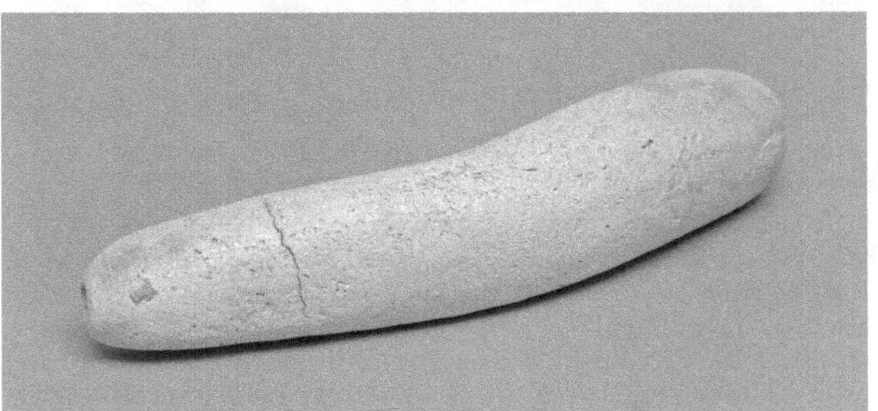

FIGURE 19. An elongate melon, also known as an Armenian cucumber (*Cucumis melo* ssp., *melo flexuosus* group), Bahar, Iran. Photograph by Shams Bahari via Wikimedia Commons (CC BY-SA 4.0).

FIGURE 20. Faience model of an elongate melon (*Cucumis melo*), Lisht, Egypt, c. 1850–1700 BCE. Metropolitan Museum of Art, New York (public domain).

FIGURE 21. An elongate melon (*Cucumis melo*) rests against a leg of beef (*far right*). Detail of a funerary banquet from the stela of the Steward Mentuwoser, Abydos, Egypt, c. 1944 BCE. Metropolitan Museum of Art, New York (public domain).

references to *sikúa* and *síkuos* could well pertain to elongate melons as well.[36] In fact, most references to *sikúa* and *síkuos,* as we shall shortly observe, would appear to describe melons rather than gourds. Polymorphism within the same cucurbit species, a rather common feature of bottle gourds (*Lagenaria siceraria*) and melons (*Cucumis melo*), was undoubtedly responsible for the sheer semantic flexibility

36. Konen 1995, 48–49, 51.

of cucurbit terminology and the diachronic confusion between various cucurbit terms.[37] Interspecific resemblances also resulted in the conflation of cucurbit-related terms. Several varieties of melon, including the snake melon or Armenian cucumber (*C. melo* ssp. *melo Flexuosus* group), resemble true cucumbers in both morphology and taste.[38] These slender cucumber-like snake melons, known as *faqqūs* in Arabic, are still familiar cultivars across much of the Middle East and North Africa, where they are consumed raw, pickled, or cooked.[39]

In light of the dangerously slippery semantics, the use of the terms *sikúa* and *síkuos* cannot serve as a useful index for the arrival of the true cucumber, since their earlier and later usages extended to include elongate varieties of the bottle gourd and the melon. Cucurbit semantics in antiquity was variously governed by local familiarity, size, morphology, taste, and culinary usage rather than any kind of systematic botanical taxonomy. Distinguishing the true cucumber from the melon and the bottle gourd will require further contextual materials read in conjunction with other strands of evidence, in particular the archaeological data.

SPLITTING HAIRS: DISTINGUISHING THE CUCUMBER FROM THE MELON

In a series of influential scientific papers, Harry Paris and Jules Janick, two horticultural scholars (along with their colleagues) have argued, on the basis of textual and iconographic sources, that the cucumber (*Cucumis sativus*) only arrived in the Mediterranean and the Middle East in late antiquity, beginning around 500 CE.[40] The present analysis of the textual and iconographic sources from both regions confirms the findings of Paris and Janick in establishing that the primary and default meaning of the Sumerian *ukuš*, Akkadian *qiššû*, Hebrew *qišūt*, Arabic *qithā*, Greek *síkuos*, and Latin *cucumis* should be melon and not cucumber. However, Paris and Janick have overlooked the archaeobotanical data which suggest an earlier westward dispersal date for the cucumber. A combined reading of textual and archaeological materials indicates that the cucumber was introduced into the Middle East and the Mediterranean by the first century CE at the latest and shared the same terminology with melons owing to morphological and gustatory resemblances.

To begin with the negative evidence, some cucumber-like fruits in Egyptian paintings and reliefs of the second millennium BCE, and Roman mosaics of the early centuries CE, are striped and furrowed.[41] These features are more character-

37. Paris 2016.

38. McCreight et al. 2013; Paris, Amar, and Lev 2012b, 24; Pitrat 2008, 284.

39. Paris 2012, 2016; Paris, Amar, and Lev 2012b, 24l.

40. Paris 2012, 2016; Janick, Paris, and Parrish 2007; Paris, Daunay, and Janick 2012; Paris and Janick 2008a, 2008b, 2010–11; Paris, Janick, and Daunay 2011.

41. Avital and Paris 2014; Konen 1995, 45; Janick, Paris, and Parrish 2007, 1449–50, 1454; Paris 2012, 33.

istic of melons than of cucumber fruits. Latin authors like Columella and Pliny, Hebrew rabbinical sources like the Mishnah and Tosefta, and medieval Arabic authors describe the fruits of *cucumis, qishut,* and *qithā* as hairy.[42] Cucumber fruits are usually glabrous or hairless, while some species of melons, especially snake melons (*C. melo* ssp. *melo Flexuosus* group) and Carosello melons (*C. melo* ssp. *melo Chate* group), are hairy when immature.[43]

The Hippocratic author of the treatise *On Regimen* (fifth/fourth century BCE) observes that the unripe *síkuos* is indigestible (II.55), a description which cannot apply to cucumbers and snake melons, which are both consumed immature.[44] As the seeds of the cucumber are underdeveloped when consumed, the comments of another Hippocratic writer who speaks of passing *síkuos* seeds in the stool should be taken to refer to a variety of melon.[45] One of the epithets provided for the melon (*qiššû*) in a Middle Babylonian manuscript (c. 1400 BCE) of the lexical series Urra = *ḫubullu* from Alalakh is "stinking."[46] Snake melons are known to have a sour taste when mature and do not keep fresh for long, suggesting that the malodorous cucurbit in question is the snake melon.[47] Columella similarly remarks of the "foul juice" of the *cucumis,* which is not characteristic of true cucumbers.[48] The Pseudo-Aristotelian author of the *Problems* (fourth to third centuries BCE) claims the *síkuos* was bitter toward the root.[49] Even as late as the seventh century CE, Isidore of Seville remarks that "*cucumis* are so called because they are sometimes bitter; they are thought to grow sweet if their seeds are steeped in honeyed milk."[50] Theophrastus claims that the Megarians protected the *síkuos* from the Etesian winds by raking dust over the fruits and hence made them sweeter.[51] By modern standards, cultivated cucumbers are neither bitter nor sweet, suggesting that these descriptions refer to the melon. The mutability of taste sensations across space and time, however, caution against mapping contemporary notions of bitterness and sweetness onto the past. As we observed earlier, a cucurbit that was described as sweet in antiquity may simply have been a non-sour or non-bitter variety.

42. E.g., Columella *Rust.* X.389–393; Mishnah, *ʿUqṣin* 2.1; Mishnah and Tosefta, *Maʿaserot* 1.5; Pliny *HN* XIX.70; Janick, Paris, and Parrish 2007, 1444–47, 1454; Nasrallah 2007, 792 (glossary entry in the text edition for Ibn Sayyār al-Warrāq's *Kitāb al-Ṭabīkh*); Paris 2012, 33; Paris, Daunay, and Janick 2012, 118, 121; Paris and Janick 2008b, 35–39, 2010–11.

43. McCreight et al. 2013; Paris 2016; Pitrat 2008, 284.

44. Hippoc. *Vict.* II.55.

45. Hippoc. *Morb.* IV, 596–598.

46. MSL 10, 36 W r iv.55.

47. Janick, Paris, and Parrish 2007, 1444.

48. Columella *Rust.* X.393.

49. Pseudo-Aristotle, *Problems* XX.25.

50. Isidore of Seville XVII.10.16; cf. Columella *Rust.* XI.51; Pliny *HN* XIX.23.65; Theophr. *Caus. pl.* III.9.4.

51. Theophr. *Hist. pl.* II.7.5.

A few other ancient remarks on plant morphology also suggest the predominance of melons among the commonly used cucurbits of the Mediterranean and the Middle East. Theophrastus's comment on the sterile or male flowers of the *síkuos* supports identification with the melon rather than the cucumber, as *Cucumis melo* typically produces pistillate (female) or hermaphroditic flowers on the first few nodes of shoots, while all apical nodes of the plant yield staminate (male) flowers.[52] Several ancient references to cucurbit fruit morphology invoke large spherical, ovoid, or even snake-like elongate shapes, which are uncharacteristic of elongate cucumbers. Among the diplomatic gifts the pharaoh Akhenaten (c. 1353–1335 BCE) sent to his Babylonian counterpart were *qiššû*-shaped gold, ivory, and stone vessels filled with aromatic oils.[53] These vessels must have been modelled after globular or ovoid bottle gourds and/or melons, or they would have been incapable of storing aromatic oils. Vergil speaks of the *cucumis* swelling into a paunch, a shape and size atypical of cucumbers.[54] Columella, on the other hand, describes a snake-like *cucumis*.[55] The Greek equivalent of Columella's "snake-melon" was the *drakontías,* which is classed as a variety of the *síkuos*.[56] These terms evidently refer to an elongate melon variety still known as the snake melon (*C. melo* var. *flexuosus*).

So far, the true cucumber has proved elusive if not nonexistent in ancient Middle Eastern and Mediterranean records. There are a handful of ancient descriptions of cucurbits where reference to cucumbers appears admissible on account of morphology and the practice of consuming the unripe fruit and its immature seeds. But a straightforward identification of these as cucumbers is impeded by the widespread use of snake melons (*C. melo* ssp. *melo Flexuosus* group) in antiquity. The latter display morphological similarities to cucumbers and are also consumed unripe, with immature seeds.

One of the earliest references to a cucurbit with immature seeds occurs in the inscriptions of the seventh-century BCE Assyrian king Sennacherib. Sennacherib describes the *pindû* stone, a fossiliferous limestone, as one whose texture is as fine-grained as a cucurbit seed (*pindû ša kīma zēr qiššê šikinšu nussuqu*), alluding to the underdeveloped seeds of a cucurbit which is consumed immature.[57] And in reference to the mutilation of his Elamite enemies in battle, he remarks: "I cut off their beards and ruined their pride and I cut off their hands like the sprout of summer cucurbits" (*sapsapāte unakis-ma baltašun abut kīma bīni qiššê simāni unakis qātišun*).[58] The term used by Sennacherib to describe this plant is *qiššu,* whose

52. Theophr. *Hist. pl.* I.13.4.
53. Rainey 2015: EA 14 i.57, iii.38, iv.5.
54. *Georgics* IV.122.
55. *Rust.* VII.2.
56. Euthydemus of Athens, ap. Ath. III.74b.
57. RINAP 3 34, 72; 49, 1; 50, 1.
58. RINAP 3 18, vi.1; 22, vi.2; 23, v.77b; 145, i.1; 230, 87b.

default meaning is melon, but in Akkadian it is also used as a generic term for other cucurbits. The references to a cucurbit in Sennacherib's inscriptions, otherwise exceedingly rare in Assyrian royal inscriptions, could refer to a new cucurbit species, perhaps the cucumber.[59] A keen interest in the collection of new plants on the part of Assyrian kings, the rarity of cucurbit appearances in royal inscriptions, and the oblique reference to soft immature seeds raise the possibility that the inscriptions of Sennacherib refer to the cucumber. Yet there is not enough contextual evidence to establish that it is not the snake melon.

Slightly later, some Greek authors refer to the consumption of cucurbits with immature seeds, often described as seedless or less-seeded cucurbits. In his fragmentary play *Odysseuses,* the fifth-century BCE Athenian comic poet Kratinos refers to ripe melons (*pépōn*) as "seed-filled *síkuos*" (*síkuos spermatías*), which suggests that less-seeded or seedless varieties of the *síkuos* were known.[60] Athenaeus, who preserves this fragment, contrasts the "seed-filled cucurbit" of Kratinos with the "sterile or seedless melons" mentioned by the Attic comic poet Plato (late fifth century BCE), who likens them to the shins of Leagros, a character held in mockery.[61] The comparison of "sterile or seedless melons" (*pépōn eunoukhías*) to shins (*knḗmē*) suggests an elongated rather than globular fruit. Phaenias (fourth century BCE), also preserved in Athenaeus, distinguishes *síkuos* and *pépōn* from gourds (*kolokúnthē*) on the basis that "gourds are inedible when raw, but are edible if stewed or baked," unlike *síkuos* and *pépōn,* which are edible "once the flesh is soft."[62] Like Phaenias, Straton of Sardis (second century CE) refers to a cucurbit consumed unripe:

> "Know the time" said one of the seven sages; for all things, Philippus, are more loveable when in their prime. A *síkuos,* too, is a fruit we honor at first when we see it in its garden bed, but after, when it ripens, it is food for swine.
>
> *Greek Anthology,* XII.197

Once again, the references to the consumption of the unripe fruit and its immature seeds, and an elongated morphology, offer no certain link with true cucumbers, since these characteristics are shared with snake melons. The availability of other names for true cucumbers in late antique and medieval texts (e.g., *khiyār* in Arabo-Persian and Turkish, *citruli* in medieval Latin, *angoúri* or *angouriá* in Byzantine and modern Greek) has been cited as evidence to argue for a later acquaintance with the true cucumber.[63]

59. Cf. *naṣṣabu* (bottle gourd) in the banquet stela of Aššurnasirpal (RIMA II A.0.101.30: 43).

60. Ath. II.68c.

61. Plato Laius, ap. Ath. II.68d-e.

62. Ath. II. 68d-e.

63. E.g., Ibn Waḥshīyah, *Al-filāḥah al-nabaṭīyah* II. 891–92; Paris, Daunay, and Janick 2012, 119–22; Paris, Janick, and Daunay 2011, 474, 481–83.

However, as we have observed, a range of ancient cucurbit terms, like the Akkadian *qiššu,* Greek *síkuos,* and Latin *cucumis,* while ordinarily indicating the melon, were semantically flexible enough to cover what a modern botanist would recognize as intraspecific and interspecific varieties. Both Theophrastus and Pliny, the doyens of Greco-Roman botany, freely admit that there were several varieties of the *síkuos* and *cucumis.*[64] Pliny reports that the *cucumis* "is composed of cartilage and a fleshy substance, while the bottle gourd (*cucurbita*) consists of rind and cartilage."[65] Both cucumber and melon fit the criteria for Pliny's *cucumis.* Similarly, in Mesopotamian lexical texts dating between the second and first millennia BCE, the Akkadian term *qiššû* was qualified with toponymic and descriptive adjectives to indicate a variety of cucurbits other than the melon, among which the cucumber could well be represented.[66] The *qiššû merarû* (bitter cucurbit) or *qiššû ša ṣumameti* (cucurbit of the desert) of the lexical tradition is, for instance, to be identified with the colocynth (*Citrullus colocynthis*), a wild relative of the watermelon, whose native range extends from the southern Mediterranean to the Middle East.[67] Other varieties of *qiššû* cited in the cuneiform lexical corpus include the finger cucurbit (*ubānu*), the Egyptian cucurbit (*muṣrītu*), and the bull's-testicle cucurbit (*iški alpi*).[68] These are no doubt references to both intraspecific and interspecific cucurbit varieties. In the absence of further contextual data or commentarial literature, much of the Mesopotamian lexical data do not allow a species-specific identification and thus offer no firm evidence for the cucumber.

Since some species of melons are polymorphic and exhibit morphological and gustatory similarities to cucumbers, it should be expected that terms relating to elongate melons (Akkadian *qiššû,* Hebrew *qišūt,* Arabic *qithā,* Greek *síkuos,* Latin *cucumis*) were applied to cucumbers when the latter spread westward. This mirrors the case of the lemon, which was not distinguished from the citron in Middle Eastern and Mediterranean contexts at least until the seventh century CE. The shared vocabulary between melons and cucumbers in earlier periods is most clearly demonstrated by Greek papyrological sources from Roman Egypt (first to third centuries CE), which attest to the local Greek dialect evolving diminutive forms of the

64. Pliny *HN* XIX.64–68; Theoph. *Hist. pl.* VII.4.6, VII.4.1.

65. Pliny *HN* XIX.61–62.

66. For a discussion and identification of various cucurbit terms in Sumero-Akkadian lexical texts, see CAD, s.v. *qiššû*; Stol 1987; Thompson 1949, 81–86 (note, however, the problematic interpretations of *qiššû* and *ukuš* as cucumber). The most important Mesopotamian lexical texts for cucurbits are the Urra = *ḫubullu* series dating from the Old Babylonian period, c. 1800 BCE onward (Tablets XVII, XXIV in Landsberger et al. 1970; MSL 10, 97–98, 115; Reiner and Civil 1974, 87, 127–78, 158) and the Murgud, an early-first-millennium BCE commentary to the Urra = *ḫubullu* (Landsberger et al. 1970, 104–06).

67. McCreight et al. 2013, 1082; Thompson 1949, 84–85. It is also known by the Akkadian names *tigillû* and *irrû.*

68. CAD, s.v. *qiššû*; Stol 1987; Thompson 1949, 81–86.

term *síkuos* (*sikúdia, sikúdion*) to distinguish the smaller true cucumber from elongate melons.[69]

THE ARCHAEOLOGICAL EVIDENCE FOR CUCUMBERS

The archaeological data on cucumbers is problematic but offers more solid evidence, in contrast to the textual sources, where the cucumber has been obscured by the shared and confused cucurbit terminology. Cucumber, as has been amply remarked, is usually consumed raw as an immature fruit along with the underdeveloped soft seeds, lowering its chances of archaeological preservation. Furthermore, in archaeological contexts cucumber seeds are not easily distinguished in morphology from melon seeds, and therefore all archaeobotanical identifications to one species or the other, particularly in older publications, must be treated with caution.[70] Still, there are a few diagnostic features pertaining to cucurbit seeds, namely seed symmetry, coat patterning, and the position of the hilum and radicle exit, which offer the prospect of species-specific identification if the samples are well preserved.[71] DNA sequencing of ancient cucurbit seeds also offers the promise of concrete identification of the cucumber in archaeological assemblages.

The Danish archaeobotanist Hans Helbaek identified two cucumber seeds (*Cucumis sativus*) dating to the last quarter of the seventh century BCE during Max Mallowan's excavations at the Neo-Assyrian capital of Kalḫu (Nimrud).[72] These seeds, taken at face value, represent the earliest reported archaeological samples of cucumber in the Middle East. But the original publication does not include photographs of the find, and the seed findings are presently untraceable, so it is impossible to confirm Helbaek's identification.[73] Cucumber seeds were also recovered from a relatively close chronological horizon at the sixth-century BCE Greek port of Marseille in France, but a later publication is less certain of the identity and suggests either *C. melo* or *C. sativus*.[74] Another early finding is the cucumber seed from the fourth-century BCE necropolis of Tauric Chersonesos in the Black Sea

69. Konen 1995, 50.

70. Samuel 2001, 465; Sebastian et al. 2010, 14269.

71. Frank and Stika 1988, 48–49; Šoštarić and Küster 2001, 229; van der Veen 2011, 155.

72. Helbaek 1966, 615; Renfrew 1987, 162.

73. I have attempted to locate these cucumber seeds to validate Helbaek's identification, but they are presently not to be found in Helbaek's collection of Nimrud materials in Copenhagen. Mette Marie Hald, who has re-examined the archaeobotanical remains from Nimrud stored at the National Museum in Copenhagen, notes that Helbaek's methodology was meticulous and his identifications are almost entirely accurate (personal communication). There is a remote possibility that the cucumber seeds were left to the National Museum of Iraq in Baghdad (Jane Renfrew, personal communication), if they were not treated as expendable after identification.

74. Marinval 2000, 186; Ruas 1996, 97.

region.[75] If accepted as cucumber, the finds at Marseille and Tauric Chersonesos might indicate that the new eastern cultivars were spreading through Greek colonial networks in the Mediterranean and the Black Sea, much like citrus fruits. But the older finds of cucumber seeds need to be re-assessed in light of more recent diagnostic criteria proposed for distinguishing the melon from the cucumber before they can be confidently accepted as cucumbers.

Presently, only more recently excavated archaeobotanical samples from the Roman Mediterranean and northern Europe can be confidently identified as cucumber. These finds establish the early Roman Imperial period as the *terminus ante quem* for the arrival of the cucumber in the Mediterranean.[76] Cucumber seeds have been identified at early Roman Egyptian sites (first to second centuries CE), including Hawara, Mons Porphyrites, Mons Claudianus, and the Red Sea ports of Myos Hormos and Berenike.[77] Interestingly, Marijke van der Veen notes with reference to the archaeobotanical assemblages at Mons Claudianus and Myos Hormos that the cucumber appears to be more common, whenever it could be distinguished from the melon.[78] Cucumber seeds have also been recovered from multiple Roman sites in the Mediterranean and northern Europe dating between the first and third centuries CE, including Vado Ligure (northern Italy), Mazières-en-Mauge (Maine-et-Loire, France), Longueil-Sainte-Marie (France), Cosne-sur-Loire (France), Toulouse (France), Oedenburg (French Alsace), Pula (Croatia), Veli Brijun (Croatia), Tongeren (Belgium), Maldegem-Vake (Belgium), Augst (Switzerland), Windisch (Switzerland), Eschenz (Switzerland), Trier (Germany), and London (map 10).[79]

It follows that for the cucumber to be so well integrated into Mediterranean and northern European agricultural regimes and diet by the first century CE, it must have had a longer genealogy in the neighboring east. In light of the spatially broad

75. Chtcheglov 1992; Marinval 2003, 24.

76. In light of the specialized terms developed for cucumber in early Roman Egypt, Konen (1995, 49–50) speculated that the cucumber was introduced into Egypt during the Greco-Roman period.

77. Hawara: Newberry 1889, 52; Germer 1985, 130 (this find needs to be re-examined in light of the age of its discovery). Myos Hormos: van der Veen 2011, 155. Mons Porphyrites (late first to second century CE): van der Veen and Tabinor 2007, 94. Mons Claudianus: van der Veen 1998, 225, 2001, 180. Berenike: Cappers 1998, 320.

78. Van der Veen 2001, 195.

79. Vado Ligure: Arobba et al. 2013. Mazières-en-Mauge: Bouby and Marinval 2004, 85. Longueil-Sainte-Marie: Marinval, Maréchal, and Labadie 2002, 261. Cosne-sur-Loire: Wiethold 2003, 274. Toulouse: Bouby and Marinval 2002, 189. Oedenburg: Vandorpe 2010, 39, 50 (illustrated). Pula: Essert et al. 2018, 959, 964. Veli Brijun: Šoštarić and Küster 2001, 229 (illustrated). Tongeren: Vanderhoeven et al. 1993, 188, 192 (illustrated). Maldegem-Vake: Bastiaens and Verbruggen 1995, 38–41. Augst: Vandorpe 2010. 92. Windisch: Vandorpe, Akeret, and Deschler-Erb 2017, 139, 142. Eschenz: Pollmann 2003. Trier: König 2001, 73. London: Willcox 1977, 270, 279 (as this report is substantially older than the other cited finds, the remains of cucumber from Roman London should be re-examined with the recently proposed criteria to distinguish melons from cucumbers).

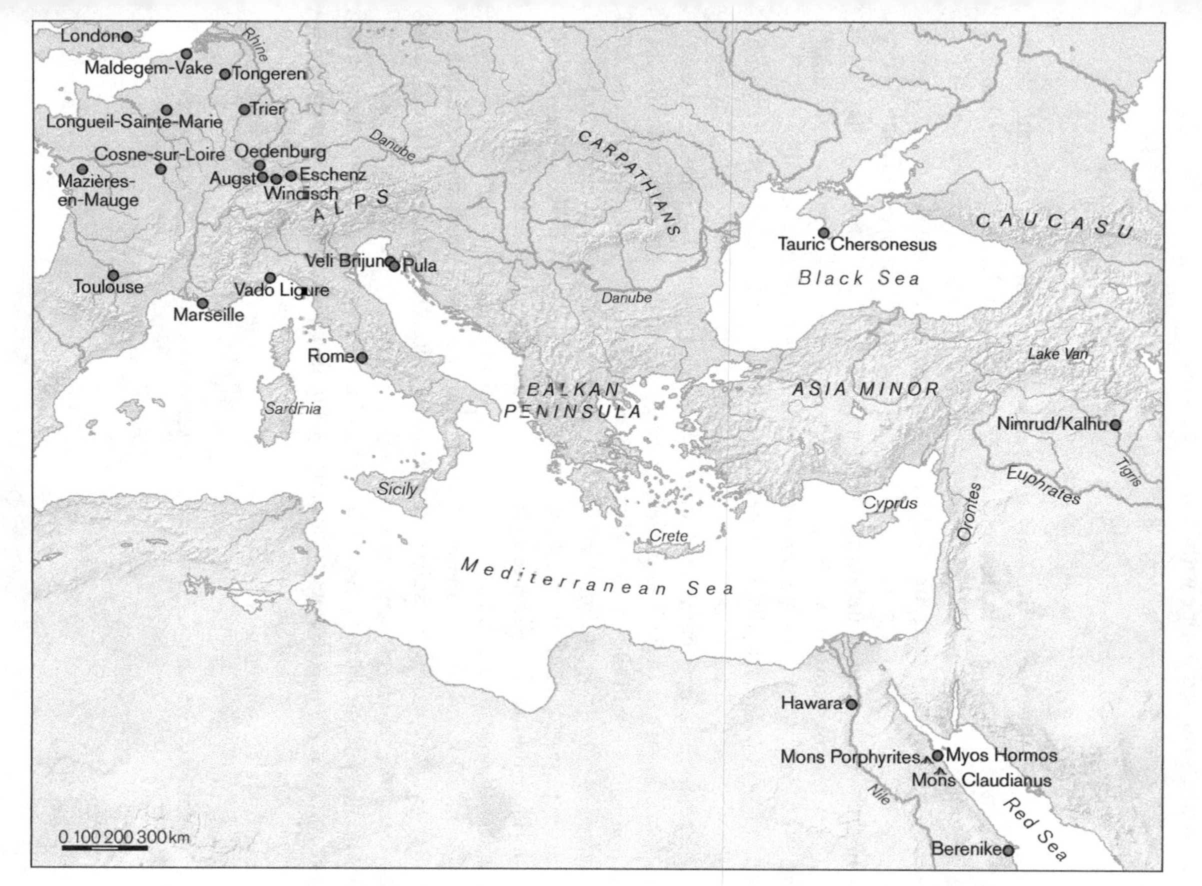

MAP 10. Ancient sites which have yielded evidence for cucumbers (*Cucumis sativus*). © Peter Palm (Berlin).

archaeological finds, there is a real possibility that the true cucumber already lurks behind some of the late-first-millennium BCE descriptions of cucurbits which are said to be consumed unripe with immature seeds.

THE LUFFA (*LUFFA AEGYPTIACA*) AND OTHER INDIAN CUCURBITS

Several much-overlooked Hellenistic Greek fragments not only explicitly designate certain cucurbits as Indian but even recognize that their seeds were brought from India. Euthydemos of Athens, in his treatise *On Vegetables* (*Perì Lakhánōn*, late fourth / early third century BCE), reports that a cucurbit known in his day was called "Indian" since the seed was brought from India.[80] Euthydemos almost certainly refers to a new or unfamiliar type of vegetable, for there would have been little need to transport the seeds of familiar crops over such long distances. A near-contemporary, Menodoros, a student of Erasistratos who was the court physician of king Seleukos I, also distinguishes a kind of cucurbit (*sikúa*) as Indian.[81] It is probable that Euthydemos of Athens and Menodoros are describing the same cucurbit, perhaps one that arrived through Seleukid diplomatic-commercial channels with Mauryan India.[82] Athenaeus, who preserves these fragments in his gastronomic treatise of the second century CE, adds that the Knidians of his day still referred to gourds (*kolokúntai*) as "Indian."[83] The use of both *kolokúnthē* and *sikúa* to describe the "Indian" cucurbit confirms not only the loose usage of these terms in antiquity but also conveys disparities in botanical taxonomies within the same cultural-linguistic horizon.

An identification of the "Indian" cucurbit with the smooth luffa or sponge gourd (*Luffa aegyptiaca*), which is still of economic and cultural significance in the Middle East and the Mediterranean, seems probable (figure 22).[84] Despite being of Indian origin, the true cucumber can be ruled out, as Menodoros's comment on the Indian cucurbits being stewed is not typical of true cucumbers, which are usually eaten raw. Instead, the description suggests a species like luffa or sponge gourd, whose immature fruit is consumed as a cooked vegetable.[85] As the modern name of the plant implies, the mature fruit of the sponge gourd is also valued for its dry, springy, fibrous interior, which is used as a cleaning scrub and as stuffing for mattresses and

80. Ath. II.58f.

81. Ath. II.59a.

82. Cf. Ath. I.18e.

83. Ath. II.59a.

84. On luffa species in India, see Decker-Walters 1999, 108–09. On luffa in the Middle East and the Mediterranean, see Chakravarty and Jeffrey 1980, 202; Charles 1987, 17.

85. Marr, Xia, and Bhattarai 2005, 137–38; Menodorus, ap. Ath. II.59a.

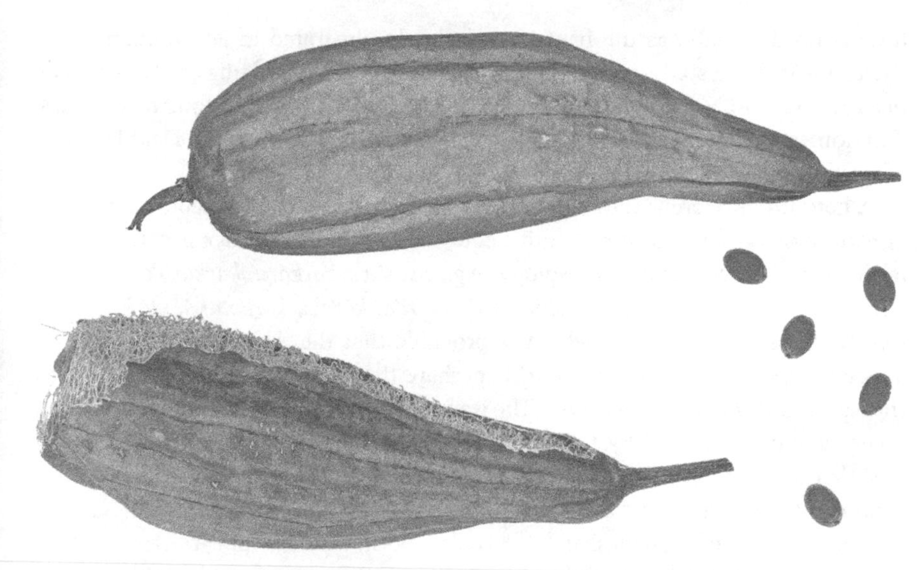

FIGURE 22. Desiccated specimen of the sponge gourd (*Luffa aegyptiaca*) from the Jardin botanique Henri Gaussen, Botanical Collection of the Muséum de Toulouse. Photograph by Roger Culos for the Muséum de Toulouse via Wikimedia Commons (CC BY-SA 3.0).

saddles.[86] But this aspect of the plant is not remarked on, which is curious should the Indian cucurbit of the Hellenistic authors turn out to be the luffa.

The early history of the luffa in Middle Eastern and Mediterranean records is far from clear. The Egyptologist Nathalie Beaux identified the luffa in reliefs in the temple of Amun at Karnak (Thebes) which commemorate the exotic flora collected by the pharaoh Thutmose III (c. 1479–1425 BCE) during his campaigns in the Levant.[87] While the reliefs undoubtedly depict an elongate and striate cucurbit broader toward the distal end, a specific identification with the luffa is not warranted, particularly since the depictions of Beaux's "luffa" bear few morphological differences from the fruits which the same scholar has identified as elongate melons.[88] These are in all likelihood elongate melons or bottle gourds. Otherwise, more secure identifications of the luffa in both iconographic and textual records are relatively late. Immanuel Löw has identified the dark-green cucurbits named *qarmulin* or *qarumalim* in the Tosefta (late third century CE) and Jerusalem Talmud (late fourth century) as *Luffa aegyptiaca*.[89] The identification of luffa in rabbinical

86. Marr, Xia, and Bhattarai 2005, 138.

87. Beaux 1990, 165–66.

88. Beaux 1990, 167–69.

89. Singular form: *qarmal*. TJ, *Shevi'it* 2:7, *Nedarim* 7:1; Tosefta, *Shevi'it* 4.12; Avital and Paris 2014, 219; Janick, Paris, and Parrish 2007, 1448; Löw 1928, vol. 1, 548–49.

texts is not in doubt, as the fruit is very clearly illustrated in late Roman–early Byzantine (fifth to sixth centuries CE) mosaics from Israel.[90] Luffa seeds have also been recovered from a near-contemporary late Roman (third to fourth centuries CE) domestic context at the site of Kellis (Ismant el-Kharab), in the Dakhla Oasis of western Egypt.[91]

A handful of references in Akkadian cuneiform texts to a vegetable named the *karkarātu* or *karkartu* have no Semitic equivalents but display strong lexical parallels with Indian cucurbit terminology (e.g., Sanskrit *karkāru, karkāruka, karkaṭī, karkaṭā, karkaśchadā, karkacirbhiṭā, kākamarda, kāravī, karkoṭaka, karkaṭikā*).[92] On the basis of lexical parallels, it is probable that the Akkadian *karkarātu* or *karkartu* represents an Indian cucurbit, perhaps the luffa if not another species like the wax gourd (*Benincasa hispida*). The wax gourd or winter melon is an important comestible in South and East Asia but is less well known in the modern Middle East.[93] The *karkarātu/karkartu*-vegetable is first attested in a late-thirteenth-century BCE tablet from Kassite Ur and then in two Neo-Assyrian texts, namely the banquet stela of Aššurnasirpal (r. 883–859 BCE), where it appears as a side dish in a royal banquet and in a medical text as an ingredient in a remedy for the *mūṣu*-disease.[94] In its first known appearance from Kassite Ur (late thirteenth century BCE), the *karkartu* plant is named twice in a fragmentary maintenance account detailing comestibles issued to some eighteen individuals:

> 2 kor of barley, 1 kor of dried *ṭurû*-plant, 3 seah of *muššu* (breast-shaped loaves?), 3 bushel(?) of *karkarātu* from the town of Hurri-būṣi, Ahu-abu and Arkāt-ili-lūmur, son of Sîn-mušēzib (received).
>
> MBLET 52 Obv. 3–5
>
> 1 seah of *muššu* (breast-shaped loaves?), 3 bushel(?) of karkartu, 3 liter of cress from the town of Sîn-karābi-išme, Imbuassu received.
>
> MBLET 52 Obv. 8–9[95]

While the identity of some plants is uncertain, the presence of barley, cress, and pomegranates in the same text suggests that the unidentified comestibles were

90. Avital and Paris 2014, 219.

91. Thanheiser 2002, 304–05.

92. *Karkāru* or *karkāruka* (wax gourd, *Benincasa cerifera*), *karkaṭī* (kind of cucumber), *karkaṭā* (Gac fruit, *Momordica mixta*), *karkaśchadā* (a luffa species), *karkacirbhiṭā* (a kind of cucurbit), *kākamarda* (gourd, colocynth), *kāravī* (a small gourd), *karkoṭaka* (Gac fruit, *Momordica mixta*), *karkaṭikā* (a kind of plant).

93. On wax gourds in India, see Chomicki, Schaefer, and Renner 2020, 1250; Decker-Walters 1999, 105–06.

94. CAD, s.v. *karkartu*; MBLET 52: 4. 1; RIMA II A.0.101.30: 129; BAM 2, 117: 3 (VAT 13910).

95. Gurney 1983, 140–42.

common enough to be included in a list of foods for regular supply. The appearance of *karkartu* as a vegetable in a royal Assyrian banquet indicates that it was valued for its taste. Unfortunately, there is simply not enough descriptive material in the cuneiform texts to establish the identity of this vegetable with any degree of certainty. Its identification as an Indian cucurbit will have to remain to remain speculative for now. Overall, the certain presence of the luffa in Mediterranean contexts by late antiquity makes it likely that it is hiding in earlier references to Indian cucurbits by Hellenistic authors or in words like the Akkadian *karkartu,* which displays lexical affinities with Indic cucurbit terminology.

THE CASE OF THE "CUCURBIT FLY"

The appearance of a parasitic faunal species, especially where it predates on specific kinds of crops, can also serve as a useful index for the arrival of a new cultivar. A Neo-Assyrian manuscript (seventh century BCE) of the lexical series Urra = *ḫubullu* (Tablet XIV) attests to a species of "fly that infests cucurbits" (*nim-ukuš*).[96] The entry for the cucurbit (cucumber and/or melon) fly stands alongside two other kinds of agricultural pests, a generic vegetable-fruit fly (*nim-nisig*) and a fly infesting ghee (*nim-inun*).[97] While the precedents of the Urra = *ḫubullu* lexical series date from as early as the Old Babylonian period (c. 1800 BCE), the entry for the "cucurbit fly" appears to be an early-first-millennium BCE updating of the lexical series, since it is absent from earlier manuscripts of the text. The specific mention of a cucurbit fly in the updated lexical series suggests that insect predation of cucurbits must have been a significant agrarian concern in the Iron Age Middle East.

Both the logographic spellings *nim-ukuš* (cucurbit fly) and *nim-nisig* (vegetable/fruit fly) are equated with the Akkadian *tambukku* (variant *tebukku*), a generic name for noxious insects, in the Neo-Assyrian manuscript of the Urra = *ḫubullu.* The *tambukku*-insect is also referred to in a handful of texts outside the lexical tradition, the oldest dating to the late second millennium BCE.[98] Incidentally, the Akkadian *tambukku* may bear some distant relationship with a type of noxious fly known as *tryambuka* in Sanskrit. The westerly origin of this word in Sanskrit is suggested by the observation of the fourth-century CE Gandhāran Buddhist scholar Vasubandhu that Persians claim "snakes, scorpions and *tryambuka*-flies should be killed because they cause harm."[99] Yaśomitra, a commentator on Vasubandhu's text, glosses it as a kind of wasp (*varaṭa*), and the same meaning is suggested by a passage

96. CT 14, pl. 1–2: K 71A, r ii 19'; text and translation in Landsberger 1934, 24–25.
97. CT 14, pl. 1–2: K 71A, r ii 18', 20'; Landsberger 1934, 24–25.
98. CAD, s.v. *tambukku.*
99. *Abhidharmakośa* 240, 25 (trans. Lindtner 1988, 440–41).

in the *Mūlasarvāstivāda-vinaya,* a corpus of monastic regulations written in Sanskrit in the early centuries CE but largely surviving in Tibetan.[100] Whatever the origins of the term, it is clear that *tambukku* was a generic name for an unpleasant insect and was used in the lexical tradition as a synonym for the "cucurbit fly."

The "cucurbit fly" of the Neo-Assyrian Urra = *ḫubullu* manuscript is to be identified with the Baluchistan melon fly (*Myiopardalis pardalina*), the melon fly (*Bactrocera cucurbitae*), or, less plausibly, with the red pumpkin beetle (*Aulacophora foveicollis*), insects well known for their near-exclusive predation on cucurbit species.[101] While the present geographical distribution of all three species covers parts of the Middle East, the Baluchistan melon fly (*Myiopardalis pardalina*) appears to be the more virulent pest of the three in Iraq and Iran. One Iraqi government report from 1962 describes the Baluchistan melon fly as the "worst pest of melon in Iraq."[102] It is also a major pest in neighboring Iran (Persian *magas-e ḵarboza,* "melon-fly").[103] While the distribution of both *Bactrocera cucurbitae* and *Aulacophora foveicollis* is more easterly and shows a distinct preference for tropical climes, *Myiopardalis pardalina* prefers temperate zones, suggesting that the "cucumber/melon fly" of the Neo-Assyrian Urra = *ḫubullu* is most likely the Baluchistan melon fly (*Myiopardalis pardalina*).[104]

The appearance of the melon fly in Mesopotamia could be associated with the landward spread of a new eastern cucurbit, although there is no certainty as to what species this might be—perhaps the *karkartu*-vegetable, if this is to be identified as an Indian cucurbit. While *Bactrocera cucurbitae* is believed to originate from the Central-South Asian frontier zone, the original range of *Myiopardalis pardalina* and *Aulacophora foveicollis* is presently unclear.[105] The severity of *Myiopardalis pardalina* and *Aulacophora foveicollis* predations in modern South Asia and the Irano-Indian borderlands might suggest a similar eastern epicenter.[106] An eastern origin for these species is also not surprising in light of the great diversity of cultivated cucurbits in South Asia. While the cuneiform lexical tradition provides a *terminus ante quem* for the presence of the melon fly in Mesopotamia (seventh century BCE), it fails to shed light on the earliest appearance of the melon fly in that region. At best, one can speculate that a new eastern cucurbit species, the *karkartu* perhaps, and a related parasitic fly arrived in the Mesopotamia sometime between the late second millennium and the early first millennium BCE.

100. *Abhidharmakośa* 240, 25 (trans. Lindtner 1988, 440–41); Silk 2008, 116–17.

101. Chakravarty and Jeffrey 1980, 198; Rivnay 1962, 283. The red pumpkin beetle is locally known in Iraq as the "little red one" (Ar. *humaira*).

102. Government of Iraq 1962, 161.

103. Abivardi 2008.

104. Atwal 1986, 262; Khan et al. 2012.

105. Virgilio et al. 2010.

106. Stonehouse et al. 2008.

THE USE AND APPEAL OF NEW CUCURBITS IN ANTIQUITY

Cucurbits rank among the most important vegetable-fruit plants cultivated globally, both today and in the past. Cucurbitaceous vegetables represent low-risk crops, since they are easily cultivated in small plots of well-drained soil in household gardens and on the margins of agrarian fields. The minimal inputs and fast-maturing nature of cucurbits made them particularly attractive to ancient cultivators. Cucumbers, for instance, start flowering in six to seven weeks.[107] Much of the Middle East and Mediterranean is well suited to cucurbit cultivation, as most Old World cucurbits grow best in warm, rainless summers.[108] The early familiarity of melons in the Middle East and the Mediterranean meant that other eastern cucurbits were easily accommodated in the agricultural and dietary repertoire when they were transmitted westward.

Cucumbers are typically eaten uncooked in the Middle East and the Mediterranean. This can range from simply salting diced cucumbers, mixing grated cucumbers in yogurt, salads, and beverages, or stuffing raw cucumbers in cooked food.[109] In this regard, the late Roman anthology of recipes attributed to Apicius (second to fourth centuries CE) probably refers to true cucumbers, which were added uncooked to a type of cold poultry salad which included pepper, mint, cheese, pine nuts, and chicken liver.[110] Cucumbers are also pickled, which extends their role as a calorific source beyond the summer months.[111] The sixth-century CE Mesopotamian Syriac author Qūthāmā, preserved in an Arabic translation, reports that snake melons (*qiththā'*) and cucumbers (*khiyār*) were kept in storage and given as birthday presents in several districts of Iraq.[112] The ability to store and preserve cucurbits in pickled form made favorable the large-scale cultivation of cucurbits in the summer season.

While most tropical and subtropical cucurbits are frost intolerant and therefore predominantly summer cultivars in the Middle East and the Mediterranean, ancient famers in both regions attempted small-scale, out-of-season cultivation through innovative frost-protection methods. Both Theophrastus and a Pseudo-Aristotelian author describe the planting of the *síkuos,* either an elongate gourd, melon, or cucumber, in baskets of soil which in winter were carried out into the

107. Paris, Daunay, and Janick 2012, 117.

108. Charles 1987, 6; Paris, Amar, and Lev 2012b, 24, 30.

109. A'lam 2011b.

110. Apicius, *De re coquinaria* IV.1.1. For similar cold poultry salad recipes from the Abbasid Middle East, see Ibn Sayyār al-Warrāq, *Kitāb al-Ṭabīkh* 31, trans. Nasrallah 2007, 163–70.

111. A'lam 2011b; cf. Pliny *HN* XIX.24 on pickling cucurbits.

112. Ibn Waḥshīyah, *Al-filāḥah al-nabaṭīyah*, 538–41; Hämeen-Anttila 2006, 233.

sun or placed near a fire for warmth.[113] Pliny and Columella likewise attest to out-of-season production of melon for the emperor Tiberius in mica or talc containers which were wheeled out on sunny days and kept indoors during the cold season.[114] In modern Iraq, cucurbits sown in winter are protected from the cold by covers made of palm fronds.[115] As the example cited by Pliny and Columella indicates, the out-of-season production of cucurbits was likely a luxury afforded by wealthier citizens.

Apart from the obvious comestible function of fleshy cucurbits, both the melon and cucumber are also valued for their oil-bearing seeds, as ornamental climbing or trailing plants, and for their medicinal applications, particularly as a diuretic and a remedy for heat exhaustion in summer.[116] Among the more outlandish non-comestible uses of cucurbits, the use of cucumbers, elongate melons, and gourds as dildos, a practice well attested in ancient South Asia, is not entirely unknown in the pre-Islamic Middle East, either.[117]

To sum up the new findings on the history of cucurbits in the ancient Middle East and the Mediterranean: the true cucumber was indeed known in the Middle East and the Mediterranean by the first century CE at the latest. It probably arrived earlier but was regularly confused and conflated with elongate melons and gourds, which were familiar in the wider region from distant antiquity. The melon itself was a mid-third-millennium BCE introduction from the east. A few other Indian cucurbit species spread westward in the first millennium BCE if not earlier, although the comparatively slight impression these other species have left on the extant sources indicate that they were not of great economic value. The sponge gourd or luffa (*Luffa aegyptiaca*), which remains important in the Middle East and the Mediterranean, was quite likely among these new cucurbits dispersed westward by anthropic networks.

113. Pseudo-Aristotle, *Problems* XX.15; Theoph. *Caus. pl.* V.6.6.

114. Columella *Rust.* XI.3.52–53; Pliny *HN* XIX.64; Paris and Janick 2008b.

115. Charles 1987, 9.

116. A'lam 2011b; Nasrallah 2007, 789 (glossary entry in the text edition for Ibn Sayyār al-Warrāq's *Kitāb al-Ṭabīkh*).

117. *Kāmasūtra* 7.2.13; Sanghadāsa, *Bṛhatkalpabhāṣya* 1050–56; TB *Megillah* 12a.

6

The Egyptian Bean: The Sacred Lotus

In this chilly winter time,
may your cooking pots be full
with paste of lotus stem and rhizome,
bright and smooth as elephant tusk,
with fritters rich in pepper,
and pieces of śakuni-fowl.

VṚIDDHI THE SCYTHIAN[1]

The genus *Nelumbo,* the sole representative of the family Nelumbonaceae, consists of two widely recognized perennial aquatic plants: *Nelumbo nucifera,* the sacred Asian or the Indian lotus (figure 23), and *Nelumbo lutea,* the American lotus. The latter, as the name implies, is a New World species native to eastern North and Central America. The Indian lotus, which is the subject of our present study, is widely distributed across eastern Eurasia as far south as northern Australia and as far north as the Amur region of the Russian Far East.[2] While the Indian lotus is found in greater abundance in the tropical and subtropical zones of Asia, it is well suited to growing in temperate zones (e.g., northern China and the Russian Far East), as long as the rhizomes do not freeze.[3]

The westernmost natural distribution of the Indian lotus presently includes disjunct wild populations in the Talysh region of the southeastern Caucasus, the south Caspian zone, and the Volga basin. These wild populations are relicts from a wider pre-Pleistocene-glaciation distribution (c. 2.5 million years ago).[4] The vast spatial distribution of *Nelumbo* species is also borne out by macrofossil records across Eurasia and North America, with the earliest samples dating back to the

1. Ap. Vallabhadeva, *Subhāṣitāvalī,* 395. The poet's dates are unknown (but no later than the seventh century CE).
2. He, Shen, and Jin 2010, 159; Li et al. 2014a, 2014b.
3. Kintaert 2010, 488.
4. Baldina et al. 1999, 175–76; Li et al. 2014a.

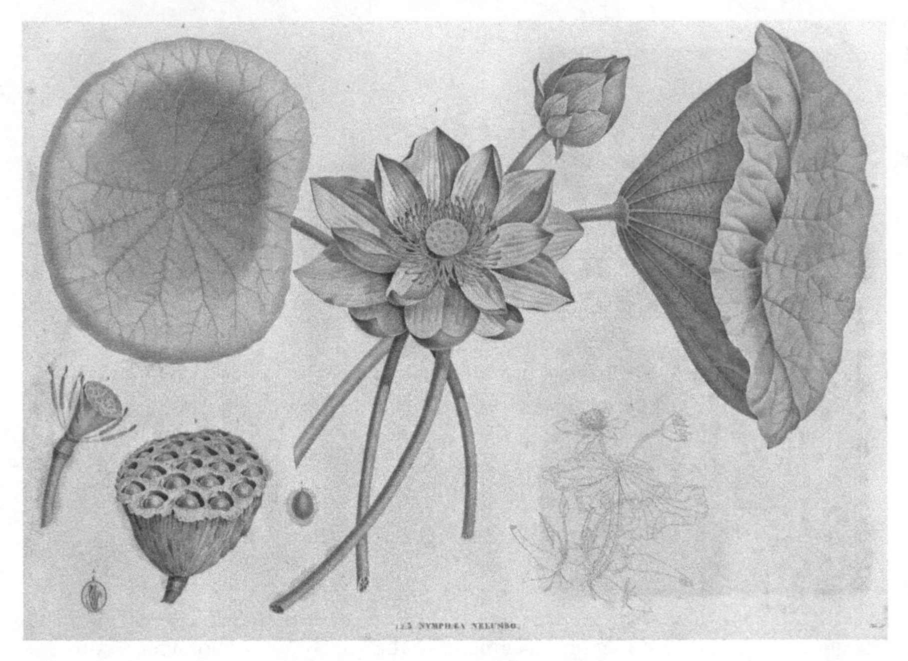

FIGURE 23. Sacred lotus (*Nelumbo nucifera*), illustration from *Description de l'Égypte, Histoire naturelle. Planches,* Vol. 2 (Paris, 1817). New York Public Library (public domain).

Late Cretaceous.[5] The Caspian lotus, sometimes distinguished as a separate species (*Nelumbo caspica*), is morphologically smaller than its East and South Asian counterparts and bears fewer edible seeds on the honeycombed fruit.[6] The phenotypic disparity between the Caspian lotus and its East and South Asian counterparts is probably the result of human selection for a larger and densely seeded variety of lotus in tropical and subtropical Asia. Small isolated wild populations and the lack of consistent economic uses for the lotus in the Caspian, Volgan, and Caucasian regions means that these lotus populations were not the source for the anthropogenic diffusion of the lotus to the Middle East and the Mediterranean in the first millennium BCE.[7]

5. He, Shen, and Jin 2010; Li et al. 2014b.

6. Koren, Yatsunskaya, and Nakonechnaya 2012, 912; Walter and Breckle 1986, 213; Zeiss and Bachmayer 1986, 116.

7. The eastern associations of the lotus in the Middle East are suggested by the use of Sanskritic terms in later Arabo-Persian botanical vocabulary for the lotus and water-lily (Nymphaceace) species, e.g., *nīlūfar*, from the Sanskrit *nīlotpala*. The earliest surviving Persian pharmaceutical text, the *Ketāb al-abnīa ʿan ḥaqāʾeq al-adwīa* of Abū Manṣūr Mowaffaq Heravī (late tenth century), describes the seeds of *Nelumbo nucifera* using another Sanskrit loanword, namely *aṭmaṭ* (A'lam 1989).

The Indian lotus is typically found in muddy, shallow, and stagnant lakes or slow-moving streams, and is cultivated as a food crop across tropical and subtropical Asia.[8] Both the acorn-like seeds (or nutlets), held in a large seedpod, and the starchy rhizomes are consumed.[9] The large, water-repellent leaves and showy and fragrant flowers, ranging in color from pink to white, also make the sacred lotus an ornamental plant *par excellence*. Apart from its role as a comestible and ornamental plant, the lotus has also acquired an all-pervasive religious and symbolic role in Asian cultures, representing in particular the ideals of beauty, non-attachment, and purity.[10] In South Asian tradition, the lotus is also the embodiment of Śrī or Lakṣmī, the goddess of sovereignty, abundance, and prosperity.

The lotus probably already held symbolic, ornamental, and comestible functions in the Harappan civilization, the earliest urban culture of South Asia. Lotus seedpod faience models were found at late-third-millennium BCE contexts at both Harappa and Mohenjo-daro, suggesting that the plant was of some cultural importance to be reproduced in artistic media.[11] The vast number of names used for the lotus in Indic languages (e.g., Skt. *nalinī, padma, aravinda, tāmarasa, kamala, sarasīruha*) also testifies to its high rank in South Asian cultures.[12]

THE EGYPTIAN BEAN: *NELUMBO NUCIFERA* IN EGYPT

> *You will laugh at this Nile vegetable and its clinging fibers when with teeth and hands you extract its obstinate threads.*
>
> MARTIAL, *EPIGRAMS* XIII.57 (TRANS. CULPEPPER STROUP 2006, 307)

While the Indian lotus (*Nelumbo nucifera*) is no longer cultivated as a food crop in modern Egypt, or anywhere along the Nile Valley for that matter, Greco-Roman literary and iconographic sketches of Nilotic landscapes invariably reserved a prominent place for the Indian lotus, which by the fourth century BCE was toponymically styled the "Egyptian bean" (Greek *kúamos Aigúptios*). The latter designation reflected both its importance as a comestible in late-first-millennium BCE Egypt and the fact that Greek visitors to Egypt had first encountered the plant growing in the Nile.

Amid the crocodiles and hippopotami, temples, and reed boats, one finds in the grand riverine vista of the late-second-century BCE Nile mosaic from the Italian

8. Small and Catling 2005.

9. Ancient sources do not lexically distinguish between root and rhizome (e.g. Greek *rhíza*). I have chosen to consistently render all such words as "rhizome," since in the case of the lotus it is the rhizome and not the root which is being consumed.

10. Cielas 2014; Garzilli 2003; Kintaert 2010, 2011–12.

11. Vats 1940, 169, 467.

12. Lienhard 2000; Schmidt 1913.

FIGURE 24. Mosaic depicting Indian lotuses, including peltate leaves and seedpods, in a Nilotic scene, Casa de Neptuno, Italica, southern Spain, late second century CE. Photograph by the author.

city of Palestrina (Praneste) a profuse representation of the Indian lotus in every stage of its flowering.[13] Mosaics, paintings, glass tiles, terra cottas, and textiles were some of the media in which the Indian lotus came to be widely represented, both in Egypt and elsewhere in the Greco-Roman Mediterranean (figure 24).[14] Kallixeinos of Rhodes, in his description of the magnificent riverine barge of Ptolemy IV Philopator of Egypt (221–204 BCE), refers to a hall supported by pillars whose capitals were adorned with representations of the Indian lotus and other Nilotic flora, a veritable mirror of the aquatic world on which the barge floated.[15] The Indian lotus, like the crocodiles and hippopotami, was identified by Greek and Roman observers as one of the diagnostic features of Nilotic landscapes. So close was the sacred lotus's association with Egypt that Alexander the Great, who sighted

13. Meyboom 1995.

14. Darby, Ghalioungui, and Grivetti 1977, 635; Daszewski 1985, 9, 68, 136; Goede 2005; Jaksch 2012, 45, 48, 53–61. 96; Kumbaric and Caneva 2014, 186; Meyboom 1995, 261; Turnheim 2002; Versluys 2002, 47, 69, 77, 86, 126, 205, 229, 232, 263.

15. Callixeinos, ap. Ath. V.206b-c.

lotuses in the Akesines River (Skt. Asiknī, modern Chenab) in India, mistakenly thought he had found the headwaters of the Nile.[16]

Herodotus (fifth century BCE) is the first ancient author who produces unambiguous evidence for the Indian lotus in Egypt. Unlike later authors, however, he does not assign a specific name to the plant and simply considers it a variant of native Egyptian lilies (*krínon*):

> When the river is in flood and flows over the plains, many lilies, which the Egyptians call lotus, grow in the water. They gather these and dry them in the sun; then they crush the poppy-like center of the plant and bake loaves of it. The rhizome of this lotus is edible also, and of a sweetish taste; it is round, and the size of an apple. Other lilies grow in the river, too, that are like roses; the fruit of these is found in a calyx springing from the rhizome by a separate stalk, and is most like a comb made by wasps; this produces many edible seeds as big as olive pits, which are eaten both fresh and dried.[17]

Here the first lotus, whose fruit is described as having a "poppy-like center," is the native Egyptian *Nymphaea lotus* or white lotus, which has an esculent rhizome like the Indian lotus.[18] The ensuing description of a rose-colored lily with a seedpod resembling a wasp's nest is unmistakably the Indian lotus. The Indian lotus has usually been considered an Achaemenid Persian introduction to Egypt on account of the lack of clear iconographic, textual, or archaeological evidence in earlier periods.[19] Herodotus's description of the Indian lotus does not, however, betray its supposedly recent foreign origins. His casual remarks on its use as a food plant in deltaic Egypt suggest that the Indian lotus was a feature of Nilotic landscapes well before the Persian conquest of Egypt in 525 BCE. Hekataios of Abdera, in the late fourth century BCE, classified the Indian lotus as one of the "naturally occurring foods" of the Nile, suggesting that its introduction could not have been so recent for its foreign origins to be overlooked.[20]

The lack of a specific name in Herodotus's account suggests that the Greeks were largely unacquainted with the plant before the fifth century BCE. The Indian lotus is first attested under the designation of "Egyptian bean" (*kúamos Aigúptios*) in the passing remark of an anonymous Hippocratic author (fifth to fourth centuries BCE), in the account of Egypt provided by Hekataios of Abdera (late fourth

16. Arr VI.1.2–6; Nearchus, ap. Strabo XV.1.25.

17. Hdt. II.92.

18. Cf. Theophr. *Hist. pl.* IV.8.9–11. On the white lotus (*Nymphaea lotus*) and blue lotus (*Nymphaea nouchali* var. *caerulea*) species of Egypt, both well attested in Pharaonic iconography and literature, see Manniche 2006; 132–35; Pommerening, Marinova, and Hendrickx 2010; Weidner 1985.

19. Germer 1985, 40; Goede 2005, 61; Kintaert 2010, 487–88; Meyboom 1995, 261; Weidner 1985, 33–35.

20. Ap. Diod. Sic. I.10.1. On Hekataios of Abdera, see Lang 2012.

century BCE), and in the lengthy survey of the plant provided by Theophrastus in his botanical enquiries (late fourth century BCE). Theophrastus says:

> The [Egyptian] bean grows in marshes and lakes. Its stem reaches a maximum of four cubits in length and a finger in thickness; it looks like a pliant reed without knots, but has interior interstices distributed over all [its diameter] like a honey-comb. This stem is surmounted by the head, which looks like a rounded wasps' nest. Each of the cells contains a slightly prominent bean; there are thirty at most. The flower is twice as large as a poppy's, and its color is like a rose, of a deep shade. The head is raised above the water. Beside each of the bean [heads] grow large leaves equal in size to a Thessalian hat; these have a stem identical to that of the bean [heads]. If one of the beans is crushed, you can clearly see the bitter part coiled up, from which the *pīlos* (seed embryo) comes. So much for the fruit. The rhizome is thicker than the thickest reed, and has interstices like the stem. They eat it raw, boiled, and roasted, and the people of the marshes make this their food. It mostly grows of its own accord; however they also sow it, after having coated it well with mud, so that it is dragged to the bottom and remains there without rotting. In this way, they [the Egyptians] prepare the bean fields. Once it takes hold, it lasts permanently.[21]

Apart from the detailed and botanically precise description, much lacking in Herodotus, the most striking observation of Theophrastus is that the Indian lotus was not simply a naturalized wild plant growing in the Nile but one that was deliberately cultivated in a riverine "bean field" (*kuamṓn*). Also, while Herodotus had only cited the consumption of lotus seeds, Theophrastus notes that lotus rhizomes also formed an essential component of the deltaic Egyptian diet.[22] The harvesting of lotus rhizomes, flowers, and seedpods was done either by wading into the shallow water or (more typically) atop a canoe, as depicted in the mosaic from Palestrina, mentioned earlier.[23] The remark that lotuses were farmed and not left to grow on their own accord is confirmed by the eyewitness account of the Augustan geographer Strabo, who says that the "bean fields" were not small subsistence plots but large commercialized holdings with the prospect of substantial profit:

> The [Egyptian] bean-fields afford a pleasant sight, and also enjoyment, to those who wish to feast there. They [the Egyptians] feast on cabin-boats, in which they enter the bean thickets and the shade of the leaves; these [the leaves] are so very large that they are used both for drinking-cups and for bowls, for these have a kind of concavity suited to this purpose; and in fact Alexandria is full of these in the workshops, where they are used as vessels; and the countryfolk have this as one source of their revenues—I mean the revenue from the [sale of the] leaves.[24]

21. Theophr. *Hist. pl.* IV.8.7–8; cf. Pliny *HN* XVIII.122, who quotes Theophrastus verbatim.
22. Cf. Diod. Sic. I.10.1, 1.34.7.
23. Meyboom 1995, 34.
24. Strabo XVII.1.15.

Pliny, like Strabo, also remarks on the plaiting of lotus leaves into vessels of various shapes in Egypt.[25] The use of the large peltate leaves of the lotus as receptacles for food and beverages or as packaging material is still well known across South, East, and Southeast Asia.[26] The literature of these regions from as early as the first millennium BCE already attests to use of lotus leaves as vessels and plates.[27] For instance, Śākuntala, the eponymous heroine of the Sanskrit playwright Kālidāsa's *Abhijñānaśākuntalam*, speaks of having sighted her paramour in a jasmine bower, where he held a lotus-leaf cup brimming with water.[28] Strabo also alludes to the use of lotus leaves as shades, a practice which has textual and iconographic parallels in the Indian subcontinent.[29]

The trade in the economically valuable parts of the lotus, especially the seeds, is suggested by the anonymous author of *Diseases of Women*, a Hippocratic gynecological treatise dating to the late fifth or early fourth century BCE. The author refers to administering a herbal application to the womb, the "size of the Egyptian bean."[30] The "size of an Egyptian bean" (2.34 grams) then becomes a standard unit of measurement for small doses of herbal applications, since later Greco-Roman medical authors quote it as well.[31] This specialized use of the edible seed of the lotus implies that it was fairly well known to medical professionals in the eastern Mediterranean from as early as the fourth century BCE.

While the Hippocratic corpus is silent on the medical applications of the lotus, Greco-Roman physicians from the Hellenistic period on describe the nutritional properties of the lotus and also recommend its use in compound remedies. Diphilos of Siphnos, a physician at the court of the Macedonian king Lysimachos (early third century BCE), remarks that the "Egyptian bean's rhizome, called a *kolokásion*, is tasty and nourishing, but is difficult to digest because it is rather astringent." He adds that the "beans produced within the pods, when green, are difficult to digest, contain little nutrition, are laxative, and produce a great deal of gas; but after they dry, they produce less gas."[32] Menophilos, a poorly known Hellenistic pharmacological author (c. second to first century BCE), prescribes the bitter embryo of the Egyptian bean (among other ingredients) for the treatment of ear ulceration.[33] Dioscorides notes that a meal of lotus seeds was recommended for those with colic and dysentery and, like Menophilos, recommends the bitter embryo of

25. Pliny *HN* XXI.87.
26. Kintaert 2010, 493–500.
27. Kintaert 2010, 493–94.
28. Kālidāsa, *Abhijñānaśākuntalam* 5.21.
29. Kintaert 2010, 496–98.
30. Hippocrates *Morb. Mul.* 181.
31. Celsus *Med.* V.23.2–3, 25.6; Galen *Antid.* 1.1, 14.3K.
32. Ap. Ath. III.73a-b.
33. Ap. Celsus *Med.* VI.7.2c.

the seed for earaches.[34] But Galen, the prolific physician-author of the second century CE, had a poor estimate of the Indian lotus, whose seed and rhizome he considered unwholesome and of little nutritive value.[35]

The Hellenistic physician Diphilos of Siphnos is incidentally the first known author to refer to the edible rhizomes of the lotus as *kolokásion*.[36] Early Latin authors applied this term, perhaps mistakenly, to the entire plant.[37] In Greek, the seedpod of the lotus was named the *kibṓrion*.[38] These specialized terms for the economically valuable parts of the Indian lotus (rhizomes and seedpods) from the Hellenistic period onward indicate a greater familiarity with the plant among Greek-speakers in the eastern Mediterranean. The cup-like shape of the honeycombed fruit of the Indian lotus (*kibṓrion*) even inspired similarly shaped metal wine cups in the Hellenistic period.[39] Apart from the seeds and rhizomes, the fibers produced by the stalk of the lotus may have been harvested for fiber production. In India, lotus-stalk fibers are spun into wicks used for lighting lamps, nowadays reserved for religious rituals.[40] Pliny notes that these fibers resembled the threads of spider silk (*araneosus*) but does not suggest an economic use for them.[41]

While the importance of the Indian lotus as a cultivar in Egypt is clear from at least the middle of the first millennium BCE, the routes and chronology of *Nelumbo nucifera*'s westward dispersal, as well as its cultivation in neighboring regions, are less well studied. There is no clear evidence for Achaemenid Persian agency in the introduction of the lotus to Egypt. Herodotus's treatment of the Indian lotus as a common feature of deltaic landscapes suggests a longer history of cultivation. It appears more likely that the lotus was introduced to Egypt under the cosmopolitan Saite rulers (664–525 BCE), who projected naval power and expanded trade in the Mediterranean and the Red Sea zones, as well as sponsoring a Phoenician-led naval expedition which attempted the circumnavigation of Africa.[42] Significantly, the earliest reference to cotton in Egypt, another eastern cultivar, occurs in the context of the Saite pharaoh Amasis's gifts to the Spartans and the Lindians of Rhodes.[43] While the Indian lotus is prominent in Greco-Roman discourses on Egypt, Egyptian sources have not been able to clarify the chronology of

34. Dioscorides *Mat. Med.* II.106; cf. Celsus *Med.* VI.7e.

35. Galen *Aliment. fac.* 39.

36. Cf. Dioscorides *Mat. Med.* II.106; Nicander, ap. Ath. III.72a-b. See chapter 7 for a discussion of the etymology of this term and its later semantic shift to mean taro.

37. Columella *Rust.* VIII.15.4; Martial XIII, 57; Pliny *HN* XXI.87; Vergil *Ecl.* IV.20.

38. Ath. III.72a-b; Diod. Sic. I.34.7; Dioscorides *Mat. Med.* II.106; Strabo XVII.2.4, XVII.1.15.

39. Athenaeus 3.72b; XI.477e-f.

40. Arundhati 1994, 29.

41. Pliny *HN* XXI.87.

42. Van de Mieroop 2011, 297–98.

43. Hdt. III.47; see chapter 2.

the Indian lotus, since it was simply taken as a variant of the native lily species. Lise Manniche notes that it is difficult to even to distinguish between the native blue (*Nymphaea nouchali var. caerulea*) and white (*Nymphaea lotus*) lotuses (*zšn, nḫb*) in Egyptian texts.[44]

THE INDIAN LOTUS BEYOND EGYPT

The Indian lotus is sporadically attested across the Mediterranean in Greco-Roman sources. Theophrastus, in the fourth century BCE, observes that the Indian lotus "also grows in Syria and throughout Cilicia."[45] Dioscorides similarly notes that "the Egyptian bean which some call Pontic, grows abundantly in Egypt and it is found both in Asia and in Cilicia, in the marshes."[46] According to the testimony of Klaudios Iolaos, a shadowy Greek author of the first century CE whose work on Phoenician history and myth survives only in fragments, the lotus, greatly valued for its healing properties, grew in the Na'aman River (Gk. Belos) near the Phoenician city of Akko (now Acre, Israel):

> Klaudios Iolaos, in Book 1 of his On Phoenician Matters says that it (the city of Ákē) was named after Herakles: "For after being ordered by Eurystheus to undertake the most difficult labor, when he was drenched by the poison of the Lernaian Hydra, he was in pain with the wounds of the stings. So the Delphic oracle commanded him to go east, until he should find a river that waters a plant resembling the hydra. After striking it down, he would be freed from his wounds. He found the river and the plant prophesized by the Pythian god, of which the stem and the rhizome were completely serpentine in their many colors and heads. For however many one cut off, new ones were immediately born. At any rate, they call the rhizome *kolokásion*, and that which grows above it *kibṓrion*; it provided the Egyptians with a pleasant sight and edible veins." It grows abundant around the Nile, but the one that grows around Belos heals wounds that are hard to cure; for, when rubbed down by the rhizome, it yields a white juice. With this juice, he says, Herakles too was healed. And so they called the city "healing" (Ákē).[47]

The lotus was also cultivated in western Anatolia, Cyprus, and Greece. Theophrastus attests to the lotus growing "around Torone in Chalkidike, in a lake of modest size."[48] Phylarchus states that the Indian lotuses growing in a swamp near the Thyamis River in Thesprotia attracted so much attention from locals, who were fond of snapping off the fruits, that Alexander II of Epirus installed a guard to

44. Manniche 2006, 132.
45. Theophr. *Hist. pl.* IV.8.8.
46. Dioscorides *Mat. Med.* II.106.
47. Klaudios Iolaos, *BNJ* 788, fr. 1 (Stephanus of Byzantium, *Ethnika*, s.v. *Ákē*).
48. Theophr. *Hist. pl.* IV.8.8.

protect the plants.[49] This anecdote suggests that in parts of the eastern Mediterranean lotuses were more of a botanical curiosity or medicinal plant than a common food crop. In Sikyon (northeast Peloponnese), the lotus was held sacred to Athena, as Athenaeus speaks of a temple to Athena Kolokasia, or "Athena of the lotus."[50] The remarks of Nicander of Colophon, a poet-physician associated with the late Attalid court in Pergamon (second century BCE), suggest that lotus rhizomes were prized as desirable elite food:

> Sow the Egyptian variety of bean, so that in the summer
> you can produce garlands from its flowers and, when the pods
> full of ripe fruit have fallen, put them into the hands
> of young men who are dining and have long been desiring them.
> As for the rhizomes, I boil them and serve them at banquets.[51]

The use of Indian lotuses to weave garlands is also noticed by the second-century CE Greek author Athenaeus, a native of Naukratis in Egypt, who says that locally they were called honey-lotus garlands, and they were "fragrant and very cooling in the hot season."[52] In another passage, Athenaeus adds that at the suggestion of the poet Pankrates of Alexandria, the rose-colored lotus (i.e., the Indian lotus) was renamed *antinóeios* in honor of Antinous, the emperor Hadrian's favorite who accompanied him to Egypt and drowned in the Nile in 130 CE.[53]

The ornamental rather than subsistence value of the lotus was probably more important in its spread further west into the Italian Peninsula, where it is attested by the first century BCE. For the Augustan poet Vergil, the Indian lotus counted among the "cradle of alluring flowers" yielded by the earth (*blandos fundet cunabula flores*).[54] Columella, in describing a pond for breeding waterfowl, recommends the growing of lotus as an ornamental plant which provides shade for fowl in the middle of the pond: "On the other hand, the middle part of the pond should be of earth, so that it may be sown with the Egyptian bean and other green stuff which lives in or near water and provides shade for the haunts of the waterfowl."[55] Pliny, who explicitly notes that the lotus grew in Italy, admired the flowering seedpod and the large peltate leaves of the plant.[56] The consumption of lotus in Italy is also suggested by Martial's scorn for its stringy rhizomes.[57]

49. Ap. Ath. III.73b-c.
50. Ath. III.72b.
51. Ath. III. 72a-b.
52. Ath. III.73a-b.
53. Ath. XV.677d-e.
54. Vergil *Ecl.* IV.20.
55. Columella *Rust.* VIII.15.4.
56. Pliny *HN* XXI.87.
57. Martial XIII.57; cf. Martial's reference to the slender stalks of the lotus in VIII.33.

While Mediterranean sources for the lotus are extraordinarily rich, the same cannot be said for ancient Mesopotamia and Iran. No convincing identifications have been proposed for the Indian lotus in cuneiform records. R. C. Thompson matched *Nelumbo nucifera* with a medicinal plant called *ankinutu* in Akkadian.[58] The attribution is spurious as the literal meaning of the Akkadian *ankinutu* (from Sumerian *ankinudi,* "reaching neither sky nor earth") indicates that the plant in question is a creeper or epiphytic plant and not the lotus.[59] An image from Sasanian Taq-i Bustan of the Indo-Iranian god Mithra standing on an open lotus suggests the lotus's sacral associations in Iran, but this image almost certainly draws inspiration from contemporary Hindu-Buddhist iconography from Central Asia and India rather than reflecting earlier Iranian attitudes to the lotus.[60]

THE ARCHAEOLOGICAL DATA FOR *NELUMBO NUCIFERA*

Not surprisingly, the lotus is amply witnessed in the archaeological records of Greco-Roman Egypt.[61] Parts of the peltate leaf of the Indian lotus were recovered from an ibis bird necropolis in Ptolemaic Saqqara.[62] A nearly complete seedpod, seeds, and flower fragments were found at Greco-Roman Hawara, in the Fayyūm Oasis.[63] Finds of lotus seeds further east in the early Roman phases (first and second centuries CE) of Mons Claudianus in the Eastern Desert and the Red Sea port of Berenike indicate local trade in lotus seeds.[64] But the earliest extant archaeobotanical finds of lotus derive not from Egypt but neighboring Cyprus. Several well-preserved carbonized nutlets of the lotus were recovered from funerary offerings in the late-fourth-century BCE necropolis of the port city of Salamis, in eastern

58. Thompson 1949, 234.

59. CAD, s.v. *ankinutu.*

60. Carter 1981.

61. There are no secure attestations of the lotus in Egyptian archaeological records before the Ptolemaic period. Renfrew tentatively reported a few lotus nutlets at the New Kingdom site of Amarna (fourteenth century BCE). Renfrew herself doubts the identification and suggests a match with the fruits of the native soapberry tree (*Balanites aegyptiaca*). This identification appears more likely as the nutlets of the lotus are spherical and not columnar like the samples published in Renfrew's (1985, 186–87) report.

62. Germer 1985, 40; Hepper 1981, 148.

63. Germer 1985, 1987, 246, 1988, 17, 19; Newberry 1889, 52. Germer (1985, 40) draws attention to another lotus receptacle preserved in Leiden which has been mislabeled as a New Kingdom find from Deir el-Bahari.

64. Three seeds of the lotus were found at Roman Mons Claudianus (van der Veen 2001, 197, 181) and two at Berenike (Cappers 2006, 101–02).

Cyprus.[65] These may represent either localized cultivation of the lotus or imports from nearby Egypt.

CHRONOLOGY AND ROUTES OF DISPERSAL

While the archaeological data confirm the presence of the Indian lotus in the Mediterranean of the late first millennium BCE, they fail to shed light on the pathways by which the lotus arrived in the Mediterranean. The specific functions acquired by the lotus in the Mediterranean (e.g., consumption of rhizomes and nutlets, use of leaves as vessels, medical and sacral uses) suggest that the lotus may have been transmitted in more direct fashion from the Indic world, where similar uses were in place. The present invisibility of the lotus in the regions between the Levant and India in antiquity could mean either that the lotus has yet to be identified in the textual and archaeological records of regions like Mesopotamia and Iran or that the sacred lotus arrived in Egypt through southerly Red Sea trading routes, bypassing much of the Middle East.

There is some doubtful iconographic evidence for an earlier Late Bronze Age dispersal of the Indian lotus. Hakon Hjelmqvist draws attention to a dagger blade from Late Bronze Age Mycenae depicting a swampy landscape with plants bearing obconical fruits interpreted as lotus seedpods.[66] A diadem from contemporary Mycenae decorated with flowers interspersed among circular "fruits" with small globules has likewise been interpreted as the seedpod of the Indian lotus.[67] An unprovenanced faience model of a seedpod, allegedly of New Kingdom date (late second millennium BCE), has also been suggested to be the Indian lotus.[68] Hjelmqvist's identifications are overly optimistic, since the depictions of the plants in both Mycenaean objects are highly stylized. The faience model of a seedpod, now at the Walters Museum of Art in Baltimore, is a more credible representation of a lotus seedpod.[69] It seems unlikely, however, that it dates to the New Kingdom. As similar models of the lotus seedpod in glass and clay are known from Greco-Roman contexts, a late-first-millennium BCE dating appears more plausible.[70]

Unlike other South Asian crops which spread to the Middle East and the Mediterranean in antiquity, the Indian lotus had no lasting impact on the diet and agriculture of either region. It disappears from cultivation in the Mediterranean and the Middle East sometime in the early medieval period. The modern botanical name of the lotus and its genus derives from the Sinhala *nelum(-ba)*. The Dravid-

65. Hjelmqvist 1973, 244–45.
66. Hjelmqvist 1973, 245.
67. Ibid.
68. Brovarski, Doll, and Freed 1982, 42.
69. Inv. no. 48.459.
70. Darby, Ghalioungui, and Grivetti 1977, 635; Goede 2005.

ian *tāmara* was also commonly encountered as a name for the lotus in nineteenth-century European botanical literature.[71] The present familiarity of the lotus in the West is the result of the European study of Indian flora in the colonial period.[72] While commonly grown as an ornamental in temperate greenhouses, the lotus has no comestible value in the modern Mediterranean or Middle East. Cooked lotus seeds and rhizomes are more likely to be encountered in East and Southeast Asian bakeries and restaurants in the modern West.

Between the mid-first millennium BCE and the mid-first millennium CE, however, the lotus was widely cultivated as an important comestible in the Nile Valley. The importance of the lotus in Egypt is also evident in its frequent representation across a range of iconographic media, both within Egypt and in foreign representations of Egyptian landscapes. The lotus was also grown elsewhere in the Mediterranean for its edible seeds and rhizomes, and as an ornamental plant. The lotus's status across much of the Middle East, on the other hand, is unfortunately unclear, owing to its invisibility in the extant textual and archaeological records, particularly with regard to Mesopotamia and Iran. Future archaeological work may rectify this lacuna in the history of the Indian lotus's westward dispersal.

The story of the lotus in the Middle East and the Mediterranean reminds us of the complexity of crop histories. Botanical translocations are not always stable or lasting events. Some crops spread as a result of singular introduction events; others, through multiple introductions with varying peregrinations. Some crops "go viral" and change the course of history, while others never make it past their centers of origin. Still others, like the lotus, diffuse and catch on for a while, but in the long run of history appear as passing fads.

71. Cuvier et al. 1828, 159; Smith 1814, 283.
72. See Desmond 1992 on early modern and modern European studies of Indian flora.

7

A Forgotten Tuber: Taro

Taro (*Colocasia esculenta*), also known as the dasheen, cocoyam, or colocasia, is a semiaquatic herbaceous vegetable crop native to South and Southeast Asia (figure 25).[1] It is found growing naturally in or near streams, lakes, and ponds. Taro is primarily valued for its starchy edible corms or the swollen underground stem. It is widely cultivated and naturalized in tropical and subtropical Asia, sub-Saharan Africa, the Mediterranean, and the Middle East.[2] But the advent of the New World potato (*Solanum tuberosum*) has reduced the importance of taro as a comestible across much of its former range since the two vegetables are cooked in similar ways.[3] For instance, fried colocasia corms, a medieval Cairene favorite, were eaten much like French fried potatoes are today.[4] The leaves and stalks of taro can also be consumed as vegetables.[5] Taro cultivars contain varying amounts of calcium oxalate, a toxic compound, which is removed by peeling and thoroughly cooking the corms. The taro plant's large peltate leaves, which earned it the alternative name of elephant's ear, have encouraged its use as an ornamental in gardens with aquatic features (figure 25).

1. Hoogervorst 2013, 43; Sanderson 2005, 70; van der Veen 2011, 95.
2. Grimaldi 2014; Matthews 1991; van der Veen 2011, 95. On taro in sub-Saharan Africa, see Boivin et al. 2013, 215, 257–59, 2014, 554; Fuller and Boivin 2009; Rangan et al. 2015, 144–51.
3. Van Wyck 2005, 150.
4. Lewicka 2011, 249; van der Veen 2011, 97.
5. Simoons 1991, 105.

FIGURE 25. Taro (*Colocasia esculenta*), illustration from Basilius Besler's *Hortus Eystettensis* (Nuremberg, 1640). Peter H. Raven Library, Missouri Botanical Garden (public domain).

COLOCASIA: FROM LOTUS TO TARO— A LATE SEMANTIC SHIFT

The earliest history and terminology pertaining to taro in the Middle East and the Mediterranean are much disputed, if not confused, in the secondary literature.[6] The vernacular names of taro (e.g., Arabic *qolqas,* modern Greek *kolokási,* Turkish *gölevez, kolokas*), as well as the modern botanical name of taro (*Colocasia esculenta*), ultimately derive from the ancient Greek *kolokásion*/*-ia* (Latin *colocasia*), which was used in antiquity for the edible rhizomes of the Indian lotus.

The claim that the Greek *kolokásion* derives from the Sanskrit *kālakacu,* one of several Indian terms for taro, is misleading.[7] The adjective *kāla,* meaning black or dark, is rarely prefixed to the usual words for taro in Sanskrit (*kacu, kacvī*). The form *kālakacu* appears to be restricted to lexicographers. Even in the modern vernaculars of North India, the Sanskrit-derived *kacu*-forms predominate in taro terminology (Assamese *kasu*, Bengali *kacu,* Hindi *kacū, kaccū*).[8]

While the precise linguistic affiliations of *kolokásion* are murky, it is certainly a term of eastern-Mediterranean rather than Indian origin. Athenaeus explicitly remarks that the Alexandrians named the lotus rhizome *kolokásia,* suggesting that the term was coined in early Hellenistic Egypt.[9] The *kolo*- prefix, still familiar to us in the word "colossal" (from the Greek *kolossós*), is attested in a few other Greek words as well (e.g., *kolokúnthē, kolókuma*), where the meaning is invariably "big."[10] The base word **kásion* is perhaps related to *kórsion* or *korsaîon,* the term used for the esculent rhizomes of the white lotus (*Nymphaea lotus*), rather than the Sanskritic *kacu.*[11] *Kolokásion* thus translates as "big lotus tuber," in contrast to the smaller rhizomes of the white lotus.

In late antiquity, the meaning of the Greek *kolokásion* and its cognates in other Mediterranean languages was extended to describe the esculent starchy underground storage organs of other wetland plants, including taro. The similar ways of consuming lotus and taro tubers were probably responsible for the application to taro of terms relating to the lotus.[12] The expanded semantic field and confused

6. Genaust 1996, 168; Löw 1928, 214; Nicolson 1987; Portères 1960, 170–74; Thiselton-Dyer 1918, 299–303; Watson 1983, 66–69.

7. Brust 2005, 350–51; De Candolle 1885, 73–74; Thiselton-Dyer 1918, 299.

8. CDIAL 2609.

9. Ath. III.72a-b.

10. Brust 2005, 350–51; Kronasser 1960; Prellwitz 1905, 234. Note also Isidore of Seville's folk etymology for the colocynth, a cucurbit species: "It is called 'colocynth' because it has a *spherical* fruit, and leaves like the common cucumber's" (*Etymologies* XVII.ix.32).

11. Diod Sic. I.10.1; Strabo XVII.2.4; Theophr. *Hist. pl.* IV.8.11.

12. In this respect, the late Roman (second to fourth centuries CE) collection of recipes attributed to Apicius (*De re coquinaria*) contains references to colocasia, whose mode of cooking (e.g., bulking out meat dishes and barley broth) could easily apply to both lotus rhizomes and taro corms (III.4.2, VI.2.5, VI.8.10, VII.15).

identity of *kolokásion* in later periods is exemplified by the comments of Aetius of Amida, an early-sixth-century CE Greek physician who curiously links the *kolokásion* with the eggplant (*manzizánion*):

> *Kolokásion* or eggplant. The strength of the root is similar to that of the turnip and the onion, its body is sticky so that it is used for cleansing and for easing the bowels.[13]

The reference to the sticky body of the *kolokásion* indicates that the plant in question is taro, whose starch-rich tubers are sticky to the touch when peeled. Elsewhere Aetius includes the *kolokásion* with other aroids, including *aron* and *dracontium*, in a list of plants with aphrodisiac properties, confirming the identification of *kolokásion* with taro.[14] While the association of *kolokásion* with the eggplant (*Solanum melogena*, Gk. *manzizánion*, from the Persian *bāḏinğān*) in Aetius is puzzling, it intriguingly attests to the presence of another Indian cultivar, whose records in the pre-Islamic Middle East and the Mediterranean are extremely patchy.[15]

By early medieval times, as the Indian lotus fell out of familiarity in the Mediterranean and the Middle East, the regional cognates of the Greek *kolokásion* came to exclusively denote taro.[16] Taro is well attested in medieval Arabic authors, and desiccated remains of taro corms were also found at the medieval levels (eleventh to thirteenth centuries) of the Egyptian Red Sea port of Quseir al-Qadim (ancient Myos Hormos).[17] As *kolokásion* was only used to describe taro from late antiquity on, it will have no relevance to the present discussion of taro's presence in the ancient Middle East and the Mediterranean.

EARLY TARO TERMINOLOGY

In the absence of early archaeological data for the taro in the Middle East and the Mediterranean, the evidence for its spread and cultivation in these regions is derived entirely from ancient textual descriptions of edible tuberous crops in Greek, Latin, and Hebrew. Yet the prospect of identifying taro in ancient records is

13. Aet. Amid. *Med.* I.210 (trans. Grimaldi).

14. Aet. Amid. *Med.* XI.35.

15. On the eggplant in the late antique and early medieval Middle East and Mediterranean, see Amichay et al. 2019, 206; Fuks, Amichay, and Weiss 2020, 7; van der Veen 2011, 93–94; Watson 1983, 70–71.

16. Woenig (1897, 45) speculates that environmental changes and agrarian practices, e.g. draining of swamps and lakes, siltation of canals, and the gradual elevation of the alluvium, could have been responsible for the disappearance of the Indian lotus in the Nile valley. Elsewhere in the Mediterranean, the lotuses reported by Phylarchus as growing in the swamp near the Thyamis River in Thesprotia lasted less than a generation. The swamp was described as having dried up within the lifetime of king Alexander II of Epirus, leaving no trace of water (ap. Ath. III.73b-c).

17. Lewicka 2011, 245, 248–49; Ibn Sayyār al-Warrāq, *Kitāb al-Ṭabīkh* 46, trans. Nasrallah 2007, 237 (see also the glossary entry at 793); Portères 1960, 172–73; van der Veen 2011, 78–9, 95–97 (Quseir al-Qadim find).

not dismal, since few edible starchy tuberous crops of a non-bitter variety were known in the ancient Mediterranean and Middle East. The present discussion will assess three terms which have been identified in the secondary literature as denoting taro: the Greek *oúïngon* and *áron*, and the Hebraic *qarqas*.

OÚÏNGON

The earliest and most credible reference to taro is found in Theophrastus's description of an aquatic tuberous crop in Egypt called the *oúïngon* (manuscript variant *oúïton*). The identification of *oúïngon* as *Colocasia esculenta* has a long genealogy in classical scholarship and goes back to the Latin translation of Theophrastus by the Greek Renaissance humanist Theodorus Gaza (c. 1400–1475).[18] While Theophrastus's description of the *oúïngon* is short, the general morphological features and uses agree with those of taro. He notes that "its leaves are large and its shoots short, while the root is long and is, as it were, the fruit. It is . . . consumed; they gather it when the river goes down by turning the clods."[19] The "fruit" of the plant—that is, its edible tuber—is described as growing underground.[20] Theophrastus also remarks that the *oúïngon* was not regarded as the root of the plant.[21] This is an accurate recognition of the edible tuber of taro as the underground stem tissue or corm of the plant and not the root.

Theophrastus's *oúïngon* or *oúïton* is reproduced in Pliny as *oetum*. But the account provided by Pliny appears to be corrupt. While Pliny follows Theophrastus in describing it as an Egyptian plant with a large esculent corm, he claims the leaves are few and small.[22] The etymology of *oúïngon* is not known. The fifth-century CE lexicographer Hesychius, remarking on *oúïton,* a variant spelling of *oúïngon,* notes that it was also called *oitón*.[23] The inconsistent orthography of *oúïngon* betrays a foreign origin for the word. Theophrastus in fact remarks of it as an Egyptian word, but no straightforward cognates have been found for *oúïngon* in the ancient Egyptian language.[24]

THE *ÁRON*

The plant family Araceae (to which taro belongs) and the genus *Arum* derive their modern botanical names from the ancient Greek *áron* and the Latin *aron* or *arum*,

18. Sharples 1989, 198; Sharples and Minter 1983, 155.
19. Theophr. *Hist. pl.* I.6.11.
20. Theophr. *Hist. pl.* I.1.7.
21. Theophr. *Hist. pl.* I.6.9.
22. Pliny *HN* XII.89.
23. Hesychius, s.v. *oúïton*.
24. Theophr. *Hist. pl.* I.1.7, 1.6.11.

which were used in antiquity to describe several wild herbaceous species with acrid and frequently poisonous rhizomes, including Solomon's lily (*Arum palaestinum*) and the cuckoo pint (*Arum maculatum*).[25] The difficulty of getting rid of the toxins found in wild aroids through cooking and the invariably small size of the rhizomes meant that the different varieties of *áron/arum* were typically consumed as famine foods or for their perceived medical benefits.[26]

On the other hand, Pliny's (first century CE) and Galen's (second century CE) descriptions of an edible non-bitter variety of *áron* from Egypt and neighboring Cyrene (Libya) appear to be references to taro.[27] But Pliny's use of the term *áron* to describe taro suggests a secondary acquaintance with the crop, limited perhaps to the edible corms, which were traded over long distances:

> Among the varieties of the bulb, too, there is the plant known in Egypt by the name of aron. In size it is very nearly as large as the squill, with a leaf like that of sorrel, and a straight stalk a couple of cubits in length, and the thickness of a walking-stick: the root of it is of a milder nature, so much so, indeed, as to admit of being eaten raw.[28]

While Pliny probably did not see the plant he describes, the account compares remarkably well with taro. The garden herb sorrel (Latin *lapathum, Rumex acetosa*) has sagittate or arrow-shaped leaves much like taro, although taro's are much larger. Likewise, the large bulb of squill (Latin *scilla, Drimia maritima*), a medicinal herb, easily matches the size of the taro's corm. Pliny's description of *áron*'s thick stalks and the corm's mild acridity also supports identification with taro. Galen's discussion of *áron* is much more extensive, with references to several species, most being bitter in taste. But the *áron* from Cyrene (Libya), which was imported to Italy, is described as less acrid and eminently edible:

> The root of this plant (*áron*) is eaten much the same as that of the turnip, but in certain regions it grows somewhat more bitter, so that it is very like the root of the edderwort. In cooking, one should pour off its first water and add more hot water, as was described in the cases of cabbage and lentils. But in Cyrene the plant is the reverse of what it is in our country. For in those parts the arum has very little pharmacological activity and very little bitterness, so that it is more useful than turnips. Because of this they also export the root to Italy, on the grounds that it can keep for a very long time without rotting or sprouting. It is clear that this sort is better as nutriment, but if one wants to cough up any of the thick, viscid fluids that accumulate in the chest and lung, the more bitter and more pharmacologically active root is better.

25. On arons, see Aristotle *Hist. an.* 611b; Dioscorides *Mat. Med.* II.197; Hippocrates *Morb.* 2.47, 3.15–16; Pliny *HN* XIX.96–7; Theophr. *Hist. pl.* 1.6.7–8, 7.9.4, 7.12.2, 7.13.2. See Löw (1928, 213–18) for the Araceae in Hebraic sources. On the genus *Arum*, see Bedalov and Küpfer 2005.

26. Dalby 2003, 28–29.

27. Grimaldi 2014.

28. Pliny *HN* XIX.96.

> When boiled in water, it is eaten with mustard or with oil, vinegar and fish sauce, and of course with other mashed dishes, especially those prepared with cheese. But it is plain that the humor distributed from it to the liver and the body as a whole, from which animals are nourished, is somehow thicker, as was mentioned in the case of turnips. This is especially the case when the roots, like those from Cyrene, have no pharmacological activity. With us in Asia, many arums are more bitter and have a medicinal property.[29]

Galen's reference to the Libyan *áron* being consumed with mashed dishes is strongly suggestive of taro, which becomes mashable when boiled. Mashed taro is well known today as a staple across the Polynesian world (e.g., Hawaiian *poi*).[30] Galen's comparison of the *áron* with the turnip is also echoed in the statement of the twelfth-century Arabic author al-Baghdādī that Egyptians split taro "like a turnip."[31] The identification of edible varieties of *áron* with taro is not a modern one but one that follows on the heels of late medieval and Renaissance interpretations of the term *áron.*[32] A marginal note under the entry for *áron* (II.197) in a fourteenth-century manuscript of Dioscorides's *Materia Medica* (Vaticanus Palatinus Graecus 77) notes, for instance, that "among the Cyprians *aron* was called *kolokasion.*"[33] In his influential commentary on Dioscorides (*I Discorsi*), the Italian Renaissance botanist Pietro Andrea Mattioli (1565) describes taro as the "arum of Egypt" (*aro d'Egitto*) and remarks that the corms are called *colocasia.*[34]

QARQAS

The crop named *qarqas* or *qeriqas* in the Mishnah, the earliest work of rabbinical literature (early third century CE), has conventionally been interpreted as taro at least from the time of Maimonides (1135–1204), based on later equations of *qarqas* with *qolqas,* the Arabo-Hebraic term for taro (*Ma'aserot* 5:8).[35] It seems that this equation was based solely on phonetic similarities between the two words. While *qolqas* is derived from the Greek *kolokásion,* the etymology of *qarqas/qeriqas* is unknown. It is said in the Mishnah to be exempt from tithing, alongside other

29. Galen *Aliment. fac.* II.61.
30. Krauss 1993, 7–8.
31. Al-Baghdādī, *Kitāb al-ifādah* 47; Lewicka 2011, 248.
32. Grimaldi 2014, chapter 6.
33. Wellmann 1907.
34. Mattioli 1565, 282.
35. Feliks 2007b; *Jerusalem Talmud,* trans. Guggenheimer 2002, 574 n112. Löw (1928, 217) suggests that the term is a confusion of taro with the lotus. This is unlikely especially in light of his own confusing interpretation of the term *kolokasia,* which he mistakenly claims was primarily applied to taro since antiquity (214).

cultivars foreign to the land of Israel (e.g., Baalbek garlic, Rikhpa onion, Egyptian lentils, and Cilician grits). Judging from the limited geographical scope of the other cultivars named in the text, *qarqas/qeriqas* could not have come from beyond the eastern Mediterranean zone. Otherwise there is little contextual information in the Mishnah regarding the identity of the plant. *Korkasi,* the name of taro in the Coptic language of Egypt, provides a remarkably close cognate to the Mishnaic *qarqas/qeriqas*, as does the Arabic orthographic variant *qorqas* instead of the regular *qolqas.*[36] The form *qorqas* is found, for instance, in the text of the twelfth-century Sevillian Ibn al-ʿAwwām's *Book of Agriculture.*[37] Overall, there is insufficient information to determine whether *qarqas/qeriqas* is indeed an older Semitic term for taro.

ROUTES OF DISPERSAL AND THE USES OF TARO

Like the Indian lotus, the earliest attestations of taro are centered on the Nile Valley and immediately adjacent regions. Taro was certainly not grown in the western Mediterranean, since Pliny is completely unaware that the Theophrastean *oetum* and his Egyptian *aron* may be one and the same. Galen notes that the *áron* from Cyrene (Libya) was exported to Italy, indicating that while taro was consumed in Italy, it was not grown there.[38] As for Mesopotamia and Iran, the taro remains invisible in native written and archaeological records, much as the Indian lotus is. This means that either taro was not grown in these regions, at least until the medieval period, or that it has yet to be identified in textual and archaeological materials.[39] Consequently, it is not unlikely that taro, like the sacred lotus, arrived in Egypt via Red Sea trade routes.

The meager references to taro in ancient textual sources do not necessarily mean that it was a rare food crop. Rather, its estimation as low-status food probably accounts for its inconspicuousness in textual sources. Among cultures dependent on cereals for staples, the taro and other tuberous crops were often regarded as "poor men's food."[40] The fourteenth-century Arabic author Ibn al-Ḥājj observes that "many of the weak, including old men, the poor and the young" ate taro.[41] Taro was likewise considered inferior to a meal of rice in premodern China.[42] In

36. Darby, Ghalioungui, and Grivetti 1977, 655; Tackhölm and Drar 1950, 377.

37. Ibn al-ʿAwwām, *Kitāb al-filāḥa* 24.

38. Galen *Aliment. fac.* II.61.

39. On taro in medieval Iraq, see Ibn Sayyār al-Warrāq, *Kitāb al-Ṭabīkh* 46, trans. Nasrallah 2007, 237 (see also the glossary entry on 793); Watson 1983, 66.

40. Matthews and Ghanem 2021, 103–4.

41. Ibn al-Ḥājj, *Al-madkhal* IV.99; Lewicka 2011, 131.

42. Simoons 1991, 105–06.

light of its perceived inferiority to cereal crops in later sources, the motivations for its westward spread remain puzzling. It could have arrived as a medicinal or ornamental plant or perhaps was simply a means to diversify cultivation in an aquatic environment. Like rice, taro could exploit swampy environments unoccupied by the traditional cereal crops of wheat and barley, thus proving an attractive supplement to cultivators who had already taken up rice.

8

Timber for God and King: Sissoo

They have a thousand uses, and civilization is impossible without them, declares Pliny of trees.[1] The plant under consideration here is a tree crop but not the usual oil- or fruit-producing type like olive or citrus. Rather it is a timber-yielding species which is, broadly speaking, weakly domesticated. While timber traveled long distances from its forest home, mostly by sea or river, it was rarer for the timber-yielding tree itself to travel and acclimatize itself in foreign lands, as sissoo did, from South Asia and the Indo-Iranian borderlands to Mesopotamia and eastern Arabia, in the first millennium BCE. Sissoo (*Dalbergia sissoo*), also known as Indian rosewood, is a tall, gray-barked, ovate-leaved deciduous tree which ranks as one of the most important timber-yielding species of South Asia. The durable, termite-resistant, dark-sheened heartwood of sissoo has long been valued as a raw material for roofing and building construction, the manufacture of furniture, industrial and agricultural tools, kitchenware, the frames and wheels of carts and chariots, and watercraft.[2]

The natural distribution of sissoo extends from the sub-Himalayan tracts and river valleys of northern India to southeastern Iran between Baluchistan, Sistan, and Kerman.[3] The archaeological data from the Indo-Iranian borderlands and northwestern India indicate that settled populations in these regions were exploiting

1. Pliny *HN* XII.5.

2. Gershevitch 1957, 318–89; Maxwell-Hyslop 1983, 67; Ratnagar 2004, 139; Tengberg and Potts 1999, 131; Tengberg, Potts, and Francfort 2008, 931–32.

3. Asouti and Fuller 2008, 107–08; Kothari 2007, 87; Maxwell-Hyslop 1983, 67; Tengberg 2002, 76; Tengberg and Potts 1999, 130–31; Tengberg, Potts and Francfort 2008, 928–29. *Dalbergia latifolia*, a closely related species, is widespread throughout the Indian subcontinent (Asouti and Fuller 2008, 107).

sissoo wood as a raw material and a fuel source from as early as the fifth millennium BCE. The sissoo charcoal fragments from a mid-fifth-millennium BCE context in Mehrgarh, and those from fourth-millennium BCE contexts in Shahi Tump and Miri Qalat in Pakistani Baluchistan, are the earliest archaeologically attested specimens of sissoo.[4] Sissoo charcoal and artifacts crafted from sissoo have also been recovered from third- and early-second-millennium BCE contexts at Konar Sandal in eastern Kerman, Shahr-i Sokhta in Sistan, Lal Shah in Baluchistan, and Greater Indus Valley localities like Harappa, Mohenjo-daro, and Sanghol.[5]

A terra cotta model of a sissoo leaf recovered from Harappa suggests ascription of some symbolic or cultural value to the tree in Harappan culture.[6] Sissoo, known in Sanskrit as *śiṃśapā, kapila, picchila, bhasmagarbhā, yugapattrikā, mahāśyāmā,* and *dhūmrikā,* among other names, is well known in early Indic textual sources. The *Ṛgveda,* the earliest decipherable document of Indian literature (mid-second millennium BCE), already attests to its use in constructing a cart or chariot.[7] A few other instances of its prominence in Indian texts may be cited. The Buddha preached the Parable of the Sissoo Leaves in a grove of sissoo trees at Kosambi.[8] Elsewhere, in a pivotal moment in the *Rāmāyaṇa* epic of pan-Indian appeal, the kidnapped heroine Sītā spots her husband's messenger, the monkey-god Hanumān, on a branch of a golden sissoo tree.[9] Among the various love potions recommended by Vātsyāyana, the author of the *Kāmasūtra,* a third-century CE treatise on erotics, is a mixture of calamus and mango butter soaked in the hollowed-out trunk of a sissoo tree.[10] Medicinal properties have been ascribed to the roots, leaves, bark, and heartwood of sissoo in India since antiquity.[11]

Although sissoo has a more westerly natural distribution than the other tropical and subtropical plants we have examined thus far, it ought to be considered alongside crops like cotton and rice, since it traveled along the same Persian Gulf trading routes to establish itself as a significant timber-yielding species in Mesopotamia and eastern Arabia. Sissoo still grows in modern Iraq, southwestern Iran (Khuzestan), and Oman, but its importance as a source of timber is much dimin-

4. Tengberg and Potts 1999, 132; Tengberg, Potts, and Francfort 2008, 930–31; Tengberg and Thiébault 2003, 27, 38.

5. Maxwell-Hyslop 1983, 71; Mashkour et al. 2013, 230; Saraswat 2014, 209 (Sanghol); Tengberg 2002, 76; Tengberg and Potts 1999, 132; Tengberg, Potts and Francfort 2008, 931; Tengberg and Thiébault 2003.

6. Vats 1940, 468.

7. *Ṛgveda* III.53.19.

8. *Saṃyuttanikāya* 56, 31.

9. Vālmīki, *Rāmāyaṇa* V.12.40, V.29.11.

10. Vātsyāyana, *Kāmasūtra* VII.1.31.

11. Warrier, Nambiar, and Ramankutty 1994, 300–04.

ished.[12] In modern Oman, sissoo mostly grows as a relict population along wadis in the mountainous region of Jebel Akhdar.[13] Sissoo wood, whether local or imported, remained familiar to Middle Eastern audiences up to the modern period. The nineteenth-century Lebanese writer Fāris al-Shidyāq casually refers in his novel *Saq ʿala al-saq* (*Leg over Leg*, 1855) to a well-appointed room with beds, alcoves, couches, and chairs made from sissoo (Arabic *saʾsam*), among other commonly used woods.[14]

SISSOO TERMINOLOGY IN THE ANCIENT MIDDLE EAST

Sissoo wood was known in Mesopotamia as an imported wood from c. 2250 BCE on, long before it was locally cultivated for its timber. The Sumerian term for sissoo, *mesmagana*, and its Akkadian equivalent *musukkannu* (variants: *mesukkannu*, *meskannu*), literally the "tree of Magan," indicate that sissoo was made familiar in Mesopotamia through the mediation of peoples living in the southern Persian Gulf. The toponym Magan referred to the region of modern-day Oman and occasionally the Persian coastline facing the Gulf of Oman (modern-day Makran of Iranian and Pakistani Baluchistan).[15] The identification of the Sumerian and Akkadian "tree of Magan" with sissoo is not in doubt, as a trilingual inscription (Akkadian-Elamite-Persian) of the Achaemenid Persian king Darius I from his palace in Susa gives the Old Persian equivalent of the tree as *yakā*, a term which survives in New Persian as *jag*.[16] Also, *šeššap*, the Elamite equivalent provided by the same inscription, is an eastern loanword related to the Sanskrit *śiṃśapā* (Prakrit *sīsavā, sīsama*) for sissoo.[17]

Sissoo may have been introduced to the region of modern Oman by the late third millennium BCE, hence the Mesopotamian designation "tree of Magan." It is also likely that Oman was a point of transit for sissoo wood deriving from eastern Iranian and Indian ports.[18] The earliest archaeobotanical finds of sissoo west of the Indo-Iranian borderlands are charcoal fragments and wooden handles of two bronze daggers from the coastal settlement of Tell Abraq, in the UAE, dating

12. Postgate 1992, 183; Talebi, Sajedi, and Pourhasemi 2014, 132; Tengberg 2002, 77; Tengberg, Potts and Francfort 2008, 929; Ubaydli 1993, 35.

13. Maxwell-Hyslop 1983, 67; Tengberg 2002, 77; Tengberg and Potts 1999, 131; Tengberg, Potts, and Francfort 2008, 929.

14. Al-Shidyāq, *Al-sāq ʿalā al-sāq*, 2.16.36.

15. Heimpel 1987–90, 196; Postgate 1992, 183; Ratnagar 2004, 75; Steinkeller 2013, 416.

16. Bleibtreu 1980, 17; Gershevitch 1957; Kent DSf 34–35, 44; Maxwell-Hyslop 1983, 67–68.

17. Witzel 1999a, 66.

18. Postgate 1992, 183.

between 2200 and 2000 BCE.[19] Margareta Tengberg and Daniel Potts contend that a local Arabian provenance for the sissoo found at Tell Abraq is not improbable.[20] But coastal eastern Arabia is largely unsuited for growing sissoo trees, which require well-drained sandy or loamy earth close to a perennial water source.[21] It was not temperature but the availability of water which limited the growing of sissoo trees in the vicinity of Tell Abraq. In this regard, it seems more probable that the earliest finds of sissoo in Bronze Age eastern Arabia were imports from eastern Iran or India.

SISSOO IN MESOPOTAMIA: LATE THIRD TO SECOND MILLENNIUM BCE

The earliest appearance of sissoo in Mesopotamia, albeit obliquely as a generic wood from Magan, is in two inscriptions of Gudea, the local ruler of Lagaš in the late twenty-second century BCE, who procured various exotic woods from the Persian Gulf and Indus region for the construction of a temple dedicated to the god Ningirsu: "Magan and Meluḫḫa (coming down) from their mountain, loaded wood on their shoulders for him, and in order to build Ningirsu's House they all joined Gudea (on their way) to his city Ĝirsu."[22]

A fragmentary eulogistic composition from twenty-first-century BCE Nippur (Ur III period) likewise exhorts the lands of Magan and Meluḫḫa to ship their timber to the royal or divine addressee.[23] As sissoo has always been identified in Mesopotamia as a product of Magan and Meluḫḫa, it seems highly probable that these early allusive passages refer to sissoo. The first explicit references to the wood derived from the "tree of Magan," sometimes even said to be deriving from Meluḫḫa (i.e., the Indus region) or Dilmun (Bahrain), are found in several Mesopotamian texts dating from the twenty-first century BCE on.[24] Textual sources from the Old Babylonian period (twentieth to eighteenth centuries BCE), ranging from dowry inventories of wealthy households to administrative dossiers, indicate that sissoo was used in the manufacture of tables, beds, chairs, eating bowls, and other items of furniture.[25] It is possible that sissoo was already imported into

19. Tengberg 2002, 75–76; Tengberg and Potts 1999, 129.

20. Tengberg 2002, 77; Tengberg and Potts 1999, 132.

21. Andersen et al. 2004, 225; Tengberg and Thiébault 2003, 28.

22. RIME III/1, E3 1.1.7.CylA, Col. xv, lines 8-10; see also E3 1.1.7.StD, Col. iv, lines 7-14: "Magan, Meluḫḫa, Gubin, and the land Dilmun—supplying him with wood—let their timber cargoes sail to Lagaš."

23. Michalowski 1988, 160, 163.

24. CAD, s.v. *musukkannu*; Hansman 1973, 556–57; Heimpel 1987–90, 198; Ratnagar 2004, 130; van de Mieroop 1992, 160.

25. Dalley 1980, 65–66, 72; Kupper 1992, 166; Maxwell-Hyslop 1983, 70.

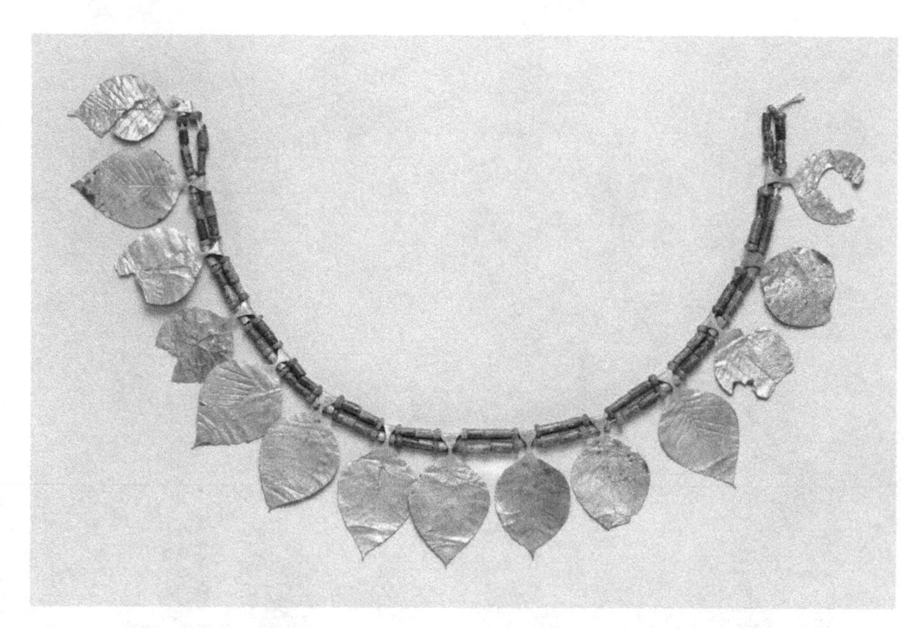

FIGURE 26. Chaplet of gold leaves (possibly modeled after *Dalbergia sissoo*) strung on a necklace of lapis lazuli and carnelian beads, "King's Grave," Royal Tombs of Ur, Iraq, c. 2600–2500 BCE. Metropolitan Museum of Art, New York (public domain).

Mesopotamia during the so-called Early Dynastic IIIa period (c. 2550–2400 BCE). Tengberg, Potts, and Francfort have identified the ovate leaf pendants with acuminate (pointed) ends in the headdresses and necklaces of queen Puabi and her retainers at the Royal Tombs of Ur as representations of sissoo leaves (figures 26-27).[26] If the identification is correct, it would suggest that the sissoo tree had some symbolic significance to the elite of Early Dynastic Ur, much as it did in the Harappan east.[27]

There is no evidence, however, for sissoo being cultivated in Mesopotamia during the late third or second millennium BCE. Its timber remained throughout this period an exotic and prestigious import from the Persian Gulf region. Even the dried leaves of sissoo, reputed to have medicinal properties, were traded across the Persian Gulf. The early-thirteenth-century BCE Babylonian physician Rabâ-ša-Marduk's prescription for headache includes dried sissoo leaves, sourced perhaps from eastern Iran or the Indus region: "If a person experiences pulsating of the

26. Tengberg, Potts, and Francfort 2008. Miller (2013, 130) disputes this view and maintains an identification with the leaves of the poplar tree (*Populus euphratica*).

27. Ibid.

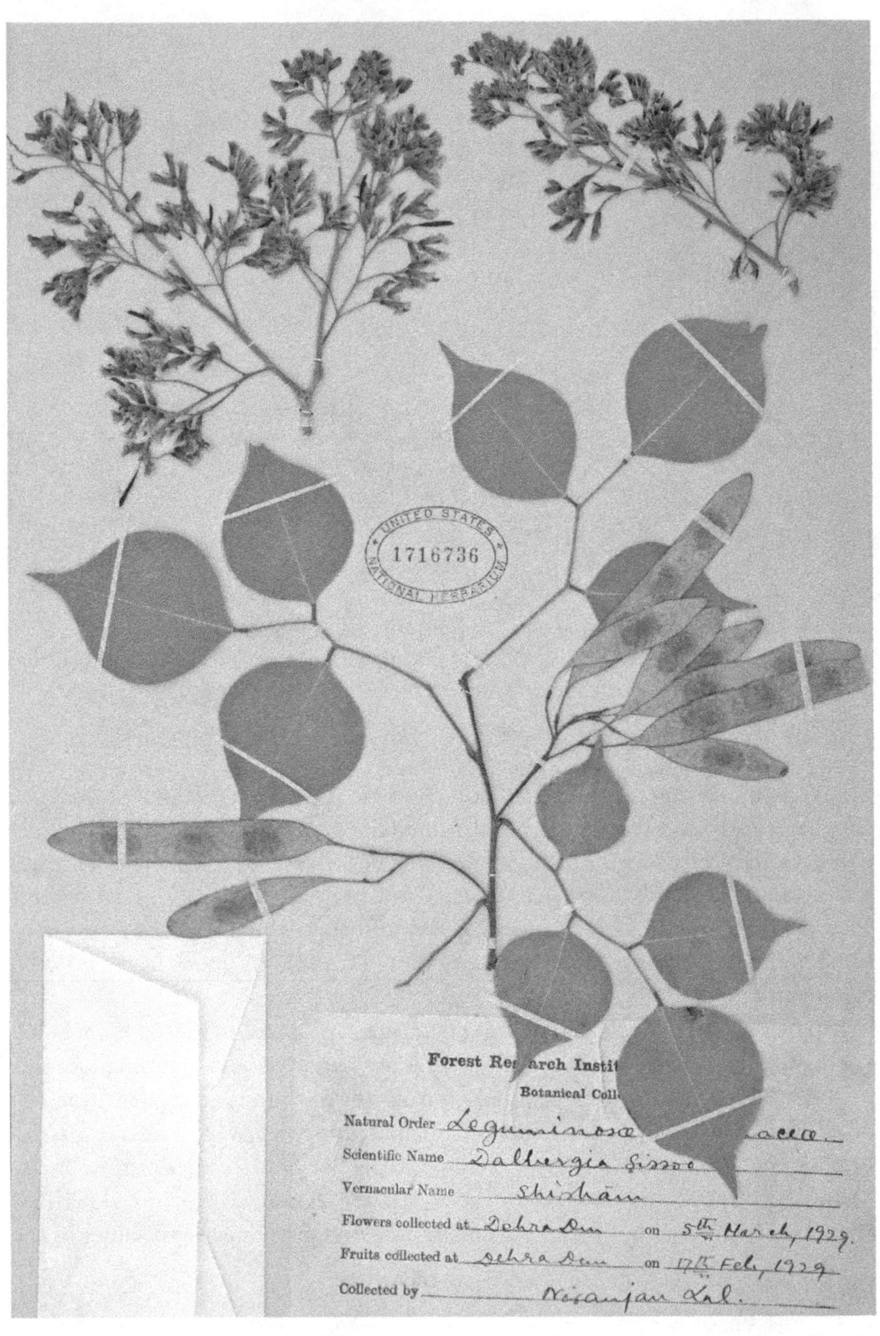

FIGURE 27. Pressed specimen of *Dalbergia sissoo* leaves collected in Dehra Dun, Uttarakhand, India, 1929. Department of Botany Collections, Smithsonian National Museum of Natural History, Washington, DC (public domain).

temples and his body hurts him, you crush (and) sift dried sissoo leaves. (You mix it with) chickpea flour, lentil flour (and) barley flour. If you continually bandage him with (it mixed) with beer dregs, he should recover."[28]

SISSOO IN ASSYRIAN SOURCES: THE EARLY FIRST MILLENNIUM BCE

It is only in the first millennium BCE that explicit references to the growing of sissoo trees in Mesopotamia are encountered in Assyrian textual sources. A clay tablet from Kalḫu containing an annalistic inscription of the Assyrian king Tiglath-Pileser III describes a plantation (*kirû*) of sissoo in southern Babylonia in connection with the siege in 731 BCE of a town belonging to the Bīt-Amukkāni, an Aramaean polity on the lower Euphrates, north of Uruk: "I confined Mukīn-zēri of (the land Bīt)-Amukkāni to Sapê (Šapīya), his royal city. I inflicted a heavy defeat upon him before his city gates. I cut down the orchards and sissoo-trees that were near his (city) wall; I did not leave a single one (standing)."[29]

The association of sissoo with southern Mesopotamia, specifically with the Aramaean chiefdom of Bīt-Amukkāni, is already made explicit in Assyrian sources of the ninth century BCE. An epigraph from a throne base of Shalmaneser III (858–824 BCE) from Kalḫu indicates that the king received sissoo wood as tribute from Mušallim-Marduk, the chieftain of Bīt-Amukkāni, and Adinu, the chieftain of Bīt-Dakkuri, during his Babylonian campaign, c. 850 BCE.[30] While Shalmaneser III's tribute report does not explicitly mention sissoo growing in the territory of the Bīt-Amukkāni and Bīt-Dakkuri, there is no doubt that sissoo was cultivated in southern Mesopotamia in the ninth century BCE, since Shalmaneser III's father, Aššurnasirpal II (883–859 BCE), reports in the "banquet stela" the growing of sissoo in his newly established capital of Kalḫu alongside other trophy trees he had collected through military campaigns and tribute missions.[31] Aššurnasirpal II's father and immediate predecessor, Tukultī-Ninurta II (890–884 BCE), received furniture crafted from sissoo wood as tribute from the governor of Suḫu, in the Middle Euphrates.[32] This suggests that either the sissoo tree was growing in the Middle Euphrates region or the governor of Suḫu had sourced the wood from localities further downstream, in southern Mesopotamia.

28. Scurlock 2014, 555–59, 11.1 (BAM 11 = KAR 188). On Rabâ-ša-Marduk and the history of this tablet (written c. 1300–1280 BCE), which was removed from Babylon and brought to Aššur by Tukultī-Ninurta I c. 1207 BCE, see Heeßel 2009, 13–28.

29. RINAP 1 47 Obv.23b.

30. RIMA III A.0.102.61; cf. A.0.102.5 vi.7.

31. RIMA II A.0.101.30, 43.

32. RIMA II A.0.100.5, 68–72.

The strong association sissoo had with southern Mesopotamia is also demonstrated by the Assyrian king Sennacherib's statement that he grew sissoo trees in his estates in Nineveh in an environment that was meant to evoke the marshlands (Akk. *agammu*) of southern Mesopotamia.[33] Sennacherib says that the locally cultivated sissoo trees were used in the construction of his Ninevite palace.[34] The king also received a tribute of large sissoo trees from Nabû-bēl-šumāti, the official in charge of the city of Ḫararatu, in northern Babylonia.[35] Sissoo is in fact amply attested in Assyrian sources of the eight and seventh centuries BCE. Wherever the provenance is specified, Babylonia is identified as the most important source of sissoo. An official writing to Sennacherib's father and predecessor, Sargon II (721–705 BCE), notes that there was not enough sissoo-wood for building construction in Babylon, revealing that sissoo orchards were present at several Babylonian localities, including Babylon:

> I [cannot] do [the work. There is] very [little] sissoo wood in Babylon for use in the work, (and) what they bring me from the city Birati [. . .] is all moist. [The king, my lord], knows that this work requires [a great deal] of sissoo wood . . . Now, [let them fetch] sissoo wood from Ki[ssik] or from wherever it is to be found.[36]

Birati (lit. "fortress"), in northern Babylonia, is a settlement near the city of Sippar which Assyrian state correspondence indicates is the same locality referred to in Sennacherib's inscriptions as Ḫararatu, where the king received a substantial tribute of sissoo logs from a local official.[37] Kissik, like Birati/Ḫararatu, has not been identified on the ground, but Mesopotamian textual sources indicate that it was in the Sealand district of southern Babylonia, close to the city of Ur.[38] Two other eighth-century Assyrian letters shed further light on the use of sissoo for building construction projects in the Assyrian heartland. An official writing to the king Sargon II asks for sissoo wood of varying sizes required in the construction of the newly established capital of Dūr-Šarrukīn: "(I need) six sissoo trees, each six cubits (3 m) long and one cubit (50 cm) in circumference; one *haluppu* fir tree, five cubits long, one cubit in circumference; ten sissoo trees, each of which are two *qû* measures thick; they may be either five or six cubits long."[39]

The source of the sissoo timber, however, is not specified in this instance. In an earlier letter from Nimrud (Kalḫu) dating to the reign of Tiglath-Pileser III (744–727 BCE), the governor of Aššur Province, writing to Šarru-dūrī, the governor of the

33. RINAP 3 8:4'; 16 viii 37; 17 viii 53; 18 viii 12"; 42: 49; 43:99; 46: 159;138 r ii' 33.
34. RINAP 3 8:7'; 16 viii 45; 17 viii 60; 18 viii 17"; 42: 50; 43: 100; 46: 160.
35. RINAP 3 2:17; 3:17; 4:15; 8: 15; 15 i 38'; 16 i 75; 17 i 67; 18 i 1"; 22 i 54; 138 i 10'; 213: 56
36. SAA 15 248 r. 2, 10.
37. See Nabû-bēl-šumāti's letters to the Assyrian king in SAA 17, 10-16.
38. Beaulieu 1992.
39. SAA 5 294: 9–12.

city of Kalḫu (*šakin Kalḫi*), who was active between 734 and 728,[40] notes that the latter had asked for sissoo-wood: "Tablet of the governor of (Aššur) province to Šarru-dūrī; health to my brother: About the sissoo-wood of which my brother wrote to me—when it has come up to me, I shall send a large quantity to my brother."[41]

It is clear that sissoo was not locally available in either case but was almost certainly sourced from southern Mesopotamia. While Aššurnasirpal II had grown sissoo in Kalḫu, it was little more than an exotic garden species in ninth-century Assyria. Only Sennacherib's inscriptions provide firm evidence for the large-scale cultivation of sissoo trees in northern Mesopotamia, although it appears that the trees were confined to well-irrigated royal estates in the capital of Nineveh. The references to sissoo in Assyrian textual sources between the ninth and seventh centuries BCE reveal that sissoo was valued as a material for producing doors, roof beams, tables, beds, and other items of furniture in both palatial and temple contexts.[42] An inscription of Aššurbanipal notes his provision of furniture crafted from sissoo for the god Marduk (Bel) at his temple in Babylon sometime in the first half of his reign (668–648 BCE):

> I had a canopy which reaches up into the heavens, made from sissoo-wood, an everlasting wood. . . . I skillfully made a bed of sissoo-wood, an everlasting wood, that is overlaid with *pašallu*-gold and studded with precious gems, as a pleasure bed for the god Bel and the goddess Beltiya.[43]

Aššur-etel-ilāni, Aššurbanipal's son and successor, likewise presented the Babylonian god Marduk with sissoo furniture, in his case an offering table inlaid with red *ṣāriru*-gold.[44] Sissoo was even esteemed as a material for the waxed writing boards (Akk. *lēʾu* or *daltu*) on which Babylonian scholars copied literary and scholastic texts destined for the royal library of Aššurbanipal in Nineveh.[45] Charred sissoo wood has in fact been recovered from a late Assyrian context in Dūr-Katlimmu (modern Tall Šēḫ Ḥamad), an important Assyrian provincial city in the lower Ḫābūr Valley of western Syria.[46] One of these charred fragments belonged to a piece of furniture with bronze fittings, vividly confirming the textual references to furniture crafted from sissoo.[47] The ample references to sissoo in Assyrian

40. On the dating of Sarru-duri's governorship, see Postgate 1973, 11.

41. ND 417, Postgate 1973, no. 189.

42. CAD, s.v. *musukkannu*; Gerisch 2013, 462; Postgate 1992, 183, 188, 190; Tengberg and Potts 1999, 131.

43. SAACT 10 31–54.

44. RIMB II 6.35.1.

45. Fincke 2003/04, 126–29; Frame and George 2005, 269, 275, 282.

46. Gerisch 2013, 443.

47. Gerisch 2013, 462.

sources suggest that its cultivation in Babylonia was actively encouraged by the Assyrian state for use in large-scale construction.

SISSOO IN BABYLONIAN, BIBLICAL, AND PERSIAN SOURCES: THE MID-TO-LATE FIRST MILLENNIUM BCE

With the demise of the Assyrian Empire in the last quarter of the seventh century, cuneiform documents from the south of Mesopotamia become the main witnesses for sissoo cultivation in the Middle East. Babylonian texts from private and temple archives, as well as royal inscriptions dating between the sixth and fifth centuries BCE, attest to the use of sissoo in the manufacture of beds (*eršu*), chairs (*kussû*), lamp stands (*bīt nûri*), stands for large vessels (*šiddatu*), writing boards (*lēʾu, daltu*), ceremonial brick molds (*nalbanu*), ploughs (*epinnu*), and even a large bucket for use in a well (*kannu*).[48] The inscriptions of the Babylonian king Nebuchadnezzar (605–562 BCE) refer to the use of sissoo for building work and furnishings at temples in Babylon and neighboring Borsippa, much as they had been used in Assyrian times.[49] By this period, furniture crafted from sissoo was accessible to wealthier private citizens in Babylonia, since it appears in marriage agreements as part of the dowry.[50] One such marriage agreement from Borsippa, dating to 493 BCE, refers to an ornate "chest of sissoo wood (adorned) with the features of a gazelle."[51] Even the leftover shavings of sissoo wood were used in Babylonian temples for burning incense.[52]

The author of Deutero-Isaiah, a Hebrew prophetic book dated to the late sixth century BCE and written in the context of the Judaean exile in Babylonia and the return under Cyrus the Great of Persia, decries idolaters and their images made of *mskn*-wood, the Hebrew equivalent of the Akkadian *musukkannu*.[53] The Masoretic vocalization accurately preserves the Akkadian variant spelling *mesukkannu*.[54] Deutero-Isaiah also describes the wood as one that will not rot, mirroring the description of sissoo as one that is "durable" or "everlasting" (Akkadian *dārê*/ *dārû*)

48. CAD, s.v. *musukkannu*; George 2005/06, 83; Holtz 2014, 48–50 (BM 32881); Joannès 1989, 60; Jursa 2003, 104, 2010, 225; MacGinnis 1996, no.3 (BM 66674); Roth 1989/90, 21–23, 28; Sherwin 2003, 147.

49. Da Riva 2009, 289, 2013a, 208, 223, 2013b, 124; George 1988, 147, 149.

50. Roth 1989, 85–86, 88, 107, 1989/90, 21–23, 28. A later parallel for the inclusion of sissoo furniture in bridal trousseaus is found in the tenth-to-eleventh-century CE documents from the Cairo Geniza archive chronicling the local Jewish community. Some of these documents refer to bedsteads of sissoo as part of dowries (Goitein 1983, 113–4).

51. Roth 1989, 86–88 (no. 24b).

52. Bongenaar 1997, 396.

53. Isaiah 40:20; Lipiński 2000, 525; Sherwin 2003, 146.

54. Sherwin 2003, 146.

in Assyrian and Babylonian royal inscriptions.[55] Deutero-Isaiah's suggestion that images of deities, undoubtedly in reference to Babylonia, were fashioned with sissoo is confirmed by an early-first-millennium BCE Mesopotamian ritual text series called the *mîs-pî,* or "washing of the mouth," which refers to a divine image crafted from sissoo, among other woods.[56] The Assyrian king Esarhaddon's inscriptions detailing his building works in Babylon also note that the pedestal and footstool of the image of the goddess Tašmētu in the temple of the god Marduk were crafted from sissoo wood.[57]

There is no doubt that the sissoo invoked in Babylonian texts continued to be sourced locally. Land lease documents from the archives of the Ebabbar temple in Sippar dating to the last quarter of the seventh century BCE refer to orchards of sissoo trees located along canals in an estate owned by the temple at Bēl-iqbi, a riverside locality south of Babylon.[58] Several late-sixth-century documents from the same archive name a sluice in a canal near Sippar as the "sluice of the sissoo-trees" (*bitqu-ša-musukkanni*), indicating that a grove of sissoo trees was located in the vicinity of the city of Sippar. The sissoo orchards in the Sippar region were no doubt of great antiquity, since the town of Birati/Hararatu in the vicinity of Sippar was already noted in Assyrian sources of the late eighth and seventh centuries BCE as a significant source of sissoo timber.

While Babylonian cuneiform documents dating to the early Achaemenid period indicate that the local sissoo orchards continued to be active in supplying domestic needs, there were apparently no sissoo trees growing in Babylonia of a size required for the construction of the monumental palace of king Darius I at Susa. The foundation charter of Darius for his palace in Susa records that the needed sissoo was sourced further afield, from its native habitat in Karmāna (modern Kerman) and Gandhāra in northwest India. Alternatively, it may have been easier to transport large logs via the maritime routes rather than the terrestrial one connecting Babylonia with Susa. The fate of sissoo plantations in Mesopotamia in later periods is unclear. Greek texts refer to the export of sissoo timber from Indian ports to the Persian Gulf. The *Periplus Maris Erythraei* witnesses the export of Indian sissoo and other timbers to the Persian port of Omana.[59] Two coffins from the site of Shakhoura in northern Bahrain, carbon-dated to the first century BCE, were made of sissoo.[60] The third-century CE Samaritan Targum, or translation of

55. E.g., Aššurbanipal, RINAP 5/1 6.i.28', 38'; Esarhaddon, RINAP 4 48 r. 91, 51 iv 9; CAD, s.v. *musukkannu.*

56. Walker and Dick 2001, 115.

57. RINAP 4 48 r. 91, 51 iv 9.

58. Da Riva 2002, 111–15 (e.g., BM 50350, BM 114781); Jursa 1995, 126–27 (BM 49930).

59. *PME* 36. The sixth-century *Christian Topography* of Cosmas Indicopleustes (11, 15) refers to the export of sissoo from the port of Kalyana, near modern Mumbai.

60. Andersen et al. 2004.

the Hebrew Bible into Samaritan, a western Aramaic dialect, uses sissoo (Samaritan *sysam*) as the equivalent of "gopher-wood," the material out of which Noah's ark was built (Gen. VI:14).[61] It seems likely that sissoo continued to be grown on a modest scale in parts of Mesopotamia and eastern Arabia to serve local needs, but large-scale construction projects invariably required the import of timbers from eastern Iran and India.

CHRONOLOGY OF DISPERSAL AND THE APPEAL OF SISSOO

The earliest Assyrian sources to document sissoo in southern Babylonia do not tell us when or by whose agency sissoo was introduced into cultivation. All that can be surmised is that sissoo was growing in Babylonia by the late tenth century BCE, if not earlier, since the Assyrian king Aššurnasirpal was able to procure sissoo saplings for his royal gardens in Nimrud in the early ninth century. The long-standing familiarity of sissoo-wood in Mesopotamia means that it is unhelpful to credit any single polity with the introduction of sissoo as a cultivar. The rise of sissoo as a plantation crop in southern Mesopotamia was probably the result of experimentation within the context of royal and temple estates. It is likely that the sissoo tree established itself in Mesopotamia in a series of reintroductions rather than being rooted in a singular event.

The paucity of good timber in the semiarid alluvial plains of southern Mesopotamia, and the ease of water transportation, are the most obvious factors motivating the introduction of sissoo as a local cultivar. Boat builders in the Persian Gulf region have relied on imports of tropical woods to construct the frame of seafaring watercraft up to the present day.[62] But the Persian Gulf and the Indus region were not the sole sources of wood used in southern Mesopotamia. Trade routes leading from the Zagros and the Caspian region in the east and from Lebanon and Anatolia in the west brought numerous other varieties of timber into Mesopotamia. The cedar of Lebanon (*Cedrus libani*) is perhaps the best-known exotic timber used for large-scale construction in ancient Mesopotamia.[63] But not all exotic trees could be grown locally. Sissoo is particularly well adapted to growing in Mesopotamia since it is drought resistant, tolerant of cold winters, and, most importantly, fast maturing.[64] Sissoo plantations already had a model in southern Mesopotamia in the form of irrigated date palm plantations. The uses of sissoo were largely limited to supplying good timber needed for the roofs of large buildings, doors, columns,

61. Gershevitch 1957, 319; Löw 1881, 65.
62. Willcox 1992, 5.
63. Willcox 1992, 7.
64. Kothari 2007, 87; Maxwell-Hyslop 1983, 67.

and high-quality furniture. It was too precious to be used as fuel. The leaves were ascribed medicinal properties in both South Asia and Babylonia and could also be used as fodder for livestock. The oil extracted from sissoo seeds is presently used as an insect repellent and a treatment for dermatological disorders.[65] It is not clear, however, whether the latter uses were known in antiquity.

While the introduction of sissoo as a local cultivar was not the doing of the Assyrians, the frenzied building activity of the Assyrian state and its immediate successors encouraged the expansion of sissoo groves in southern Mesopotamia and, for a brief moment in the seventh century BCE, in the Assyrian capital of Nineveh as well. The non-royal uses for sissoo appear to be limited in the Assyrian period, perhaps suggesting a royal monopoly. Alternatively, the prominence of sissoo in royal contexts may be an artifact of the way the surviving evidence is skewed toward elite cultures. The later history of sissoo plantations in Mesopotamia and eastern Arabia is presently unclear, but the timber of this species remained familiar to Persian Gulf societies and the wider Middle Eastern world till recent times, testifying to its high estimation since antiquity.

65. Kothari 2007, 87; Tengberg and Potts 1999, 131; Tengberg, Potts, and Francfort 2008, 932.

9

How to Turn Tropical

For most crops, the earliest introductions are largely invisible processes. It is a general truism in history and archaeology that we rarely uncover, in detail, the very first instances of phenomena as complex as crop transfers. By the time a crop is embedded in the archaeological and historical record, it is likely that it was brought to that moment by a certain regularity of exchange or localized cultivation. Yet there is no need to assume that these processes were prolonged. Some introductions may have been relatively quick, others protracted. Some were singular events guided by imperial agency; others were multiple introductions led by varied social groups through a multiplicity of routes. On the basis of the evidence collected for the crops discussed in this book, four stages may be distinguished in the introduction of easterly tropical and subtropical crops to the Middle East and the Mediterranean: familiarization, experimentation, routinization, and finally indigenization. Crop movers were a diverse group, and the motivations for crop translocations cannot be reduced to subsistence values. The crop adoption trajectory needs to be read "within the sphere(s) of competition, emulation, negotiation, performance and communication."[1]

FAMILIARIZATION

The first stage naturally involves the movement and consumption of a plant or, more commonly, its associated storable botanical product, like grain, fiber, oil, timber, or desiccated or preserved fruits or flowers. The knowledge of foreign crops

1. Sherratt 1999, 27.

could also be provided by finished goods involving vegetal materials like textiles and handicrafts. These would have been conveyed through either transit trade or direct long-distance networks. Some of these exchanges may have begun as an unplanned barter of the merchant's own foodstuffs and personal belongings at the completion of a journey. At a very basic level, the need to move a crop may have arisen as a response to risky and costly long-distance trade. But the role of the merchant, which is amply discussed in chapter 1, should not be overemphasized. Plants and their products could and did arrive through non-market processes like gifting, tribute, dowry, loot, and war indemnities.

A whole range of other mobile peoples functioned as movers of crops, fibers, foodstuffs, and commodities of vegetal origin in the ancient Middle East and the Mediterranean. These included slaves, deportees, refugees, soldiers, messengers, diplomats, bureaucrats, colonists, road surveyors, porters, and guides; professionals like cooks, physicians, scribes, and augurs; and even elite brides and grooms traveling long distances with their large entourages.[2] The multiethnic character of Achaemenid and Hellenistic Babylonia, which was host to various occupational and ethnic groups, including Scythians, Carians, Lydians, Egyptians, and Indians who settled there in exchange for labor and military service, probably generated a vibrant gastronomic culture. This is unfortunately invisible in the relatively insular cuneiform tradition of the late first millennium BCE.

By the mid-first millennium BCE, a competitive culinary tradition striving for innovation also led to a widening repertoire of foods consumed by elites in the Middle East and the Mediterranean.[3] The emergence of gastronomic treatises and cooks as a professional, albeit servile, class are best attested in Greek sources from the late Classical period on.[4] A cook's address to his companion in a fragmentary comedy by Baton (third century BCE) expresses, no doubt with exaggeration, the burdens placed on cooks to innovate and perfect the dishes attributed to eminent cook-gastronomes: "Good for us, Sibyne, that we don't sleep at night or even lie down. Instead, a lamp stays lit, and there are books in our hands, and we puzzle over what Sophon's left behind, or Semonaktides of Chios, or Tyndarichos of Sikyon, or Zopyrinos."[5]

The situation was probably no different in the Middle East, although we have little by way of culinary texts in the cuneiform tradition of the first millennium

2. Boivin, Fuller, and Crowther 2012, 455. See Beckman (2013) for a discussion of foreigners in ancient Middle Eastern societies; Henkelman (2017, 68–78) on the mobile peoples needed to maintain road networks in the Achaemenid Empire; and Livarda (2008, 106–110, 2011, 156–57) for the role of the Roman army in the movement of new crops to northern Europe.

3. Dalby 2003, 98–99, 102–03, 157–58.

4. Berthiaume 1982; Dalby 2003, 97–99; Nadeau 2015; cf. Columella (*Rust.* I.5), who speaks of professional training schools (*officina*) for cooks.

5. Ap. Ath. XIV.662c-d.

BCE.[6] Greek sources on the Persian king's table hint at the enthusiasm for the consumption of new and exotic foods.[7] Xenophon claims that "men travel through the Persian king's entire territory for him, trying to find wines he would enjoy drinking (and) countless people produce foods he might like to eat."[8] Strabo stresses that the Persian king only consumed the finest foods, whether wheat from Assos, wine from Syria, or even water fetched from the Eulaios in southwestern Iran.[9] While there is certainly a moralizing element to Greek descriptions of Persian gastronomic luxury, it does not detract from the cosmopolitan and neophilic nature of contemporary elite food culture, which widened exposure to a range of foreign vegetal materials.

Temporary and permanent population movements, both coerced and voluntary, must have played a crucial role in widening food choices and fostering a cosmopolitan outlook. The archaeological finds of citrus and eastern cucurbits at Greek and Phoenician colony-sites like Carthage, Cumae, and Marseille in the Mediterranean and Tauric Chersonesos in the Black Sea indicate seaborne distribution of these cultivars through colonial networks (see chapters 4 and 5). As with the merchant bartering personal foodstuffs, some of the earliest foreign crop acquaintances may have been the unplanned result of food shared or left behind in the course of a journey.[10]

EXPERIMENTATION

The Cultivator

The second stage involves small or large-scale experimental cultivation. The scale depended on who the cultivator was. Cultivators were not a homogeneous group but display variations in social status, cultivation capacity, and intent. For a middling farmer, a new crop could simply fulfill calorific needs and perhaps help with resilience to environmental stresses and the resulting risk of variable annual production, crop failure, and food shortages. Diodorus Siculus, drawing on several Hellenistic sources on India, including Onesicritus and Megasthenes (late fourth century BCE), claimed that famine never visited India since the regular monsoonal rains combined with winter–summer double cropping ensured a secure

6. There is no doubt that a formalized culinary tradition existed in the Middle East, since elaborate recipes for food preparation are already attested in cuneiform texts dating to c. 1700 BCE (Bottéro 1995; Milano 2004).

7. E.g., Aristoxenus of Tarentum, ap. Ath. 545d; Clearchus, ap. Ath. 529d; Plut. *Quaest. Conv.* I.4. See also the discussion of Persian elite food in Briant 2002, 286–92.

8. Xenophon, *Agesilaus* 9.3; cf. Dinon, ap. Ath. 14.652b.

9. Strabo XV.3.22.

10. Boivin, Fuller, and Crowther 2015, 350.

food supply.[11] While this is evidently a romanticized and patently false sketch of Indian agriculture, it draws attention to an undercurrent of food insecurity in a Mediterranean context.

For those with surplus labor and a larger farmstead, calorific needs might go hand in hand with considerations of the crop as a marketable summer surplus. In both cases, the new crops represent a case of "innovative intensification."[12] The lack of taxation and tithing traditions for new crops, particularly non-grain crops, may have enhanced their production value. But this also means that new crops either leave an impressionistic paper trail, as we have observed throughout this book, or are completely invisible in textual records.

Crop experiments are opportunistic events involving a willingness to exploit plant resources for both calorific and non-calorific ends. Subsistence—that is, the production of food resources at the bare minimum to support existence over a long period—cannot account for all crop movements. The archaeologist Andrew Sherratt pertinently remarks that subsistence "is actively constructed in opposition to an accurate depiction of everyday reality. Subsistence is, in short, a rhetorical rather than a scientific term, a utopian representation of a world without ostentation and cupidity."[13] Beyond nutritive functions, crops could also simultaneously fulfill a range of other purposes, including the provision of fodder for animals, fuel, fiber, medicine, ritual substances, cosmetics, and raw materials for construction and manufacturing. Some, like citruses, cotton, and sissoo, are most obviously not subsistence crops. The ornamental aspects of a crop are not mutually exclusive with its economical contributions, either. Ancient gardens hosted agrarian crops and may well have been the starting point of the cultivation of many an exotic cultivar.

Negotiating Climate and Environment

The biggest barrier faced by the ancient Middle Eastern–Mediterranean cultivator experimenting with subtropical and tropical crops was climate and environment. The general pattern of weather across the Mediterranean and the Middle East is one of hot dry summers and mild wet winters—the eponymous Mediterranean climate. Farming strategies involving foreign crops had to adapt to a range of climatic and environmental stresses, including variable day length, temperature, soil conditions, water supply, altitude, pests, and plant pathogens. The paleoclimatic data for the Iron Age eastern Mediterranean and the Middle East, determined by proxies like ancient pollen, maritime and terrestrial sediments, and speleothems, cannot be reliably correlated with the migration of new crops. Notwithstanding local microclimatic variations, the general impression provided by palaeoclimatological analyses

11. Diod. Sic. II.36.5–7; Muntz 2012.
12. Horden and Purcell 2000, 201–24.
13. Sherratt 1999, 12.

of late-second- and first-millennium BCE southern Eurasia is one of an aridifying trend interspersed with wet and humid phases like the second quarter of the first millennium BCE.[14] It would appear that thirsty tropical and subtropical crops were adopted in spite of long-term fluctuations in pluviality. Significantly, the landscape archaeologist Dan Lawrence and his colleagues observe that in the long term there is no correlation between atmospheric moisture and settlement patterns after 2000 BCE.[15] This cautions against environmentally determinist approaches which ignore human and plant adaptability.

Most of these *longue durée* climatic developments would, of course, have been imperceptible to the ancient cultivator, for whom interannual variability in temperature and precipitation was more tangible. As the bioarchaeologist Arkadiusz Sołtysiak remarks, "the amplitude of the Holocene climatic changes is usually lower than regular inter-annual variability."[16] It was the interannual variability in rainfall, above all, which rendered the Middle East and the Mediterranean a risk-laden landscape for the cultivator.[17] The ancient cultivator, as the Roman agricultural author Columella remarks, had to expect interannual climate variability: "He must observe the behavior of the current weather and season, for they do not always wear the same habit as if according to a fixed rule; summer and winter do not come every year with the same countenance; the spring is not always rainy or the autumn moist."[18]

Precipitation was less important for summer cultivation in the Middle East and the Mediterranean, which depended on irrigation for reliable growth. But irrigation was a laborious affair and had to be carefully regulated, especially in the spring and early summer, when large rivers like the Tigris and Euphrates were prone to floods generated by Anatolian snowmelt. The spring and early-summer flow accounts for nearly 50 percent of the annual runoff in the Tigris-Euphrates basin.[19] In a region of hot dry summers with high evapotranspiration rates, groundwater was generally a negligible source of moisture for cultivators. Southern Mesopotamia and southwestern Iran proved excellent hosts for thirsty and labor-intensive crops like cotton and rice, where they were grafted onto a pre-existing irrigation infrastructure supported by a relatively large population.[20] The sheer productive potential of the Mesopotamian alluvium is articulated in the *Genesis Rabah*, a late

14. Altaweel et al. 2019; Avnaim-Katav et al. 2019; Gurjazkaite et al. 2018; Kaniewski et al. 2008; Kuzucuoğlu et al. 2011, 183, 186; Langgut et al. 2015; Lückge et al. 2001; Psomiadis 2018; Schilman et al. 2001; Sorrel and Mathis 2016.

15. Lawrence et al. 2016.

16. Sołtysiak 2016, 393.

17. Samuel 2001, 351.

18. Columella, *On Agriculture*, I.23.

19. Cullen and deMenocal 2000, 856.

20. On irrigation in Mesopotamia, see the various articles in the fourth volume of the *Bulletin of Sumerian Agriculture* (1988) and Wilkinson and Rayne 2010.

antique Hebrew exegetical text, by none other than the personified Euphrates, who declares: "If someone plants a plant near me, it brings forth produce in thirty days. If someone sows seed by me, it comes up in three days."[21] But a reliable summer supply of water for crops was not uniformly present across the Middle East and the Mediterranean, and hence tropical and subtropical crops spread unevenly across the region, thriving predominantly in flat and fertile plains watered by perennial streams and rivers. And an oversupply of water without adequate drainage modifies pedological conditions, particularly in arid regions where the accumulation of salts and minerals in the upper layers of the soil, coupled with stagnant water, increases soil salinity to the detriment of crops.[22] In general, neutral soils with pH values between 6 and 7 are ideal for most agronomic crops.

Annual variability in production was prompted not just by variable weather and natural disasters but also by conflict, which siphoned off manpower and damaged agricultural fields. Seasonal labor scheduling necessitated by building projects and canal construction also subtracted labor from summer cropping and may have hampered the spread of labor-intensive crops like rice and cotton. Diodorus Siculus explicitly contrasts his idealized Indian cultivator, mentioned earlier, with the "rest of mankind," who harassed agriculturists and cut down orchards in times of conflict.[23]

Crop Features

Crops whose cultivation was slow and required large capital investments presented another impediment to experimentation. In Petronius's *Satyricon*, citrons, which were grown in the ancient Mediterranean, are included in a list of otherwise outlandish, if not impossible, farm produce emanating from the estates of the *nouveau riche* character Trimalchio. The estate produce included pepper, which does not grow in the Mediterranean, and "poultry milk."[24] Petronius further lampoons the rich Roman farmstead lifestyle by remarking that Trimalchio had written for a shipment of mushroom spores from India.[25] The inclusion of citron in this context indicates that even in Roman Italy, where the archaeobotanical evidence for citruses is relatively dense, the cultivation of citron was perceived as a difficult enterprise and probably the preserve of wealthier denizens. Arboriculture in general, whether involving fruit or timber-yielding species, was a long-term and costly investment motivated by non-subsistence values. Citrus trees have a long juvenile phase and only achieve maximum productivity in the second decade of growth.[26]

21. *Genesis Rabah* 16.3 (trans. Neusner 1985, 172).
22. Samuel 2001, 352–54.
23. Diod. Sic. II.36.6.
24. Petronius, *Satyricon* 38.
25. Petronius, *Satyricon* 38.
26. Cooke and Fuller 2015, 453; Zohary, Hopf, and Weiss 2012, 114.

The growing of tree and cash crops like cotton is an expression of calculated long-term production and profit-oriented trade strategies. This was frequently beyond the reach of the average householder.

Conversely, crops whose features include quick maturation, high yield, ease of harvest, disease resistance, and storability would have appealed to cultivators, and these must have been important factors in adoption or rejection.[27] With the notable exception of cotton and tree crops like citruses and sissoo, most summer crops in the ancient Middle East and the Mediterranean were quick-maturing annual varieties which were planted in spring or early summer and harvested in autumn. Storage, a key component of food security, would have played a significant role in the experimentation phase for comestible crops. Rice can be stored as grain or processed into flour, much like wheat and barley. Taro corms can be stored for up to three months in underground pits or cool, well-ventilated storage areas. Citruses and some cucurbits can be pickled or desiccated for later use. In rural India, some varieties of melon are preserved for weeks by suspending them from the ceiling with coir ropes.[28]

Pre-existing food processing and cooking technologies also significantly influenced the choice of new comestible crops, which were expected to be processed and consumed in ways that were familiar.[29] Rice in the early Middle East was processed and consumed as bread, cake, and porridge rather than as boiled grain as was the norm across the rest of Asia. The morphological and gustatory resemblance of early varieties of the cucumber to non-sweet, vegetable-type melons encouraged its spread and confusion with the melon.

Finally, the plants themselves are not as passive as one imagines them to be in the processes of domestication and translocation. Many plants are capable of adjusting to their new environment through adaptive mechanisms involving phenological changes, particularly with regard to features like germination, flowering, seed dispersal, and seasonal dormancy.[30] Genomic studies of modern rice crops growing in Italy and Japan indicate that artificial selection pressures and adaptive mechanisms favored crops with reduced sensitivity to photoperiod (day length), allowing them to flower in the longer summer days of the northern hemisphere.[31] This phenological adaptation was likely a feature of the early South Asian rice crops moving into the Middle East as well. A variety of rice grown in the Al-Aḥsā' Oasis of eastern Saudi Arabia is not only less photoperiod sensitive but also

27. Jones et al. 2011; Smith 2006, 481.

28. Swamy 2017.

29. Boivin, Fuller, and Crowther 2012, 454.

30. Borrell et al. 2020, 7.

31. Fujino and Sekiguchi 2005; Gómez-Ariza et al. 2015.

highly tolerant of drought and salinity, features which otherwise are not associated with rice.[32]

The Cultivator-King

At the far end of the cultivation spectrum we find large-scale experimentation with exotica involving large labor and capital inputs on the part of non-agrarian, typically urban, elites. The motivations for experimentation may be more complex here. In honoring religious establishments with the finest produce, claiming symbolic victory over foreign landscapes—as Sennacherib does with his replication of the southern Babylonian marshscape in Nineveh—or marking status through conspicuous consumption, the costlier experiments were ultimately more about stratification than subsistence. They could be singular or recurring events. But the cultivars which eventually proved successful are more likely the result of a series of reintroductions rather than a single introduction event.[33]

Political actors, unsurprisingly, played a major role as agents of crop diffusion. Political elites across history were fond of transplanting exotic flora as a means of displaying their political clout and their territorial and diplomatic reach.[34] Exotic flora, particularly those with aesthetic, medico-magical, or spiritual value, functioned as an extension of royal paraphernalia, and their possession conferred social prestige.[35] In some cases, the royal collecting of plants was simply motivated by pleasure in novelties and sheer intellectual curiosity.[36] Ancient sources mention a strong personal interest on the part of elites in horticultural activities.[37] In Xenophon's (c. 430–354 BCE) *Oikonomikos,* a treatise on household management, the Persian prince Cyrus the Younger, much to the surprise of his Spartan guest Lysander, actively engages in horticulture and even ranks it alongside warfare and competitive activities:

> Cyrus personally showed him (Lysander) around his paradise at Sardis. Now Lysander admired the beauty of the trees there, the accuracy of their spacing, the straightness of the rows, the regularity of the angles, and the multitude of sweet scents that clung around them as they walked; and in amazement he exclaimed, "Cyrus, I really do admire all this loveliness, but I am far more impressed with your agent's skill in measuring and arranging everything so exactly." Cyrus was delighted to hear this and replied, "Well, Lysander, the whole of the measurement and arrangement is my own work, and I did some of the planting myself." "What did you say, Cyrus?" exclaimed Lysander, looking at him, and noting the beauty and perfume of

32. Zhang et al. 2012.
33. Boivin, Fuller, and Crowther 2012, 456.
34. Boivin et al. 2012; Jones et al. 2011; Pollard 2009; Totelin 2012.
35. Boivin and Fuller 2009, 165.
36. Boivin, Fuller, and Crowther 2012, 454.
37. Totelin 2012, 126–41.

his robes, and the splendor of the necklaces and bangles and other jewels that he was wearing; "Did you really plant part of this with your own hands?" "Does that surprise you, Lysander?" asked Cyrus in reply. "I swear by Mithras that I never yet sat down to dinner when in sound health, without first working up a sweat at some task of war or agriculture, or exerting myself in some sort of competition."[38]

Plants could also arrive by way of diplomatic gifts, tribute, expeditions, or even wartime loot. The literary records of the ancient Middle East and the Mediterranean are replete with instances of royally commissioned plant translocation.[39] The Old Kingdom pharaoh Sahure's (c. 2458–2446 BCE) transplanting of myrrh trees from the land of Punt (Eritrea–Ethiopia corridor, Djibouti, and eastern Somalia) to his royal garden was celebrated in relief in the king's pyramid complex.[40] This is among the earliest depictions of deliberate plant transplantation in the ancient world.[41] It also substantially predates the better-known Egyptian maritime expeditions to Punt and the procurement of incense trees under the New Kingdom queen Hatshepsut (c. 1473–1458 BCE).[42] Possessing rare products and bestowing them on others could sustain patronage networks and enhance the prestige of the giver. The family of Senedjemibinty, a deceased Old Kingdom official, gratefully advertised the Pharaoh Djedkare's (c. 2381–2353 BCE) provision of myrrh to embalm Senedjemibinty's body in the latter's funerary biographical inscription with the hyperbolic remark that such an honor "never happened alike to any man before."[43]

In the Early Iron Age, the best-attested examples of royal botanical collections are those of the Assyrian kings.[44] From the end of the twelfth century BCE on, the Assyrian royal garden took on a more pronounced political dimension than its Mesopotamian precursors.[45] Drawing on the botanical resources of the empire and beyond, it functioned as a microcosm of the known world. King Aššurnasirpal II (883–859 BCE) boasts in the "banquet stela" of collecting plants from various regions of his empire:

38. IV.20–24, trans. Marchant and Henderson 2013.

39. Amrhein 2015; Foster 1998, 320; Wiseman 1983. On the Hellenistic period, see Schneider 2012; Secord 2016.

40. El Awady 2009, 155–73; 253–57. The location of Punt was recently clarified through strontium and oxygen isotope analysis of ancient baboon tissue (Dominy et al. 2015, 2020).

41. The earlier Old Kingdom pharaoh Sneferu (c. 2613–2589 BCE) may have outdone Sahure in acquiring myrrh trees, since an inscribed relief from his pyramid temple at Dahshur depicts him inspecting exotic trees, one of which is named as myrrh (*ʿnt.w*). It is not clear, however, whether these trees were acquired via intermediaries or through direct contact with peoples in the Horn of Africa (Creasman and Yamamoto 2019, 350–01; Edel 1996, 200–03).

42. Creasman 2014.

43. El Awady 2009, 254.

44. Dalley 2013; Foster 1998; Reade 2004; Thomason 2005; Wiseman 1983.

45. Amrhein 2015, 92

> I irrigated the meadows of the Tigris (and) planted orchards with all kinds of fruit trees in its environs. . . . In the lands through which I marched and the highlands which I traversed, the trees (and) plants which I saw (and collected) were: cedar, cypress, *šimiššalû*, *burāšu*-juniper . . . *daprānu*-juniper, almond, date, ebony, sissoo, olive, *ṣuṣūnu*, oak, tamarisk, *dukdu*, terebinth and *murrānu*, *meḫru* . . . *tīatu*, Kaniš oak, *ḫaluppu*, *ṣadānu*, pomegranate, *šallūru*, fir, *ingirašu*, pear, quince, fig, grapevines, *angašu*-pear, *ṣumlalû*, *titipu*, *ṣippūtu*, *zanzaliqqu*, swamp apple, *ḫambuququ*, *nuḫurtu*, *urzīnu* and *kanaktu*. The canal cascades from above into the gardens. Fragrance pervades the walkways.[46]

Royal gardens like those of Aššurnasirpal II and his successors were effectively "warehouses of biodiversity," much like modern botanical gardens.[47] Crop experimentation within the confines of royal gardens could be the starting point for the wider dissemination of foreign crops, especially fruit trees and aromatics. Assyrian gardens were practical as much as they were political. The gardens supplied the royal table with fruits and vegetables, the perfume industry and temples with aromatics, and physicians with pharmaceuticals.[48] The inscriptions of the Assyrian king Sennacherib make unambiguous that cotton cultivation met the needs of a palace textile industry, while sissoo trees in the royal estates were felled for palatial construction in Nineveh (see chapters 2 and 8). Botanical experimentation in royal estates could also become a significant source of revenue for the royal household. Profit may have motivated the introduction of balsam trees (which are native to the Arabian Peninsula and East Africa) to Ein-Gedi, in the Dead Sea region, by King Josiah of Judah (640–609 BCE).[49] The cultivation of balsam in this region remained a lucrative source of income for royal estates down to the Hellenistic and Roman periods.[50]

Failures and Limited Successes

The history of cultivation failures and limited successes, few of which make it into the historical record, must be lengthy indeed. Agricultural innovation, in some cases, was a sluggish and incremental process which probably consisted of multiple experimental croppings. There could be a long time lag between experimental cultivation of a crop and its adoption as a regular cultivar.[51] Some cultivars were moved great distances but the environment posed an insurmountable barrier to

46. RIMA II A.0.101.30.

47. Boivin, Fuller, and Crowther 2012, 456; Boivin et al. 2014, 550.

48. Foster 1998, 322; Wiseman 1983, 142.

49. TB, *Keritot* 5b; Hirschfeld 2007, 29–35. Josephus *Ant.* VIII.174 anachronistically attributes it to Solomon.

50. Pliny *HN* 12.111–24; Strabo XVI.2.41; Theophr. *Hist. pl.* 9.6.1–4; Dalby 2003, 43; Safrai 1994 83–87; Totelin 2012, 122–25.

51. Boivin, Fuller, and Crowther 2012, 455–56; Liu and Jones 2014.

local cultivation. This was the case with most South Asian spices, like pepper, cardamom, and cinnamon, which simply could not grow outside the tropical belt. Pliny describes the failed experiments of Seleukos I, who vainly attempted the cultivation of amomum, nard, and cinnamon in Syria.[52] The cotton attested in the Assyrian heartland between the eighth and seventh centuries BCE was probably a limited palace project as well, since early varieties of tree cotton were not suited to the cold winters of northern Mesopotamia.

Cultivation failures, however, could inspire the growing of local substitutes. Pliny remarks that pepper was so popular that the berries of other plants, including juniper and myrtle, were passed off as pepper.[53] Rabbinical sources (early to mid-first millennium CE) claim that pepper was grown in Palestine.[54] This almost certainly refers to the cultivation of a substitute for true pepper, most likely Ethiopian or kimba pepper (*Xylopia aethiopica*), which is extracted from the pods of a tree native to sub-Saharan Africa.[55]

ROUTINIZATION

The third stage is marked by regularized and sustainable cultivation. Summer cropping and irrigation would have been seamlessly integrated into the cycle of the local agricultural year. The cultivator's knowledge of environmental requirements and weed and pest control would be acute. But overall, the crop would still remain a niche product, for it is temporally restricted to summer and spatially restricted to localities that fulfill the thermal, pedological, and pluvial requirements discussed in regard to the experimentation phase. This mode of production, however, worked to the advantage of tropical and subtropical crops in the Middle East and the Mediterranean since they exploited spatial and temporal zones unoccupied by the traditional winter-sown calorific staples.[56] None of the crops under consideration here grew at the expense of the traditional staples of barley and wheat. Rice may have been adopted in southern Mesopotamia as a means to reclaim marginal swamplands unsuitable for barley cultivation. Herodotus notes that the edible Indian lotus growing in the Nile among other edible aquatic flora made the acquisition of food more economical.[57] The examples of rice and lotus

52. Pliny *HN* XVI.135–36.

53. Pliny *HN* XII.29.

54. *Ecclesiastes Rabah* 2.5, 2.8; *Midrash Hagadol* on *Deut.* 8.9; Löw 1924, vol. 3, 51–52; Safrai 1994, 83, 91.

55. Löw 1924, 52.

56. The traditional suite of Middle Eastern–Mediterranean crops, most notably wheat, barley, lentil, and chickpea, were planted with the winter rains, between November and December, and harvested in spring. See Zohary, Hopf, and Weiss 2012 on the traditional suite of Middle Eastern crops.

57. Hdt. II.92.

exemplify the efficient agrarian use of previously marginal cultivable habitats, whether aquatic or terrestrial.

Not only were the new cultivars limited by environmental constraints, but also their cultivation, circulation, and consumption were dampened by long-standing cultural preferences and foodways. The availability of new subsistence options did not automatically translate to immediate and widespread adoption. Even the biological consequences of the later Columbian Exchange, although a relatively quicker event, were not immediately perceptible. The large-scale cultivation of New World crops in Europe took off only in the late seventeenth century.[58] Northern European peasants reviled the now-ubiquitous potato (*Solanum tuberosum*) as an unwholesome cultivar even as late as the eighteenth century.[59] The Central American tomato (*Solanum lycopersicum*) was described by the Italian herbalist Pietro Mattioli as early as 1544 but long remained an ornamental plant owing to the perception that its fruits were unpleasant or even dangerous to eat.[60] Its widespread use in southern European cuisine only dates to the eighteenth century—and in northern Europe, the nineteenth century.[61]

The impediments to the instantaneous adoption of new crops in antiquity were manifold. Ancient cultivators, particularly those in insular regions, preferred predictability to change. Food habits vary with age, gender, and class as well as across space and time. The palate of rural, aged, and non-elite groups was often markedly neophobic. Plutarch attests to entrenched foodways among the elderly, who were said to be averse to the consumption of exotic foods from India: "We know that many older people still cannot eat ripe cucumber, citron or pepper."[62] The Alexandrian physician Khrusermos (first century BCE) was even said, undoubtedly with great exaggeration, to be liable to a heart attack if he ever consumed pepper.[63] Consumables and consumption practices were inherently conservative because they constituted an essential expression of group membership, whether it be ethnic affiliation, gender construct, or social class.[64] And even after they were adopted, the new crops were often consumed in familiar ways, as we have observed with rice.

While conservative foodways and agricultural regimes hampered the spread of new crops, they also meant that when populations moved, voluntarily or involuntarily, they brought their familiar crops with them. In Hellenistic Egypt, Greek settlers transplanted Aegean varieties of crops, including cabbage from Rhodes,

58. Hawkes 1998; Heywood 2012.

59. Salaman 1985, 114–15.

60. Hawkes 1998, 152–53; Heywood 2012, 76.

61. Gentilcore 2010, 45-68; Heywood 2012, 76.

62. Plutarch *Quaest. Conv.* VIII.9.3 (733).

63. Sextus Empiricus, *Outlines of Pyrrhonism* I.14 (84).

64. Appadurai 1981, 494, 496, 1986; Fernández-Armesto and Sacks 2012; Gumerman 1997; Livarda 2008, 8–25; Smith 2006, 480–81; Thomas 2012, 174–75.

chickpeas from Byzantium, figs from Chios, and garlic from Tlos in Lycia.[65] Egypt was not lacking in such crops, but the Egyptian varieties were held to be inferior in taste by Greek settlers. The import of cabbage from Rhodes, for instance, was said to be necessary on account of the bitter taste of Egyptian cabbage to Greeks.[66] Thus, ironically, where demographic movements are concerned, conservative foodways could facilitate rather than hamper the spread of crops.

There could also be sumptuary distinctions during the earliest periods of a crop's cultivation. Cotton is first attested in temple and palatial contexts in Mesopotamia before becoming more widely available in the mid-first millennium BCE, suggesting controlled access to the material in its earliest phase. Food crops, as we have observed, were markers of culture, identity, and class and invested with cultural and social meaning beyond their nutritive value.[67] Differential access to foods based on appearance, taste, rarity, cost, and reputation may have created a bottleneck in the adoption of new crops among peasant-cultivators. The processes of social emulation and peer–polity interaction, however, eventually overcame elite monopolies to reproduce metropolitan crop consumption trends on a local scale.

INDIGENIZATION

The fourth and final stage involves wider spatial diffusion and nativization. The crop at this point is commoditized and integrated into profit-oriented local and regional trade networks. It may also be incorporated into state rationing, tithing, and taxing regimes, as was the case with rice and cotton. More crucially, it may be said to have gone native. The parameters for indigenization may be defined as follows. It would lack associations with the exotic, but owing to environmental constraints it might still be produced on a smaller scale. It is likely to be available to a wider spectrum of consumers, including those of lesser means outside urban centers. Most importantly, the social, symbolic, and cultural meanings gained by the crops, typically not found in their place of origin, are a hallmark of indigenization.

Mythological constructs and religious beliefs had a significant influence on food acceptability and crop usage. The citron's sacred associations in Judaism, in particular with the feast of Sukkot, ensured a permanent foothold for this tree-crop in the gardens of the Levant and beyond. Rice gained medical and magical properties in both the Middle East and the Mediterranean. The former function probably explains its rapid spread beyond the Alps with Roman legionaries in the first century CE. The 196 charred grains of rice from Roman Novaesium (Neuss

65. Crawford 1979, 139–40; Secord 2016, 215–18.
66. Diphilos of Siphnos, ap. Ath. IX 369f.
67. Appadurai 1981, 494; Gumerman 1997.

am Rhein) in Germany were in fact recovered from a building identified as a military hospital (*valetudinarium*).[68]

The process of indigenization is long-drawn and extends beyond the lifetimes of individual cultivators. The cultivated melon and sesame are also good examples of tropical-subtropical crops gone native, particularly since they were transmitted westward by the late third millennium BCE. Both were widely grown and well integrated into Middle Eastern and Mediterranean culture. Sesame's role as the premier Mesopotamian oil crop is reflected in its listing in the Babylonian Astronomical Diaries (c. 650–60 BCE), which provide prevailing market rates for the leading agricultural crops of the day, including barley, dates, cuscuta (dodder), cress, and wool.[69] But its Indian origins were not forgotten. Two millennia after its introduction to the Middle East and the Mediterranean, Pliny the Elder could still remark that sesame comes from India (*sesama ab Indis venit*).[70]

EPILOGUE

Southern Eurasia's tropical turn was a dynamic, constantly evolving, and long-drawn process. By the end of the first millennium BCE, gastronomic, agrarian, and civilizational frontiers were dissolving across this region. The long-distance anthropogenic movement of crops from South Asia to the Middle East and the Mediterranean enriched local agrobiodiversity and lent some degree of hemispheric unity to the agricultural repertoire. Foodways and agrarian landscapes of southern Eurasia now contained elements which were familiar across the entire region.

In retrospect, the tropical turn of the Middle East and the Mediterranean can be identified as an important step in a globalizing process. In time, the movement of crops across Eurasia transformed not only the agricultural landscapes, labor regimes, and cultures of the Old World but also those of the New World. Slavery in the Americas is intricately tied to the cultivation of labor-intensive tropical crops like cotton, sugarcane, and rice. Whether dictating fashions from Sulawesi to Scotland (think paisleys) or sparking industrializing processes in Britain, which was flooded with Indian cottons before the eighteenth century (much to the dismay of local producers), the cotton textile industry of India dominated and shaped global trade networks for at least two and a half millennia.[71] This changed, of course, with industrialization in Britain and the attendant rise of the East India Company's political fortunes, which effectively curtailed the lucrative export trade of South Asia in favor of mass-produced Lancashire cottons. As Karl Marx astutely

68. Knörzer 1966, 433–43, 1970, 13, 28.
69. Pirngruber 2017.
70. *HN* XVIII.96; cf. Strabo XV.13.
71. See essays in Fee 2020; Parthasarathi and Riello 2012; Riello and Roy 2009.

observed in 1853, "Till 1813, India had been chiefly an exporting country, while it now became an importing one."[72] Industrialization in North America also tied in with cotton production, the first manifestation of this process being the New Englander Eli Whitney's cotton gin (1793). What is significant to us is that these seminal events of a later age are connected to the ancient crop exchange under consideration here. The acquaintance of later Europeans with rice, cotton, or sugarcane traces its genealogy back to the crop movements within ancient Eurasia.

Tropical and subtropical crops in the Middle East and the Mediterranean underwent a long acculturation sequence, in which the earliest period is typically characterized by small-scale cultivation and adaptation to pre-existing agricultural strategies and food technologies. But this did not imply that agricultural landscapes remained static. The introduction of water-intensive summer crops like rice and cotton called for rigorous water management. In southern Mesopotamia's case this translated to the extension of pre-existing irrigation canals. Fruit orchards and the growing of ornamental plants would have also transformed the aesthetic appearance of cultivation landscapes. As crops arrived in piecemeal fashion, agrarian changes were incremental and spatially uneven. Great metropoles like Nineveh, Babylon, Susa, and Athens likely consumed a far wider spectrum of marketable botanical produce than their distant rural hinterlands. The pedological and hydrological heterogeneity of the Middle East and the Mediterranean also meant that the impact and distribution of new crops was asymmetrical. Some crops, particularly spice-bearing plants, did not survive past the experimental stage owing to environmental incompatibility or the lack of sufficient capital and labor inputs.

The environmental archaeologist David Harris rightly observes that the "great diversity of available plant products is a very recent phenomenon, [and] the beginnings of the process lie far back in the prehistoric past."[73] To modern observers accustomed to a huge variety of comestibles and botanical produce, the plants described in this book may appear numerically limited, but it must be borne in mind that the traditional repertoire of founder crops used by Neolithic and Bronze Age sedentary societies in the Middle East and the Mediterranean was equally limited and certainly narrower than the gamut of wild foods and fibers consumed by their hunter-gatherer predecessors.[74] The first waves of plant domestication essentially led to a contraction in dietary diversity. For those not in the elite, plant-derived food was a fairly monotonous affair dominated by a handful of cereal, pulse, vegetable, and fruit crops. In ancient Mesopotamia, barley, which produced both bread and beverage, was the unchanging staple of life, and an inferior version of the same was to be expected in the afterlife. In an early-first-millennium BCE

72. Marx 1853.
73. Harris 1998, 85; cf. Khoury et al. 2016.
74. Heywood 2013, 39–40.

recension of the Akkadian poem "Ištar's Descent to the Netherworld," the underworld goddess Ereškigal rhetorically asks whether she should share the fate of the dead: "Shall I eat a loaf of clay for bread, shall I drink dirty water for beer?"[75] Over time, the addition of summer-growing crops from adjacent biogeographic regions to two-cropping cycles served as a basis for food security, dietary diversity, nutritional enrichment, and demographic growth. As foods and fibers, new crops also had a key role in the construction of cultural identities across Afro-Eurasia. But the extension of the agricultural year with summer cropping meant less time for other leisurely pursuits and may have depressed the general quality of life for farmhands, some of whom were doubtless dependent or unfree laborers.

The crop exchange between tropical Asia, the Middle East, and the Mediterranean continued long after the period considered here, into late antiquity, the medieval period, and beyond. In the age when Baghdad and Cordoba flourished as the twin poles of the Middle East and the Mediterranean, respectively, the cultivation of tropical flora was commonplace and extended to include a range of other tropical species, including lime, bitter orange, shaddock, spinach, okra, banana, tamarind, indigo, mung bean, mango, coconut, jasmine, and brown mustard.[76] The incredible botanical diversity of the tropics meant that many other plant species consumed locally were unknown outside tropical Asia until recent times.[77] Nonetheless, the gradual tropicalization of Middle Eastern and Mediterranean agricultural landscapes from the Iron Age on meant that their visual, olfactory, and gustatory worlds were increasingly closer to our globalized world than to the earliest agricultural societies.

75. Foster 2005, line 33.

76. Amar and Lev 2017, 232–33; Boivin et al. 2014, 562–63; Ruas et al. 2015; Ubaydli 1993, 34–37; Varisco 1994, 181–201, 2002.

77. See Khoury et al. 2016 on contemporary patterns of foreign crop consumption. For a survey of underutilized tropical crop resources, see Ulian et al. 2020.

REFERENCES

PRIMARY SOURCES

Sumerian, Akkadian, Elamite, Hittite, Egyptian, and Old Persian Texts

Abusch, T., and D. Schwemer. 2011. *Corpus of Mesopotamian Anti-witchcraft Rituals*, Vol. 1. Leiden: Brill.

Beaulieu, P.-A. 1989. "Textes administratifs inédits d'époque hellénistique provenant des archives du Bīt Rēš." *Revue d'Assyriologie* 83: 53–87.

Beckman, G. 1999. *Hittite Diplomatic Texts*. Atlanta: Society of Biblical Literature.

Black, J. A., G. Cunningham, J. Ebeling, E. Flückiger-Hawker, E. Robson, J. Taylor, and G. Zólyomi. 1998–2006. *The Electronic Text Corpus of Sumerian Literature*. Faculty of Oriental Studies, University of Oxford (http://etcsl.orinst.ox.ac.uk/).

Bottéro, J. 1995. *Textes culinaires mésopotamiens*. Winona Lake, IN: Eisenbrauns.

Clay, A. T. 1912. *Business Documents of Murashû Sons of Nippur Dated in the Reign of Darius II*. Babylonian Section, University Museum, University of Pennsylvania.

Clay, A. T., and H. V. Hilprecht. 1898. *Business Documents of Murashû Sons of Nippur Dated in the Reign of Artaxerxes I*. Department of Archaeology and Palaeontology, University of Pennsylvania.

Collombert, P. 2002. "Le conte de l'hirondelle et de la mer." In K. Ryholt (ed.), *Acts of the 7th International Conference of Demotic Studies*, 59–76. Copenhagen: Museum Tusculanum.

Dalley, S., and J. N. Postgate. 1984. *The Tablets from Fort Shalmaneser*. London: British School of Archaeology in Iraq.

Dietrich, M. 2003. *The Babylonian Correspondence of Sargon and Sennacherib*. State Archives of Assyria 17. Helsinki: Helsinki University Press.

Dougherty, R. P. 1933. *Archives from Erech, Neo-Babylonian and Persian Periods*. Goucher College Cuneiform Inscriptions 2. New Haven: Yale University Press.

Durand, J-M. 1983. *Archives royales de Mari XXI: Textes administratifs des salles 134 et 160 du Palais de Mari.* Paris: Librairie Orientaliste Paul Geuthner.

Ebeling, E. 1919–23. *Keilschrifttexte aus Assur: religiösen Inhalts.* Leipzig: J. C. Hinrichs.

Foster, B. R. 2005. *Before the Muses: An Anthology of Akkadian Literature,* 3rd ed. University Press of Maryland.

Frame, G. 1995. *Rulers of Babylonia: From the Second Dynasty of Isin to the End of Assyrian Domination, 1157–612 BC.* Royal Inscriptions of Mesopotamia, Babylonian Periods, Vol. II. Toronto University Press.

Frayne, D. R. 1993. *Sargonic and Gutian Periods 2334–2113 BC.* Royal Inscriptions of Mesopotamia, Early Periods, Vol. II. University of Toronto Press.

———. 1997. *Ur III Period 2112–2004 BC.* Royal Inscriptions of Mesopotamia, Early Periods, Volume III/2. University of Toronto Press.

———. 2008. *Presargonic Period 2700–2350 BC.* Royal Inscriptions of Mesopotamia, Early Periods, Vol. I. University of Toronto Press.

Fuchs, A., and S. Parpola. 2001. *The Correspondence of Sargon II, Part III: Letters from Babylonia and the Eastern Provinces.* State Archives of Assyria 15. Helsinki University Press.

George, A. R. 2003. *The Babylonian Gilgamesh Epic: Introduction, Critical Edition and Cuneiform Texts.* Oxford University Press.

Grayson, A. K. 1987. *Assyrian Rulers of the Third and Second Millennia BC.* Royal Inscriptions of Mesopotamia, Assyrian Periods, Vol. I. University of Toronto Press.

———. 1991. *Assyrian Rulers of the Early First Millennium BC I, 1114–859 BC.* Royal Inscriptions of Mesopotamia, Assyrian Periods, Vol. II. University of Toronto Press.

———. 1996. *Assyrian Rulers of the Early First Millennium BC II, 858–745 BC.* Royal Inscriptions of Mesopotamia, Assyrian Periods, Vol. III. University of Toronto Press.

Grayson, A. K., and J. Novotny. 2012. *The Royal Inscriptions of Sennacherib, King of Assyria (704–681 BC), Part 1.* Royal Inscriptions of the Neo-Assyrian Period 3/1. Winona Lake, IN: Eisenbrauns.

———. 2014. *The Royal Inscriptions of Sennacherib, King of Assyria (704–681 BC), Part 2.* Royal Inscriptions of the Neo-Assyrian Period 3/2. Winona Lake, IN: Eisenbrauns.

Gurney, O. R. 1983. *The Middle Babylonian Legal and Economic Texts from Ur.* London: British School of Archaeology in Iraq.

Hallock, R. T. 1969. *Persepolis Fortification Tablets.* University of Chicago Press.

Henkelman, W., C. E. Jones, and M. W. Stolper. 2006. *Achaemenid Elamite Administrative Tablets 2: The Qaṣr-i Abu Naṣr tablet.* Arta, 3, 1–20.

Hunger, H. 1992. *Astrological Reports to Assyrian Kings.* State Archives of Assyria 8. Helsinki University Press.

Hunger, H., and E. von Weiher. 1976–98. *Spätbabylonische Texte aus Uruk.* Berlin: Gebr. Mann and von Zabern.

Kataja, L., and R. Whiting. 1995. *Grants, Decrees and Gifts of the Neo-Assyrian Period.* State Archives of Assyria 12. Helsinki University Press.

Kennedy, D. A. 1968. *Cuneiform Texts from Babylonian Tablets in the British Museum 49: Late Babylonian Economic Texts.* London: British Museum.

Kent, R. G. 1953. *Old Persian: Grammar, Texts, Lexicon.* New Haven: American Oriental Society.

King, L. W. 1912. *Babylonian Boundary-Stones and Memorial Tablets in the British Museum.* London: British Museum.

Köcher. F. 1955. *Keilschrifttexte zur assyrisch-babylonischen Drogen- und Pflanzenkunde.* Berlin: Akademie.

Kuhrt, A. 2007. *The Persian Empire: A Corpus of Sources from the Achaemenid Period.* Abingdon: Routledge.

Kwasman, T. 2015. "A Neo-Assyrian Royal Funerary Text." *Studia Orientalia Electronica* 106, 111–26.

Landsberger, B., and O. R. Gurney. 1957. "Practical Vocabulary of Assur." *Archiv für Orientforschung* 18, 328–41.

Landsberger, B., et al. 1957–74. *The Series ḪAR-ra = ḫubullu. Materials for the Sumerian lexicon (MSL) 5. 6, 7, 9, 10 and 11.* Rome: Pontificium Institutum Biblicum.

Lanfranchi, G. B., and S. Parpola. 1990. *The Correspondence of Sargon II, Part II: Letters from the Northern and Northeastern Provinces.* State Archives of Assyria 5. Helsinki University Press.

Lauinger, J. 2012. "Esarhaddon's Succession Treaty at Tell Tayinat: Text and Commentary." *Journal of Cuneiform Studies* 64, 87–123.

Leichty, E. 2011. *The Royal Inscriptions of Esarhaddon, King of Assyria. 680–669 BC.* Royal Inscriptions of the Neo-Assyrian Period 4. Winona Lake, IN: Eisenbrauns.

Luukko, M. 2012. *The Correspondence of Tiglath-Pileser III and Sargon II from Calah/Nimrud.* State Archives of Assyria 19. Helsinki University Press.

MacGinnis, J. 1987. "A Neo-Assyrian Text Describing a Royal Funeral." *State Archives of Assyria Bulletin* 1.1, 1–13.

———. 1996. "Letters from the Neo-Babylonian Ebabbara." *Mesopotamia: rivista di archeologia, epigrafia e storia orientale antica* 31, 99–160.

Mayer, W. 2013. *Assyrien und Urarṭu I: Der Achte Feldzug Sargons II. im Jahr 714 v. Chr.* Münster: Ugarit.

Novotny, J. 2014. *Selected Royal Inscriptions of Assurbanipal: L³, L⁴, LET, Prism I, Prism T, and Related Texts.* State Archives of Assyria Cuneiform Texts 10. Winona Lake, IN: Eisenbrauns.

Oberhuber, K. 1960. *Sumerische und akkadische Keilschriftdenkmäler des Archäologischen Museums zu Florenz.* Innsbrucker Beiträge zur Kulturwissenschaft. Institut für Sprachwissenschaft der Universität Innsbruck.

Parpola, S., and K. Watanabe. 1988. *Neo-Assyrian Treaties and Loyalty Oaths.* State Archives of Assyria 2. Helsinki University Press.

Pinches, T. G. 1896. *Cuneiform Texts from Babylonian Tablets in the British Museum 2.* London: British Museum.

———. 1982. *Cuneiform Texts from Babylonian Tablets in the British Museum 55.* London: British Museum.

———. 1982. *Cuneiform Texts from Babylonian Tablets in the British Museum 56.* London: British Museum.

Pomponio, F., G. Visicato, and A. Westenholz. 2006. *Le Tavolette Cuneiformi di Adab delle Collezioni della Banca d'Italia.* Rome: Banca d'Italia.

Postgate, J. N. 1973. *The Governor's Palace Archive.* London: British School of Archaeology in Iraq.

Rainey, A. 2015. *The El-Amarna Correspondence,* 2 vols. Leiden: Brill.

Reiner, E., and M. Civil. 1974. *The Series ḪAR-ra = ḫubullu: Tablets XX-XXIV (MSL 11).* Rome: Pontificium Institutum Biblicum.

Roth, M.T. 1989. *Babylonian Marriage Agreements: 7th–3rd centuries BC.* Neukirchen-Vluyn: Neukirchener.

Salvini, M. 1998. "I Testi Cuneiformi delle Campagne 1989 e 1993 a Tell Barri/Kaḫat." In P.E. Pecorella (ed.), *Tell Barri/Kaḫat II*, 187–98. Rome: CNR, Istituto per gli studi micenei ed egeo-anatolici.

Scurlock, J. 2014. *Sourcebook for Ancient Mesopotamian Medicine.* Atlanta: Society of Biblical Literature.

Smith, S. 1923. *Cuneiform Texts from Babylonian Tablets in the British Museum 37.* London: British Museum.

Strassmaier, J.N. 1887. *Inschriften von Nabonidus, König von Babylon.* Leipzig: Pfeiffer.

———. 1890. *Inschriften von Cambyses, König von Babylon.* Leipzig: Pfeiffer.

———. 1897. *Inschriften von Darius, König von Babylon.* Leipzig: Pfeiffer.

Tadmor, H., and S. Yamada. 2011. *The Royal Inscriptions of Tiglath-Pileser III and Shalmaneser V, Kings of Assyria.* Winona Lake, IN: Eisenbrauns.

Thureau-Dangin, F. 1921. *Rituels accadiens.* Paris: Ernest Leroux.

Walker, C.B.F., and M. Dick. 2001. *The Induction of the Cult Image in Ancient Mesopotamia: The Mesopotamian Mīs-Pî Ritual.* State Archives of Assyria Literary Texts 1. Helsinki: Neo-Assyrian Text Corpus Project.

Westenholz, J.G. 1997. *Legends of the Kings of Akkade.* Winona Lake, IN: Eisenbrauns.

Zawadzki, S. 2013. *Garments of the Gods: Studies on the Textile Industry and the Pantheon of Sippar according to the Texts from the Ebabbar Archive. Vol. II: Texts.* Academic Press Fribourg.

Greek and Latin Texts

Aelian, *De natura animalium,* trans. A. Zucker. Paris: Belles Lettres, 2001–02.

Aelian, *Varia historia,* trans. G. Nigel. Harvard University Press, 1997.

Aeschylus, *Suppliants,* trans. A.H. Sommerstein. Harvard University Press, 2009.

Aetius of Amida, *Libri medicinales,* ed. A. Olivieri. Berlin: Akademie, 1935–50.

Agatharchides of Knidos, *On the Erythraean Sea,* trans. S.M. Burstein. London: Hakluyt Society, 1989.

Ammianus Marcellinus, *Res gestae,* trans. J.C. Rolfe. Harvard University Press, 1939–50.

Antatticista (e cod. Coislin. 345), ed. I. Bekker. Berlin: Nauck, 1814.

Apicius, *De re coquinaria,* trans. C. Grocock, and S. Grainger. Totnes: Prospect Books, 2006.

Archigenes of Apamea, *Medical Fragments,* ed. C. Brescia. Naples: Libreria Scientifica Editrice, 1955.

Aretaeus of Cappadocia, *On Therapy of Acute and Chronic Diseases,* ed. C. Hude. Berlin: Akademie, 1958.

Aristotle, *Historia animalium,* trans. A.L. Peck and D.M. Balme. Harvard University Press, 1965–91.

Arrian, *Anabasis Alexandri,* trans. P.A. Brunt. Harvard University Press, 1976–83.

Arrian, *Indica,* trans. P.A. Brunt. Harvard University Press, 1983.

Athenaeus, *Deipnosophistai (The Learned Banqueters),* trans. S.D. Olson. Harvard University Press, 2007–12.

Celsus, *On Medicine,* trans. W.G. Spencer. Harvard University Press, 1935–38.

Columella, *On Agriculture,* trans. H.B. Ash et al. Harvard University Press, 1941–55.

Cosmas Indicopleustes, *Christian Topography,* trans. W. Wolska-Conus. Paris: Cerf, 1968–73.
Ctesias of Cnidus, *Indica,* trans. D. Lenfant. Paris: Belles Lettres, 2004.
Curtius Quintus, *History of Alexander,* trans. J.C. Rolfe. Harvard University Press, 1946.
Diocletian's Edict on Prices, ed. S. Lauffer. Berlin: De Gruyter, 1971.
Diodorus Siculus, *Library of History,* trans. C.H. Oldfather et al. Harvard University Press, 1933–67.
Dioscorides, *De materia medica,* trans. L.Y. Beck. Hildesheim: Georg Olms, 2011.
Eusebius, *Praeparatio evangelica,* trans. E.H. Gifford. Oxford: Clarendon Press, 1903.
Florus, *Epitome of Roman History,* trans. E.S. Forster. Harvard University Press, 1929.
Galen, *Antidotes,* ed. C.G. Kühn. Leipzig: Cnoblochius, 1827.
Galen, *Composition of Medicines according to Places,* ed. C.G. Kühn. Leipzig: Cnoblochius, 1826–27.
Galen, *Glossarium,* ed. C.G. Kühn. Leipzig: Cnoblochius, 1830.
Galen, *On the Powers of Simple Medicines,* ed. C.G. Kühn. Leipzig: Cnoblochius, 1833.
Galen, *On the Properties of Foodstuffs,* trans. O. Powell. Cambridge University Press, 2003.
Geoponica, trans. A. Dalby. Totnes: Prospect Books, 2011.
Greek Anthology, trans. W.R. Paton. Harvard University Press, 1916–18.
Herodotus, *The Histories,* trans. P-E. Legrand. Paris: Belles Lettres, 1932–54.
Hesiod, *Theogony,* trans. G.W. Most. Harvard University Press, 2007.
Hesychius, *Lexicon,* ed. M. Schmidt. Jena: Sumptibus F. Maukii, 1858–68.
Hippocrates, *Complete Works,* trans. É. Littré. Paris: Baillière, 1839–61.
Historia Augusta, trans. D. Magie. Harvard University Press, 1921–32.
Homer, *Odyssey,* trans. A.T. Murray and G.E. Dimock. Harvard University Press, 1995.
Horace, *Satirae,* trans. H.R. Fairclough. Harvard University Press, 1926.
Isidore of Seville, *Etymologiae,* trans. S.A. Barney et al. Cambridge University Press, 2006.
Josephus, *Jewish Antiquities,* trans. H. St. J. Thackeray. Harvard University Press, 1930–65.
Justinus, *Epitome of the Philippic History of Pompeius Trogus,* trans. J.C. Yardley. Atlanta, GA: Scholars Press, 1994.
Macrobius, *Saturnalia,* trans. R.A. Kaster. Harvard University Press, 2011.
Martial, *Epigrams,* trans. D.R.S. Bailey. Harvard University Press, 1993.
Metz Epitome, ed. P.H. Thomas. Leipzig: Teuber, 1966.
Nicander of Colophon, *Alexipharmaca,* trans. A.S.F. Gow and A.F. Scholfield. Cambridge University Press, 1953.
Nicolaus of Damascus, *De plantis,* trans. H.J. Drossaart-Lulofs and E.L.J. Poortman. Amsterdam: North-Holland, 1989.
Nonius Marcellus, *De compendiosa doctrina,* ed. W.M. Lindsay. Leipzig: Teubner, 1903.
Oribasius, *Collectiones medicae,* ed. J. Raeder. Leipzig: Teubner, 1928–33.
Oribasius, *Synopsis ad Eustathium,* ed. J. Raeder. Leipzig: Teubner, 1926.
Orosius, *Historiarum adversum paganos libri VII,* trans. A.T. Fear. Liverpool University Press, 2010.
Palladius, *Opus agriculturae,* trans. J.G. Fitch. Totnes: Prospect Books, 2013.
Pausanias, *Description of Greece,* trans. W.H.S. Jones. Harvard University Press, 1918–35.
Periplus Mari Erythraei, trans. L. Casson. Princeton University Press, 1989.
Petronius, *Satyricon,* trans. M. Heseltine. Harvard University Press, 1913.
Philostratus, *Life of Apollonius of Tyana,* trans. C.P. Jones. Harvard University Press, 2005.

Phrynichus, *Praeparatio sophistica,* ed. I. de Borries. Leipzig: Teubner, 1911.
Pliny, *Natural History,* trans. J. André et al. Paris: Belles Lettres, 1947–2015.
Plutarch, *Life of Alexander,* trans. B. Perrin. Harvard University Press, 1919.
Plutarch, *Life of Sulla,* trans. B. Perrin. Harvard University Press, 1916.
Plutarch, *Table Talk,* trans. P. A. Clement et al. Harvard University Press, 1961–69.
Pollux, *Onomasticon,* ed. E. Bethe. Leipzig: Teubner, 1900–31.
Pomponius Mela, *Chorography,* trans. F. E. Romer. University of Michigan Press, 1998.
Posidippus of Pella, *Lithiká,* trans. C. Austin and G. Bastianini. Milan: Edizioni Universitarie di Lettere Economia Diritto, 2002.
Pseudo-Aristotle, *On Marvellous Things Heard,* trans W. S. Hett. Harvard University Press, 1936.
Pseudo-Aristotle, *Problems,* trans. R. Mayhew. Harvard University Press, 2011.
Ptolemy, *Geography,* trans. A. Stückelberger and G. Graßhoff. Basel: Schwabe, 2006.
Scholia to Aristophanes, ed. W. J. W. Koster and D. Holwerda. Groningen: Bouma's Boekhuis and Egbert Forsten, 1960–2007.
Servius, *Commentaries on Vergil,* ed. G. Thilo and H. Hagen. Leipzig: Teubner, 1878–1902.
Sextus Empiricus, *Outlines of Pyrrhonism,* trans. R. G. Bury. Harvard University Press, 1933.
Solinus, *Collectanea rerum mirabilium,* trans. K. Brodersen. Darmstadt: Wissenschaftliche Buchgesellschaft, 2014.
Stephanus of Byzantium, *Ethnika,* ed. M. Billerbeck. Berlin: De Gruyter, 2008–17.
Strabo, *Geography,* trans. S. Radt. Göttingen: Vandenhoeck and Ruprecht, 2002–11.
Suda, trans. D. Whitehead et al., 2000–16. Suda On Line (http://www.stoa.org/sol/).
Suetonius, *Lives of the Caesars,* trans. J. C. Rolfe. Harvard University Press, 1914.
Theodorus Priscianus, *Theodori Prisciani Euporiston Libri III,* ed. V. Rose. Leipzig: Teubner, 1894.
Theophrastus, *Enquiry into Plants,* trans. S. Amigues. Paris: Belles Lettres, 1988–2006.
Theophrastus, *De causis plantarum,* trans. B. Einarson and G. K. K. Link. Harvard University Press, 1976–90.
Vergil, *Ecologues,* ed. R. Coleman. Cambridge University Press, 1977.
Vergil, *Georgics,* ed. R. F. Thomas. Cambridge University Press, 1988.
Vinidarius, *Excerpta Apicii,* trans. C. Grocock, and S. Grainger. Totnes: Prospect Books, 2006.
Xenophon, Agesilaus, trans. E. C. Marchant and G. W. Bowersock. Harvard University Press, 1968.
Xenophon, *Cyropaedia,* trans. M. Bizos. Paris: Belles Lettres, 1971–78.
Xenophon, *Oikonomikos,* trans. E. C. Marchant and J. Henderson. Harvard University Press, 2013.
Zosimus, *New History,* trans. F. Paschoud. Paris: Belles Lettres, 1971–89.

Greek and Latin Epigraphic and Papyrological Sources

Bernard, P., G. J. Pinault, and G. Rougemont. 2004. "Deux nouvelles inscriptions grecques de l'Asie centrale." *Journal des savants* 2.1, 227–356.
Bowman, A. K., and J. D. Thomas. 1994. *The Vindolanda Writing-Tablets.* London: British Museum.
Corpus Inscriptionum Latinarum IV: Inscriptiones parietariae Pompeianae, Herculanenses, Stabianae, ed. C. Zangemeister et al. Berlin: Georg Reimer and De Gruyter, 1871-2011.
Diocletian's Edict on Prices, trans. S. Lauffer. Berlin: De Gruyter, 1971.

Lehoux, D. 2005. "The Parapegma Fragments from Miletus." *Zeitschrift für Papyrologie und Epigraphik* 152, 125–40.
Rougemont, G. 2012. *Inscriptions grecques d'Iran et d'Asie central.* London: School of Oriental and African Studies.

Hebrew and Aramaic Texts

Babylonian Talmud, trans. I. Epstein. London: Soncino, 1935–48.
Ecclesiastes Rabah, trans. A. Cohen. London: Soncino Press, 1939.
Genesis Rabah, trans. J. Neusner. Atlanta: Scholars Press, 1985.
Goitein, S. D., and M. Friedman. 2008. *India Traders of the Middle Ages: Documents from the Cairo Geniza "India Book."* Leiden: Brill.
Hebrew Bible, ed. K. Elliger et al. Stuttgart: Deutsche Bibelstiftung, 1968–76.
Jerusalem Talmud, trans. H. W. Guggenheimer. Berlin: De Gruyter, 1999–2015.
Leviticus Rabah, trans. J. Israelstam and J. Slotki. London: Soncino Press, 1939.
Midrash Hagadol, Deuteronomy, ed. S. Fisch. Jerusalem: Mosad Harav Kook, 1972.
Mishnah, trans. H. Danby. Oxford: Oxford University Press, 1933.
Naveh, J., and S. Shaked. 2012. *Aramaic Documents from Ancient Bactria.* London: Khalili Family Trust.
Targum Onkelos, Leviticus, trans. I. Drazin and S.M. Wagner. Jerusalem: Gefen Publishing.
Targum Song of Songs, trans. P.S. Alexander. Collegeville, MI: Liturgical Press, 2003.
Tosefta, trans. J. Neusner. New York: Ktav, 1977–86.

Arabic Texts

al-Baghdādī, 'Abd al-Laṭīf, *Kitāb al-ifādah,* trans. K. F. Zand et al. London: George Allen and Unwin, 1965.
al-Dīnawarī, Abū Ḥanīfa, *Kitāb al-nabāt,* trans. M. Hamidullah. Cairo: Institut Français d'Archéologie Orientale, 1973.
al-Hamaḏānī, Ibn al-Faqīh, *Mukhtaṣar Kitāb al-buldān,* trans. H. Massé. Damascus: Institut Français de Damas, 1973.
al-Iṣṭakhrī, *Kitāb al-masalik wa-l-mamālik,* ed. I. Afshar. Tehran: Bungāh-i Tarjumah va Nashr-i Kitāb, 1961.
al-Jāḥiẓ, 'Amr ibn Baḥr, *Kitāb al-bukhalā',* ed. T. Hajiri. Cairo: Dār al-Kātib al-Miṣrī, 1948.
al-Mas'ūdī, *Murūj al-dhahab wa-ma'ādin al-jawhar,* ed. C. Pellat. Beirut: al-Jāmi'a al- Lubnāniyya, 1966–79.
al-Muqaddasī, *Aḥsan al-taqāsīm fī ma 'rifat al-aqālīm,* trans. B. A. Collins. Reading: Garnet, 1994.
al-Nuwayrī, *Nihāyat al-arab fī funūn al-adab.* Cairo: Dār al-Kutub al-Miṣriyya, 1923–55.
al-Shidyāq, Ahmad Fāris, *Al-sāq 'alā al-sāq,* trans. H. Davies. New York University Press, 2013-14.
al-Tha'ālibī, *Ta'rīkh ghurar al-siyar,* trans. H. Zotenberg. Paris: Imprimerie Nationale, 1900.
Ibn al-'Awwām, *Kitāb al-filāḥa,* ed. J. Clément-Mullet. Paris: Maisonneuve, 1864–66.
Ibn al-Ḥājj, *Al-madkhal,* ed. H. Ahmad. Beirut: Al-Maktaba l-'Aṣriyya, 2005.
Ibn Baṭṭūṭa, *al-Riḥla,* trans. H. Gibb and C. Beckingham. Cambridge University Press, 1958–2000.
Ibn Ḥawqal, *Kitāb ṣūrat al-arḍ,* ed. J. H. Kramers. Leiden: Brill, 1938–39.

Ibn Qutaybah, *'Uyūn al-akhbār.* Cairo: Dār al-Kutub, 1925–30.
Ibn Sayyār al-Warrāq, *Kitāb al-Ṭabīkh,* trans. N. Nasrallah. Leiden: Brill, 2007.
Ibn Waḥshīyah, *Al-filāḥah al-nabaṭīyah,* ed. T. Fahd. Damascus: Institut Français de Damas, 1993–98.
Yāqūt al-Ḥamawī, *Mu'jam al-buldān,* ed. F. Wustenfeld. Leipzig: F. A. Brockhaus, 1866–73.

Persian Texts

Ferdowsī, *Šāh-nāma,* trans. D. Davis. New York: Penguin, 2016.
Ḥudūd al-'ālam, trans. V. Minorsky. London: Luzac & Co., 1937.
Niẓāmī, *Makhzan al-Asrār,* trans. G. H. Darab. London: Arthur Probsthain, 1945.
Sa'dī, *Golestān,* trans. W. M. Thackston. Bethesda, MD: Ibex, 2008.

Sanskrit Texts

Amarasiṃha, *Nāmaliṅgānuśāsanam,* ed. V. Jhalakikar. New Delhi: Eastern Book Linkers, 2012.
Āśvalāyanaśrautasūtram, ed. R. Vidyaratna. Calcutta: Asiatic Society of Bengal, 1874.
Atharvaveda Saṃhitā, trans. R. T. H. Griffith. Benares: E. J. Lazarus & Co, 1916.
Baudhāyanaśrautasūtram, trans. C. G. Kashikar. New Delhi: Motilal Banarsidass, 2003.
Gītagovinda, trans. B. S. Miller. Columbia University Press, 1977.
Gobhilagṛhyasūtram, ed. C. Bhattacharya. New Delhi: Munshiram Manoharlal, 1982.
Hemacandra, *Sthavirāvalīcarita,* trans. C. C. Fynes. Oxford University Press, 1998.
Jaiminīyagṛhyasūtram, trans. W. Caland. Lahore: Punjab Sanskrit Series, 1922.
Kālidāsa, *Abhijñānaśākuntalam,* trans. S. Vasudeva. New York University Press, 2006.
Kālidāsa, *Mālavikāgnimitram,* trans. M. R. Kale. New Delhi: Motilal Banarsidass, 1960.
Kauṭilya, *Arthaśāstra,* trans. P. Olivelle. Oxford University Press, 2013.
Lāṭyāyanaśrautasūtram, trans. H. G. Ranade. New Delhi: Motilal Banarsidass, 1998.
Mahābhārata, ed. V. S. Sukthankar et al. Pune: Bhandarkar Oriental Institute, 1933–66.
Matsya-Purāṇa, trans. K. Joshi. New Delhi: Parimal, 2007.
Pāṇini, *Aṣṭādhyāyī,* trans. S. D. Joshi and J. A. F. Roodbergen. New Delhi: Sahitya Akademi, 1991–2006.
Ṛgveda, trans. S. W. Jamison and J. P. Brereton. Oxford University Press, 2014.
Suśruta, *Suśruta-Saṃhitā,* trans. P. V. Sharma. Varanasi: Chaukhambha Visvabharati, 1999–2001.
Taittirīya Saṃhitā of the Black Yajurveda, trans. A. B. Keith. Harvard University Press, 1914.
Vallabhadeva, *Subhāṣitāvalī,* trans. A. N. D. Haksar. New Delhi: Penguin, 2007.
Vālmīki, *Rāmāyaṇa,* ed. G. H. Bhatt and U. P. Shah. Baroda: Oriental Institute, 1960–75.
Vātsyāyana, *Kāmasūtra,* trans. W. Doniger and S. Kakar. Oxford University Press, 2002.
Viṣṇu-Purāṇa, ed. J. Vidyasagara. Calcutta: Saraswati Press, 1882.
Viṣṇuśarman, *Pañcatantra,* trans. P. Olivelle. New York University Press, 2006.

Prākrit Texts

Apadāna, ed. M. E. Lilley. London: Pali Text Society, 1925–27.
Āyāraṅga-sutta, trans. H. Jacobi. Oxford: Clarendon Press, 1884.
Bloch, J. 2007. *Les inscriptions d'Asoka.* Paris: Belles Lettres, 2007.
Dīghanikāya, ed. T. W. Rhys Davids and J. E. Carpenter. London: Pali Text Society, 1890–1911.

Jātaka, trans. E. B. Cowell et al. Cambridge University Press, 1895–1907.
Mahāniddesa, ed. L. L. Vallée Poussin and E. J. Thomas. London: Pali Text Society, 1916–17.
Mahāvagga, trans. I. B. Horner. London: Pali Text Society, 1993.
Mahāvaṃsa, trans. W. Geiger. London: Pali Text Society, 1912.
Nāyādhammakahāo, trans. A. Muni. Delhi: Padma Prakashan, 1996.
Saṃyuttanikāya, trans. B. Bodhi. Somerville, MA: Wisdom, 2000.
Sanghadāsa, *Bṛhatkalpabhāṣya*, ed. W. B. Bollée. Stuttgart: Franz Steiner, 1998.
Theragāthā, trans. K. R. Norman. London: Pali Text Society, 2007.

Tamil Texts

Akanānūru, ed. R. Jayabal. Chennai: New Century Book House, 2007.

Chinese Texts

Sima Qian, *Shiji*, trans. W. Nienhauser et al. Indiana University Press, 1994–.

SECONDARY WORKS

Abivardi, C. 2008. "Pests, Agricultural." *Encyclopaedia Iranica*, online edition (http://www.iranicaonline.org/articles/pests-agricultural).

Achtman, M. 2017. "Multiple Time Scales for Dispersals of Bacterial Disease over Human History." In N. Boivin, R. Crassard, and M. Petraglia (eds.), *Human Dispersal and Species Movement: From Prehistory to the Present*, 454–76. Cambridge University Press.

Ahmadinia, R., and A. Shishegar. 2019. "Jubaji, a Neo-Elamite (Phase IIIB, 585–539 BC) Tomb in Ramhurmuz, Khuzestan." *Iran* 572, 142–74.

A'lam, H. 1989. "Botanical Studies on Iran. I: The Greco-Islamic Tradition." *Encyclopedia Iranica*, 4.4, 390–401 (http://www.iranicaonline.org/articles/botanical-studies-on-iran-).

———. 2011a. "Citrus Fruits." *Encyclopaedia Iranica*, 5.6, 637–46 (http://www.iranicaonline.org/articles/citrus-fruits-).

———. 2011b. "Cucumber." *Encyclopaedia Iranica*, 6.5, 450–52 (http://www.iranicaonline.org/articles/cucumber-cucumis-sativus-l).

Albala, K. 2002. *Eating Right in the Renaissance*. University of California Press.

Albenda, P. 2005. *Ornamental Wall Painting in the Art of the Assyrian Empire*. Leiden: Brill.

Albiani, M. G. 2006. "Myrinus." *Brill's New Pauly*, online edition.

Allchin, F. R., et al. 1995. *The Archaeology of Early Historic South Asia: The Emergence of Cities and States*. Cambridge University Press.

Almathen, F., et al. 2016. "Ancient and Modern DNA Reveal Dynamics of Domestication and Cross-Continental Dispersal of the Dromedary." *Proceedings of the National Academy of Sciences* 11324, 6707–12.

Alt, K. W., et al. 2003. "Climbing into the Past: First Himalayan Mummies Discovered in Nepal." *Journal of Archaeological Science* 30, 1529–35.

Altaweel, M., and A. Squitieri. 2018. *Revolutionizing a World: From Small States to Universalism in the Pre-Islamic Near East*. London: UCL Press.

Altaweel, M., et al. 2019. "New Insights on the Role of Environmental Dynamics Shaping Southern Mesopotamia: From the Pre-Ubaid to the Early Islamic Period." *Iraq* 81, 23–46.

Altheim, F., and R. Stiehl. 1964. *Die Araber in der alten Welt.* Vol. 1. Berlin: De Gruyter.
Alvarez-Mon, J. 2010. *The Arjān Tomb: At the Crossroads between the Elamite and the Persian Empires.* Leuven: Peeters.
———. 2015. "The Introduction of Cotton into the Near East: A View from Elam." *International Journal of the Society of Iranian Archaeologists* 1.2, 43–54.
Amar, Z., and E. Lev. 2017. *Arabian Drugs in Medieval Mediterranean Medicine.* Edinburgh University Press.
Amichay, O., D. Ben-Ami, Y. Tchekhanovets, R. Shahack-Gross, D. Fuks, and E. Weiss. 2019. "A Bazaar Assemblage: Reconstructing Consumption, Production and Trade from Mineralised Seeds in Abbasid Jerusalem." *Antiquity* 93, 199–217.
Amigues, S. 2007. "L'exploitation du monde végétal en Grèce classique et hellénistique. Essai de synthèse." *Topoi* 15.1, 75–125.
Amrhein, A. 2015. "Neo-Assyrian Gardens: A Spectrum of Artificiality, Sacrality and Accessibility." *Studies in the History of Gardens and Designed Landscapes* 35.2, 91–114.
Andersen, S. F., H. Strehle, M. Tengberg, and M. I. Salman. 2004. "Two Wooden Coffins from the Shakhoura Necropolis, Bahrain." *Arabian Archaeology and Epigraphy* 15.2, 219–28.
André, J. 1981. *L'Alimentation et la cuisine à Rome.* Paris: Belles Lettres.
Andrés-Toledo, M. A. 2013. "Textiles in Zoroastrianism." In C. A. Giner, J. O. García, and M. J. M. García (eds.), *Luxury and Dress: Political Power and Appearance in the Roman Empire and its Provinces*, 23–30. University of Valencia.
Andrews, A. C. 1956. "Melons and Watermelons in the Classical Era." *Osiris* 12, 368–75.
———. 1961. "Acclimatization of Citrus Fruits in the Mediterranean Region." *Agricultural History* 35.1, 35–46.
Angelakıs, A. N., et al. 2020. "Irrigation of World Agricultural Lands: Evolution through the Millennia." *Water* 12(5), 1–50.
Anthony, D. W. 2007. *The Horse, the Wheel, and Language: How Bronze-Age Riders from the Eurasian Steppes Shaped the Modern World.* Princeton University Press.
Anthony, D. W., and D. R. Brown. 2011. "The Secondary Products Revolution, Horse-Riding, and Mounted Warfare." *Journal of World Prehistory* 24, 131–60.
Appadurai, A. 1981. "Gastro-Politics in Hindu South Asia." *American Ethnologist* 8.3, 494–511.
———. 1986. "Introduction: Commodities and the Politics of Value." In A. Appadurai (ed.), *The Social Life of Things: Commodities in Cultural Perspective*, 3–63. Cambridge University Press.
Arfa'i, A. 1999. "La grande route Persépolis-Suse: Une lecture des tablettes provenant des Fortifications de Persépolis." *Topoi* 9, 33–45.
Arias, B. A., and L. Ramón-Laca. 2005. "Pharmacological Properties of Citrus and Their Ancient and Medieval Uses in the Mediterranean Region." *Journal of Ethnopharmacology* 97, 89–95.
Arobba, D., F. Bulgarelli, C. Siniscalco, and R. Caramiello. 2013. "Roman Landscape and Agriculture on the Ligurian Coast through Macro and Microremains from a Vada Sabatia Well (Vado Ligure, Italy)." *Environmental Archaeology* 18.2, 114–31.
Arundhati, P. 1994. *Royal life in Mānasollāsa.* Delhi: Sundeep Prakashan.
Aruz, J. 1992. "Figure of a Seated Monkey" (catalogue entry). In P. O. Harper, J. Aruz, and F. Tallon (eds.), *The Royal City of Susa: Ancient Near Eastern Treasures in the Louvre*, 97–98. New York: Metropolitan Museum of Art.

———. 2003. "Art and Interconnections in the Third Millennium BC." In J. Aruz and R. Wallenfels, *Art of the First Cities: The Third Millennium BC from the Mediterranean to the Indus*, 239–50. New York: Metropolitan Museum of Art.

Ash, T. 1978. *An Introduction to the Archaeology, Ethnography and History of Ras al-Khaimah.* National Museum of Ras al-Khaimah.

Asher, D. 2015. *The Art of Natural Cheesemaking.* White River Junction, VT: Chelsea Green.

Asouti, E., and D.Q. Fuller. 2008. *Trees and Woodlands of South India.* Walnut Creek, CA: Left Coast Press.

Aston, B.G., J.A. Harrell, and I. Shaw. 2000. "Stone." In P.T. Nicholson and I. Shaw (eds.), *Ancient Egyptian Materials and Technology*, 5–77. Cambridge University Press.

Atwal, A.S. 1986. *Agricultural Pests of India and South-East Asia.* New Delhi: Kalyani.

Aubaile-Sallenave, F. 1984. "L'agriculture musulmane aux premiers temps de la conquête: apports et emprunts, à propos de Agricultural innovation in the early Islamic world de Andrew M. Watson." *Journal d'agriculture traditionnelle et de botanique appliquée* 31(3–4), 245–56.

———. 1988. "Bādenjān 'eggplant, aubergine.'" *Encyclopedia Iranica* 3, 366–68 (https://iranicaonline.org/articles/badenjan-egeplant-aubergine).

Avital, A., and H.S. Paris. 2014. "Cucurbits Depicted in Byzantine Mosaics from Israel, 350–600 CE." *Annals of Botany* 114, 203–22.

Avnaim-Katav, S., A. Almogi-Labin, A. Schneider-Mor, O. Crouvi, A.A. Burke, K.V. Kremenetski, and G.M. MacDonald. 2019. "A Multi-Proxy Shallow Marine Record for Mid-to-Late Holocene Climate Variability, Thera Eruptions and Cultural Change in the Eastern Mediterranean." *Quaternary Science Reviews* 204, 133–48.

Avni, G. 2018. "Early Islamic Irrigated Farmsteads and the Spread of Qanats in Eurasia." *Water History* 10(4), 313–38.

Ayyar, V.R., and K.P. Aithal. 1964. "*Kārpāsa* cotton: its origin and spread in Ancient India." *Adyar Library Bulletin* 28, 1–40.

Badian, E. 1998. "The King's Indians." In W. Will (ed.), *Alexander der Grosse: Eine Welteroberung und ihr Hintergrund*, 205–24. Bonn: Rudolf Habelt.

Bagg, A.M. 2017. "Watercraft at the Beginning of History: The Case of Third-Millennium Southern Mesopotamia." In P. De Souza and P. Arnaud (eds.), *The Sea in History: The Ancient World*, 127–37. Woodbridge: Boydell Press.

Bagnall, R.S. 2008. "SB 6.9025, Cotton, and the Economy of the Small Oasis." *Bulletin of the American Society of Papyrologists* 45, 21–30.

Bailey, H.W., ed. 1976. *Indo-Scythian Studies: Khotanese Texts VI: Prolexis to the Book of Zambasta.* Cambridge University Press.

———. 1979. *Dictionary of Khotan Saka.* Cambridge University Press.

Baldina, E.A., J. De Leeuw, A.K. Gorbunov, I.A. Labutina, A.F. Zhivogliad, and J.F. Kooistra. 1999. "Vegetation Change in the Astrakhanskiy Biosphere Reserve (Lower Volga Delta, Russia) in relation to Caspian Sea Level Fluctuation." *Environmental Conservation* 26.3, 169–78.

Balland, D. 1989. "Barley." *Encyclopedia Iranica,* 3, 802–05 (http://www.iranicaonline.org/articles/barley).

Balogh, C. 2011. *The Stele of YHWH in Egypt: The Prophecies of Isaiah 18–20 concerning Egypt and Kush.* Leiden: Brill.

Barber, E. J. W. 1991. *Prehistoric Textiles: The Development of Cloth in the Neolithic and Bronze Ages with Special Reference to the Aegean.* Princeton University Press.

Barjamovic, G. 2011. *A Historical Geography of Anatolia in the Old Assyrian Colony Period.* Copenhagen: Museum Tusculanum Press.

———. 2018. "Interlocking Commercial Networks and the Infrastructure of Trade in Western Asia during the Bronze Age." In K. Kristiansen, T. Lindkvist, and J. Myrdal (eds.), *Trade and Civilisation: Economic Networks and Cultural Ties from Prehistory to the Early Modern Era*, 113–42. Cambridge University Press.

Barker, G. 2006. *The Agricultural Revolution in Prehistory.* Oxford University Press.

Barkley, N. A., M. L. Roose, R. R. Krueger, and C. T. Federici. 2006. "Assessing Genetic Diversity and Population Structure in a Citrus Germplasm Collection Utilizing Simple Sequence Repeat Markers (SSRs)." *Theoretical and Applied Genetics* 112.8, 1519–31.

Barkova, L., and N. V. Polosmak. 2005. *Kostjum i tekstil' pazyrykcev Altaja (IV-III vv. do n.è.).* Novosibirsk: InFolio Press.

Barrett, H. C., and A. M. Rhodes. 1976. "A Numerical Taxonomic Study of Affinity Relationships in Cultivated Citrus and Its Close Relatives." *Systematic Botany* 1.2, 105–36.

Bar-Yosef, O. 1998. "On the Nature of Transitions: The Middle to Upper Palaeolithic and the Neolithic Revolution." *Cambridge Archaeological Journal* 82, 141–63.

Bastiaens, J., and C. Verbruggen. 1995. "Archeobotanisch onderzoek van het Romeinse kamp van Maldegem-Vake (Oost-Vlaanderen, België). Macroresten van de opgravingscampagnes 1986 en 1987." *Handelingen der Maatschappij voor Geschiedenis en Oudheidkunde te Gent* 49.1, 33–44.

Bates, J. 2019. "Oilseeds, Spices, Fruits and Flavour in the Indus Civilisation." *Journal of Archaeological Science: Reports* 24, 879–87.

———., C. A. Petrie, and R. N. Singh. 2017. "Approaching Rice Domestication in South Asia: New Evidence from Indus Settlements in Northern India." *Journal of Archaeological Science* 78, 193–201.

Baumgarten, A. I., S. D. Sperling, and S. Sabar. 2007. "Scroll of Esther." In M. Berenbaum and F. Skolnik (eds.), *Encyclopedia Judaica* 16, 2nd ed., 215–20. Detroit: Macmillan Reference USA.

Bayer, R. J., et al. 2009. "A Molecular Phylogeny of the Orange Subfamily (Rutaceae: Aurantioideae) using Nine cpDNA Sequences." *American Journal of Botany* 96.3, 668–85.

Beaulieu, P. A. 1992. "Kissik, Dūru and Udannu." *Orientalia* 61.4, 400–24.

———. 1994. "Antiquarianism and the Concern for the Past in the Neo-Babylonian Period." *Bulletin of the Canadian Society for Mesopotamian Studies* 28, 37–42.

———. 2013. "Mesopotamian Antiquarianism: From Sumer to Babylon." In A. Schnapp et al. (eds.), *World Antiquarianism: Comparative Perspectives*, 121–39. Los Angeles, CA: Getty.

Beaux, N. 1990. *Le cabinet de curiosités de Thoutmosis III: plantes et animaux du "Jardin botanique" de Karnak.* Leuven: Peeters.

Becker, C. 2008. "Die Tierknochenfunde aus Tall Seh Hamad/Dur-Katlimmu: eine zoogeographisch- haustierkundliche Studie." In H. Kühne (ed.), *Umwelt und Subsistenz der assyrischen Stadt Dur-Katlimmu am unteren Habur, Berichte der Ausgrabung von Tall she Hamad / Dur-Katlimmu 8*, 61–131. Wiesbaden: Harrassowitz.

Beckman, G. 2013. "Foreigners in the Ancient Near East." *Journal of the American Oriental Society* 133.2, 203–16.

Bedalov, M., and P. Küpfer. 2005. "Studies on the Genus *Arum* (Araceae)." *Bulletin de la Société Neuchâteloise des Sciences Naturelles* 128, 43–70.

Bedigian, D. 2003. "Sesame in Africa: Origin and Dispersals." In K. Neumann, A. Butler, and S. Kahlheber (eds.), *Food, Fuel and Fields: Progress in African Archaeobotany*, 17–36. Cologne: Heinrich Barth Institut.

Begley, V., and R. D. De Puma, eds. 1991. *Rome and India: the Ancient Sea Trade.* University of Wisconsin Press.

Bellina, B., et al. 2014. "The Development of Coastal Polities in the Upper Thai-Malay Peninsula in the Late First Millennium BCE." In N. Revire and S. A. Murphy (eds.), *Before Siam: Essays in Art and Archaeology*, 69–89. Bangkok: River Books.

Bellinger, L., and R. Pfister. 1962. "Textiles." In H. Colt et al. (eds.), *Excavations at Nessana (Auja Hafir, Palestine)*, 91–105. London: British School of Archaeology in Jerusalem.

Bellwood, P. 2004. *First Farmers: The Origins of Agricultural Societies.* Oxford: Blackwell.

Ben-Sasson, R. 2012. "Botanics and Iconography: Images of the Lulav and the Etrog, Ars Judaica." *Bar-Ilan Journal of Jewish Art* 8, 7–22.

Berthiaume, G. 1982. *Les rôles du mágeiros: études sur la boucherie, la cuisine et le sacrifice dans la Grèce ancienne.* Leiden: Brill.

Berthon, R., M. Mashkour, P. Burger, and C. Çakırlar. 2020. "Domestication, Diffusion and Hybridization of the Bactrian Camel." In D. Agut-Labordère and B. Redon (eds.), *Les vaisseaux du désert et des steppes: Les camélidés dans l'Antiquité*, 21–26. Lyon: MOM.

Betro, M. 1999. "Asoka in un testo letterario demotico?" *Studi Ellenistici* 12, 115–25.

Betts, A., K. van der Borg, A. de Jong, C. McClintock, and M. van Stryndonck. 1994. "Early Cotton in North Arabia." *Journal of Archaeological Science* 21, 489–99.

Bewley, J. D., M. Black, and P. Halmer. 2006. *The Encyclopedia of Seeds: Science, Technology and Uses.* Wallingford: CABI.

Bisht, I. S., K. V. Bhat, S. P. S. Tanwar, D. C. Bhandari, K. Joshi, and A. K. Sharma. 2004. "Distribution and Genetic Diversity of *Cucumis sativus* var. *hardwickii* (Royle) Alef. in India." *Journal of Horticultural Science and Biotechnology* 79, 783–91.

Biswas, A. K. 1993. "The Primacy of India in Ancient Brass and Zinc Metallurgy." *Indian Journal of History of Science* 28.4, 309–30.

Blackman, M. J., and S. Méry. 1999. "Les importations de céramiques harapéennes en Arabie orientale: état de la question." *Proceedings of the Seminar for Arabian Studies* 29, 7–28.

Blazek, V. 1999. "Elam: A Bridge between Ancient Near East and Dravidian India." *Archaeology and Language* 4, 48–87.

Bleibtreu, E. 1980. *Die Flora der neuassyrischen Reliefs: eine Untersuchung zu den Orthostatenreliefs des 9.-7. Jahrhunderts v. Chr.* Vienna: Verlag des Institutes für Orientalistik der Universität Wien.

Blench, R. 2003. "The Movement of Cultivated Plants between Africa and India in Prehistory." In K. Neumann, A. Butler, and S. Kahlheber (eds.), *Food, Fuel and Fields: Progress in African Archaeobotany*, 273–92. Cologne: Heinrich Barth Institut.

———. 2008. "Re-evaluating the Linguistic Prehistory of South Asia." In T. Osada and A. Uesugi (eds.), *Linguistics, Archaeology and the Human Past*, 159–78. Kyoto: Institute for Humanity and Nature.

Bloch, J. 1925. "Le nom du riz." In G. van Ouest (ed.), *Etudes Asiatiques*, Vol. 1, 37–47. Paris: L'École française d'Extrême-Orient.

Böck, B. 2011. "Sourcing, Organising and Administering Medicinal Ingredients." In K. Radner and E. Robson (eds.), *The Oxford Handbook of Cuneiform Culture*, 690–705. Oxford University Press.

———. 2013. *The Healing Goddess Gula: Towards an Understanding of Ancient Babylonian Medicine.* Leiden: Brill.

———. 2015. "Shaping Texts and Text Genres: On the Drug Lore of Babylonian Practitioners of Medicine." *Aula Orientalis* 33.1, 21–37.

Bodson, L. 1998. "Contribution à l'étude des critères d'appréciation de l'animal exotique dans la tradition grecque ancienne." In L. Bodson (ed.), *Les animaux exotiques dans les relations internationals: espèces, fonctions, significations*, 139–212. Université de Liège.

———. 1999. "Ancient Greek Views on the Exotic Animal." *Arctos* 32, 61–85.

Boisacq, E. 1938. *Dictionnaire étymologique de la langue grecque.* Paris: Klincksieck.

Boivin, N. 2017. "Proto-Globalisation and Biotic Exchange in the Old World." In N. Boivin, R. Crassard, and M. Petraglia (eds.), *Human Dispersal and Species Movement: From Prehistory to the Present*, 349–408. Cambridge University Press.

Boivin, N., R. Blench, and D. Q. Fuller. 2009. "Archaeological, Linguistic and Historical Sources on Ancient Seafaring: A Multidisciplinary Approach to the Study of Early Maritime Contact and Exchange in the Arabian Peninsula." In M. Petraglia and J. Rose (eds.), *The Evolution of Human Populations in Arabia: Paleoenvironments, Prehistory and Genetics*, 251–78. Dordrecht: Springer.

Boivin, N., A. Crowther, R. Helm, and D. Q. Fuller. 2013. "East Africa and Madagascar in the Indian Ocean World." *Journal of World Prehistory* 26.3, 213–81.

Boivin, N., A. Crowther, M. Prendergast, and D. Q. Fuller. 2014. "Indian Ocean Food Globalisation and Africa." *African Archaeological Review* 31.4, 547–81.

Boivin, N., and D. Q. Fuller. 2009. "Shell Middens, Ships and Seeds: Exploring Coastal Subsistence, Maritime Trade and the Dispersal of Domesticates in and around the Ancient Arabian Peninsula." *Journal of World Prehistory* 22.2, 113–80.

Boivin, N., D. Q. Fuller, and A. Crowther. 2012. "Old World Globalization and the Columbian Exchange: Comparison and Contrast." *World Archaeology* 44, 452–69.

———. 2015. "Old World Globalization and Food Exchanges." In M. C. Beaudry and K. B. Metheny (eds.), *Archaeology of Food: An Encyclopedia*, 350–56. Lanham, MD: Rowman and Littlefield.

Bonavia, E. 1894. *The Flora of the Assyrian Monuments and Its Outcomes.* Westminster: A. Constable.

Bongenaar, A. C. V. M. 1997. *The Neo-Babylonian Ebabbar Temple at Sippar: Its Administration and Its Prosopography.* Nederlands Historisch-Archeologisch Instituut te Istanbul.

Bonora, G. L. 2021. "The Oxus Civilization and the Northern Steppes." In B. Lyonnet and N. A. Dubova (eds.), *The World of the Oxus Civilization*, 734–75. Abingdon: Routledge.

Borger, R. 1971. "Getreide." In F. Ebeling et al. (eds.), *Reallexikon der Assyriologie und Vorderasiatischen Archäologie,* Vol. III, 308–11. Berlin: Walter de Gruyter.

Borgongino, M. 2006. *Archeobotanica: Reperti vegetali da Pompei e dal territorio vesuviano.* Rome: L'Erma di Bretschneider.

Borowski, O. 1998. *Every Living Thing: Daily Use of Animals in Ancient Israel.* Walnut Creek, CA: Rowman Altamira.

Borrell, J. S., et al. 2020. "The Climatic Challenge: Which Plants Will People Use in the Next Century?" *Environmental and Experimental Botany* 170, 1–14.

Bottema, S., and A. Sarpaki. 2003. "Environmental Change in Crete: A 9000-Year Record of Holocene Vegetation History and the Effect of the Santorini Eruption." *Holocene* 13(5), 733–49.

Bouby, L., and P. Marinval. 2002. "Les données encore partielles sur l'alimentation à Vieille-Toulouse." In J. M. Pailler (ed.), *Tolosa: Nouvelles recherches sur Toulouse et son territoire dans l'Antiquité*, 187–90. Ecole française de Rome.

———. 2004. "Fruits and Seeds from Roman Cremations in Limagne (Massif Central) and the Spatial Variability of Plant Offerings in France." *Journal of Archaeological Science* 31.1, 77–86.

Bouchaud, C., A. Clapham, C. Newton, G. Tallet, and U. Thanheiser. 2018. "Cottoning on to cotton (*Gossypium* spp.) in Arabia and Africa during Antiquity." In A. M. Mercuri et al. (eds.), *Plants and People in the African Past: Progress in African Archaeobotany.* Cham: Springer.

Bouchaud, C., J. Morales, V. Schram, and M. van der Veen. 2017. "The Earliest Evidence for Citrus in Egypt." In V. Zech-Matterne and G. Fiorentino (eds.), *AGRUMED: Archaeology and History of Citrus Fruit in the Mediterranean.* Naples: Centre Jean Bérard (https://books.openedition.org/pcjb/2179).

Bouchaud, C., M. Tengberg, and P. D. Prà. 2011. "Cotton Cultivation and Textile Production in the Arabian Peninsula during Antiquity: The Evidence from Madâ'in Sâlih (Saudi Arabia) and Qal'at al-Bahrain (Bahrain)." *Vegetation History and Archaeobotany* 20, 405–17.

Bouchaud, C., E. Yvanez, and J. P. Wild. 2019. "Tightening the Thread from Seed to Cloth. New Enquiries in the Archaeology of Old World Cotton. A Case for Inter-Disciplinarity." *Revue d'ethnoécologie* 15, 1–30.

Breniquet, C., and C. Michel, eds. 2014. *Wool Economy in the Ancient Near East and the Aegean: From the Beginnings of Sheep Husbandry to Institutional Textile Industry.* Oxford: Oxbow.

Briant, P. 2002. *From Cyrus to Alexander: A History of the Persian Empire*, trans. P. T. Daniels. Winona Lake, IN: Eisenbrauns.

———. 2009. "The Empire of Darius III in Perspective." In W. Heckel and L. A. Tritle (eds.), *Alexander the Great: A New History*, 141–70. Oxford: Blackwell.

Brigand, J. P., and P. Nahon. 2016. "Gastronomy and the Citron Tree (*Citrus medica* L.)." *International Journal of Gastronomy and Food Science* 3, 12–16.

Brinkman, J. A. 2017. "Babylonia under the Kassites: Some Aspects for Consideration." In A. Bartelmus and K. Sternitzke (eds.), *Karduniaš: Babylonien zur Kassitenzeit*, 1–44. Berlin: Walter de Gruyter.

Brite, B. B., and J. M. Marston. 2013. "Environmental Change, Agricultural Innovation, and the Spread of Cotton Agriculture in the Old World." *Journal of Anthropological Archaeology* 32, 39–53.

Brovarski, E., S. K. Doll, and R. E. Freed. 1982. *Egypt's Golden Age: The Art of Living in the New Kingdom, 1558–1085 B.C.* Boston: Museum of Fine Arts.

Brown, T., and K. Brown. 2011. *Biomolecular Archaeology: An Introduction.* Chichester: Wiley-Blackwell.

Brust, M. 2005. *Die indischen und iranischen Lehnwörter im Griechischen.* Institut für Sprachen und Literaturen der Universität Innsbruck.

Buccellati, G., and M. Kelly-Buccellati. 2002. "Tar'am-Agade, Daughter of Naram-Sin, at Urkesh." In L. Al-Gailani Werr et al. (eds.), *Of Pots and Plans: Papers on the Archaeology and History of Mesopotamia and Syria Presented to David Oates in Honour of his 75th Birthday*, 11–31. London: NABU.

Buchholz, H-G. 1988. "Archäologische Holzfunde aus Tamassos, Zypern." *Acta Praehistorica et Archaeologica* 20, 75–157.

Buckland, P.C., and E. Panagiotakopulu. 2001. "Rameses II and the Tobacco Beetle." *Antiquity* 75, 549–556.

Buell, P.D., E.N. Anderson, M. de Pablo Moya, and M. Oskenbay. 2020. *Crossroads of Cuisine: The Eurasian Heartland, the Silk Roads and Food.* Leiden: Brill.

Burke, B. 2012. "Looking for Sea-Silk in the Bronze Age Aegean." In M-L. Nosch and R. Laffineur (eds.), *Kosmos: Jewellery, Adornment and Textiles in the Aegean Bronze Age. Proceedings of the 13th International Aegean Conference, University of Copenhagen, Danish National Research Foundation's Centre for Textile Research*, 171–76. Leuven: Peeters.

Burkill, I.H. 1997. *The Useful Plants of West Tropical Africa*, 2nd ed., Vol. 4. London: Royal Botanic Gardens, Kew.

Burstein, S.M. 2012. "Agatharchides of Knidos (86)." *Brill's New Jacoby*, online edition.

Çakırlar, C., and S. Ikram. 2016. "'When Elephants Battle, the Grass Suffers.' Power, Ivory and the Syrian Elephant." *Levant* 482, 167–83.

Calabró, G.L. 1961. "Frammenti inediti di Archigene." *Bolletino del comitato per la preparazioni dell'edizione nationale dei classici greci e latini* 9, 67–71.

Caldwell, R. 1856. *A Comparative Grammar of the Dravidian or South-Indian Family of Languages.* London: Williams and Norgate.

Calligaro, T. 2005. "The Origin of Ancient Gemstones Unveiled by PIXE, PIGE and μ-Raman Spectrometry." In M. Uda, G. Demortier, and I. Nakai (eds.), *X-rays for Archaeology*, 101–12. Dordrecht: Springer.

Cameron, J. 2010. "The Archaeological Textiles from Ban Don Ta Phet in Broader Perspective." In B. Bellina et al. (eds.), *50 Years of Archaeology in Southeast Asia: Essays in Honour of Ian Glover*, 141–51. Bangkok: River Books.

———. 2015. "Iron and Cotton in the Indian Ocean Littoral: New Data from ThaKae, Central Thailand." In S. Tripati (ed.), *Maritime Contacts of the Past: Deciphering Connections amongst Communities*, 198–207. Delhi: Kaveri Book Service.

Canard, M. 1959. "Le riz dans le Proche Orient aux premiers siècles de l'Islam." *Arabica* 6, 113–31.

Cappers, R.T.J. 1998. "Archaeobotanical Remains." In S.E. Sidebotham and W.Z. Wendrich (eds.), *Berenike 1996: Report of the 1996 Excavations at Berenike (Egyptian Red Sea Coast) and the Survey of the Eastern Desert*, 289–330. Leiden: Research School, CNWS.

———. 2006. *Roman Footprints at Berenike.* Cotsen Institute of Archaeology, University of California, Los Angeles.

Carbonell-Caballero, J., R. Alonso, V. Ibañez, J. Terol, M. Talon, and J. Dopazo. 2015. "A Phylogenetic Analysis of 34 Chloroplast Genomes Elucidates the Relationships between Wild and Domestic Species within the Genus Citrus." *Molecular Biology and Evolution* 32, 2015–35.

Carter, H. 1923. "An Ostracon Depicting a Red Jungle-Fowl (The Earliest Known Drawing of the Domestic Cock)." *Journal of Egyptian Archaeology* 9.1–2, 1–4.

Carter, M.L. 1981. "Mithra on the Lotus: A Study of the Imagery of the Sun-god in the Kushano-Sasanian Era." In J. Duchesne-Guillemin (ed.), *Monumentum Georg Morgenstierne* I, 74–98. Leiden: Brill.

Carter, R. 2001. "Saar and Its External Relations: New Evidence for Interaction between Bahrain and Gujarat during the Early Second Millennium BC." *Arabian Archaeology and Epigraphy* 12.2, 183–201.

———. 2013. "Sumerians and the Gulf." In H. Crawford (ed.), *The Sumerian World*, 579–99. Abingdon: Routledge.

Casanova, M. 2001. "Le lapis-lazuli, la pierre précieuse de l'Orient ancien." *Dialogues d'histoire ancienne* 27.2, 149–70.

Casson, L. 1984. "Cinnamon and Cassia in the Ancient World." In L. Casson, *Ancient Trade and Society*, 225–46. Detroit: Wayne University Press.

Castillo, C.C., B. Bellina, and D.Q. Fuller. 2016. "Rice, Beans and Trade Crops on the Early Maritime Silk Route in Southeast Asia." *Antiquity* 90(353), 1255–69.

Castillo, C., and D.Q. Fuller. 2010. "Still Too Fragmentary and Dependent upon Chance? Advances in the Study of Early Southeast Asian Agriculture." In B. Bellina et al. (eds.), *50 Years of Archaeology in Southeast Asia: Essays in Honour of Ian Glover*, 91–111. Bangkok: River Books.

Celant, A., and G. Fiorentino, G. 2017. "Macroremains of Citrus Fruit in Italy." In V. Zech-Matterne and G. Fiorentino (eds.), *AGRUMED: Archaeology and History of Citrus Fruit in the Mediterranean*. Naples: Centre Jean Bérard (https://books.openedition.org/pcjb/2194).

Chaisuwan, B. 2011. "Early Contacts between India and the Andaman Coast." In P-Y. Manguin, A. Mani, and G. Wade (eds.), *Early Interactions between South and Southeast Asia: Reflections on Cross-Cultural Exchange*, 83–112. Singapore: Institute of Southeast Asian Studies.

Chakrabarti, D.K. 1999. *India: An Archaeological History.* Oxford University Press.

———., ed. 2004. *Indus Civilization Sites in India: New Discoveries.* Mumbai: Marg.

Chakraborti, S., and R.K. Bari. 1991. *Handicrafts of West Bengal.* Calcutta: Institute of Art and Handicraft.

Chakravarty, H.L., and C. Jeffrey. 1980. "Cucurbitaceae." In C.C. Townsend, E. Guest, and S.A. Omar, *Flora of Iraq, Vol. IV: Cornaceae to Resedaceae*, 191–208. Baghdad: Ministry of Agriculture.

Chambers, J.D., and G.E. Mingay. 1966. *The Agricultural Revolution 1750–1880.* London: Batsford.

Charles, M.B. 2008. "Alexander, Elephants and Gaugamela." *Mouseion: Journal of the Classical Association of Canada* 8.1, 9–23.

———. 2010. "Elephants, Alexander and the Indian Campaign." *Mouseion: Journal of the Classical Association of Canada* 10.3, 327–53.

Charles, M.P. 1987. "Onions, Cucumbers, and the Date Palm." *Bulletin of Sumerian Agriculture* 3, 1–21.

Chen, S., et al. 2010. "Zebu Cattle Are an Exclusive Legacy of the South Asia Neolithic." *Molecular Biology and Evolution* 27.1, 1–6.

Childe, V.G. 1936. *Man Makes Himself.* London: Watts.

Choi, J. Y., A. E. Platts, D. Q. Fuller, R. A. Wing, and M. D. Purugganan. 2017. "The Rice Paradox: Multiple Origins but Single Domestication in Asian rice." *Molecular Biology and Evolution* 34(4), 969–79.

Choi, J. Y., and M. D. Purugganan. 2018. "Multiple Origin but Single Domestication Led to Oryza sativa." G3: *Genes, Genomes, Genetics* 8(3), 797–803.

Chomicki, G., H. Schaefer, and S. S. Renner. 2020. "Origin and Domestication of Cucurbitaceae Crops: Insights from Phylogenies, Genomics and Archaeology." *New Phytologist* 226, 1240–55.

Chowdhury, K. A., and G. M. Buth. 1971. "Cotton Seeds from the Neolithic in Egyptian Nubia and the Origin of Old World Cotton." *Biological Journal of the Linnean Society* 3, 303–12.

———. 2005. "Plant Remains from Excavation of Terraces of the Nile at Afyeh, Nubia and Egypt." *Purattatva* 35, 154–59.

Chowdhury, M. P., S. Campbell, and M. Buckley. 2021. "Proteomic Analysis of Archaeological Ceramics from Tell Khaiber, Southern Iraq." *Journal of Archaeological Science* 132, 105414.

Chtcheglov, A. 1992. *Polis et chora: Cité et territoire dans le Pont-Euxin.* Paris: Belles Lettres.

Ciancaglini, C. A. 2008. *Iranian Loanwords in Syriac.* Wiesbaden: Reichert.

Ciaraldi, M. 2007. *People and Plants in Ancient Pompeii: A New Approach to Urbanism from the Microscope Room. The Use of Plant Resources at Pompeii and in the Pompeian Area from the 6th Century BC to AD 79*. London: Accordia Research Institute.

Cielas, H. 2014. "The Eight-Petalled Lotus Flower Pattern in Sanskrit Figurative Poetry." *Pandanus* 13 7/1, 73–88.

Cierny, J., and G. Wiesgerber. 2003. "Bronze Age Tin Mines in Central Asia." In A. Giumlia-Mair and F. Lo Schiavo (eds.), *Le probléme de l'étain à l'origine de la métallurgie*, 23–31. Oxford: Archaeopress.

Cimino, R. M., ed. 1994. *Ancient Rome and India: Commercial and Cultural Contacts between the Roman World and India.* New Delhi: Munshiram Manoharlal.

Civáň, P., and T. A. Brown. 2017. "Origin of Rice (*Oryza sativa* L.) Domestication Genes." *Genetic Resources and Crop Evolution* 64(6), 1125–32.

———. 2018. "Role of Genetic Introgression during the Evolution of Cultivated Rice (*Oryza sativa* L.)." *BMC Evolutionary Biology* 18.1, 1–11.

Civáň, P., H. Craig, C. J. Cox, and T. A. Brown. 2015. "Three Geographically Separate Domestications of Asian Rice." *Nature Plants* 1.11, 1–5.

Civáň, P., et al. 2019. "Origin of the Aromatic Group of Cultivated Rice (*Oryza sativa* L.) Traced to the Indian Subcontinent." *Genome Biology and Evolution* 11(3), 832–43.

Clapham, A. J. 2019. "The Archaeobotany of Nubia." In D. Raue (ed.), *Handbook of Ancient Nubia,* Vol. 1, 83–101. Berlin: De Gruyter.

Clapham, A. J., and P. Rowley-Conwy. 2006. "Rewriting the History of African Agriculture." *Planet Earth,* Summer 2006, 24–26.

———. 2007. "New Discoveries at Qasr Ibrim, Lower Nubia." In R. Cappers (ed.), *Fields of Change: Proceedings of the 4th International Workshop for African Archaeobotany.* Groningen: Barkhuis.

———. 2009. "The Archaeobotany of Cotton (*Gossypium* sp. L) in Egypt and Nubia with Special Reference to Qasr Ibrim, Egyptian Nubia." In E. Weiss and A. S. Fairburn (eds.), *From Foragers to Farmers: Papers in Honor of Gordon Hillman*, 244–53. Oxford: Oxbow Books.

Clayden, T. 1989. *Aspects of the Early History of the Kassites and the Archaeology of the Kassite Period in Iraq (c. 1600–1150 BC).* Doctoral dissertation, University of Oxford.

———. 2009. "Eye-Stones." *Zeitschrift für Orient-Archäologie* 2, 36–86.

Cleuziou, S. 1992. "The Oman Peninsula and the Indus Civilization: A Reassessment." *Man and Environment* 17.2, 93–103.

Cleuziou, S. 2003. "Early Bronze Age Trade in the Gulf and the Arabian Sea: The Society behind the Boats." In D. T. Potts, H. Al Naboodah, and P. Hellyer (eds.), *Archaeology of the United Arab Emirates: Proceedings of the First International Conference of the U.A.E.*, 133–49. London: Trident Press.

Cleuziou, S., and S. Méry. 2002. "In Between the Great Powers: The Bronze Age Oman Peninsula." In S. Cleuziou, M. Tosi, and J. Zarins (eds.), *Essays on the Late Prehistory of the Arabian Peninsula*, 273–316. Rome: Istituto italiano per l'Africa e l'Oriente.

Cleuziou, S., and M. Tosi. 2000. "Ra's al-Jinz and the Prehistoric Coastal Cultures of the Ja'alan." *Journal of Oman Studies* 11, 19–73.

Cobb, M. A. 2015. "Balancing the Trade: Roman Cargo Shipments to India." *Oxford Journal of Archaeology* 34.2, 185–203.

———. 2018a. *Rome and the Indian Ocean Trade from Augustus to the Early Third Century CE.* Leiden: Brill.

———. 2018b. "From the Ptolemies to Augustus: Mediterranean Integration into the Indian Ocean Trade." In M. A. Cobb (ed.), *The Indian Ocean Trade in Antiquity: Political, Cultural and Economic Impacts*, 17–51. London: Routledge.

Cockburn, E. 1986. "Cotton in Ancient Egypt." In R. A. David (ed.), *Science in Egyptology*, 469–73. Manchester University Press.

Collard, C. 2005. "Colloquial Language in Tragedy: A Supplement to the Work of P.T. Stevens." *Classical Quarterly* 55, 350–86.

Collins, P. 2010. "Attending the King in the Assyrian Reliefs." In A. Cohen and S. E. Kangas (eds.), *Assyrian Reliefs from the Palace of Ashurnasirpal II: A Cultural Biography*, 181–97. Hanover: University Press of New England.

Collon, D. 1996. "Mesopotamia and the Indus: The Evidence of the Seals." In J. Reade (ed.), *The Indian Ocean in Antiquity*, 209–25. London: Kegan Paul International.

———, ed. 2008. "Nimrud Treasures: Panel Discussion." In J. E. Curtis et al. (eds.), *New Light on Nimrud*, 105–18. London: British Institute for the Study of Iraq.

Coltherd, J. B. 1966. "The Domestic Fowl in Ancient Egypt." *Ibis* 108.2, 217–23.

Compagnoni, B., and M. Tosi. 1978. "The Camel: Its Distribution and State of Domestication in the Middle East during the Third Millennium BC in Light of Finds from Shahr-i Sokhta." In R. H. Meadow and M. A. Zeder (eds.), *Peabody Museum Bulletin* 2, 91–103. Harvard University Press.

Cooke, M., and D. Q. Fuller. 2015. "Agricultural Continuity and Change during the Megalithic and Early Historic Periods in South India." In K. K. Basa, R. K. Mohanty, and S. B. Ota (eds.), *Megalithic Traditions in India: Archaeology and Ethnography*, 445–76. Delhi: Aryan Books International.

Corbino, C. A., J. D. G. Mazzorin, C. Minniti, and U. Albarella. 2022. "The Earliest Evidence of Chicken in Italy." *Quaternary International* 626–27, 680–86 (https://doi.org/10.1016/j.quaint.2021.04.006).

Cornille, A., et al. 2012. "New Insight into the History of Domesticated Apple: Secondary Contribution of the European Wild Apple to the Genome of Cultivated Varieties." *PLoS Genetics* 8.5, 1–13.

Cornille, A., et al. 2019. "A Multifaceted Overview of Apple Tree Domestication." *Trends in Plant Science* 24(8), 770–82.

Costantini, L. 1981. "Paleoethnobotany at Pirak." In H. Hartel (ed.), *South Asian Archaeology 1979*, 271–78. Berlin: Dietrich Reimer.

Costantini, L., and L. C. Biasini. 1985. "Agriculture in Baluchistan between the 7th and 3rd Millennium B.C." *Newsletter of Baluchistan Studies* 2, 16–37.

Coubray, S., V. Zech-Matterne, and A. Mazurier. 2010. "The Earliest Remains of a Citrus Fruit from a Western Mediterranean Archaeological Context? A Microtomographic-Based Re-assessment." *Comptes Rendus Palevol* 9.6, 277–82.

Cousin, L. 2020. "Le dromadaire (*Camelus dromedarius*) dans le Proche-Orient ancien au Ier millénaire av. J.-C." In D. Agut-Labordère and B. Redon (eds.), *Les vaisseaux du désert et des steppes: Les camélidés dans l'Antiquité*, 65–79. Lyon: MOM.

Crawford, D. J. 1979. "Food: Tradition and Change in Hellenistic Egypt." *World Archaeology* 11:2, 136–46.

Crawford, H. E. 1996. "Dilmun, Victim of World Recession." *Proceedings of the Seminar for Arabian Studies* 26, 13–22.

———. 1998. *Dilmun and its Gulf Neighbours.* Cambridge University Press.

Creasman, P. P. 2014. "Hatshepsut and the Politics of Punt." *African Archaeological Review* 31.3, 395–405.

Creasman, P. P., and K. Yamamoto. 2019. "The African Incense Trade and Its Impacts in Pharaonic Egypt." *African Archaeological Review* 36(3), 347–65.

Crosby, A. W. 1972. *The Columbian Exchange: Biological and Cultural Consequences of 1492.* Westport, CT: Greenwood Press.

Crüsemann, N., M. van Ess, M. Hilgert, and B. Salje. 2013. *Uruk: 5000 Jahre Megacity.* Petersberg: Michael Imhof.

Cullen, H. M., and P. B. Demenocal. 2000. "North Atlantic Influence on Tigris–Euphrates Streamflow." *International Journal of Climatology* 20(8), 853–63.

Culpepper Stroup, S. 2006. "Invaluable Collections: The Illusion of Poetic Presence in Martial's *Xenia* and *Apophoreta*." In R. R. Nauta, H-J. van Dam, and J. J. L. Smolenaars (eds.), *Flavian Poetry*, 299–313. Leiden: Brill.

Cunliffe, B. W. 2015. *By Steppe, Desert, and Ocean: The Birth of Eurasia.* Oxford University Press.

Cuno, J. 2011. *Who Owns Antiquity? Museums and the Battle over Our Ancient Heritage.* Princeton University Press.

Curk, F., F. Ollitrault, A. Garcia-Lor, F. Luro, L. Navarro, and P. Ollitrault, P. 2016. "Phylogenetic Origin of Limes and Lemons Revealed by Cytoplasmic and Nuclear Markers." *Annals of Botany* 117, 565–83.

Cuvier, M. F., et al. 1828. *Dictionnaire des sciences naturelles,* Vol. 52. Strasbourg: Levrault.

Da Riva, R. 2002. *Der Ebabbar-Tempel von Sippar in frühneubabylonischer Zeit (640–580 v. Chr.).* Münster: Ugarit.

———. 2009. "The Nebuchadnezzar Inscription in Nahr el-Kalb." In A-M. Afeiche (ed.), *Le Site du Nahr el-Kalb*, 255–302. Beirut: Ministry of Culture, Republic of Lebanon.

———. 2013a. "Nebuchadnezzar II's Prism (EŞ 7834): A New Edition." *Zeitschrift für Assyriologie und vorderasiatische Archäologie* 103.2, 196–229.

———. 2013b. *The Inscriptions of Nabopolassar, Amel-Marduk and Neriglissar.* Berlin: de Gruyter.

Dafni, A., and B. Böck. 2019. "Medicinal plants of the Bible—Revisited." *Journal of Ethnobiology and Ethnomedicine* 15.1, 1–14.

D'Agostino, A. 2009. "The Assyrian-Aramaean Interaction in the Upper Khabur." *Syria* 86, 17–41.

Dahl, J. L., C. A. Petrie, and D. T. Potts. 2013. "Chronological Parameters of the Earliest Writing System in Iran." In C. A. Petrie (ed.), *Ancient Iran and its Neighbours: Local Developments and Long-range Interactions in the Fourth Millennium BC*, 353–78. Oxford: Oxbow Books.

Dalby, A. 1996. *Siren Feasts: A History of Food and Gastronomy in Greece.* London: Routledge.

———. 2003. *Food in the Ancient World from A to Z.* London: Routledge.

Dales, G. F. 1968. "Of Dice and Men." *Journal of the American Oriental Society* 88.1, 14–23.

Dalley, S. 1980. "Old Babylonian Dowries." *Iraq* 42, 53–74.

———. 1991. "Ancient Assyrian Textiles and the Origins of Carpet Design." *Iran* 29, 117–35.

———. 2013. *The Mystery of the Hanging Garden of Babylon: An Elusive World Wonder Traced.* Oxford University Press.

Dandamaev, M. A. 1992. *Iranians in Achaemenid Babylonia.* Costa Mesa: Mazda.

Darby, W. J., P. Ghalioungui, and L. Grivetti. 1977. *Food: The Gift of Osiris,* 2 vols. London: Academic Press.

Daressy, G. 1922. "Le riz dans l'Egypte antique." *Bulletin de l'Institut d'Egypte* 4, 35–37.

Daryaee, T. 2006–07. "List of Fruits and Nuts in the Zoroastrian Tradition: An Irano-Hellenic Classification." *Nāme-ye Irān-e Bāstān: The International Journal of Ancient Iranian Studies* 11/12, 75–84.

Daszewski, W. A. 1985. *Corpus of Mosaics from Egypt: Hellenistic and Early Roman Period.* Mainz: Philipp von Zabern.

Davies, G. I. 2001. "Introduction to the Pentateuch." In J. Barton and J. Muddiman (eds.), *The Oxford Bible Commentary*, 12–38. Oxford University Press.

de Candolle, A. 1885. *Origin of Cultivated Plants.* New York: Appleton.

de Caylus, A. C. 1752. *Recueil d'antiquitès egyptiennes, etrusques, grecques et romaines,* Vol. I. Paris: Chez Desaint et Saillant.

De Romanis, F. 2020. *The Indo-Roman Pepper Trade and the Muziris Papyrus.* Oxford University Press.

De Romanis, F., and A. Tchernia, eds. 1997. *Crossings: Early Mediterranean Contacts with India.* New Delhi: Manohar.

Decker, M. 2009. "Plants and Progress: Rethinking the Islamic Agricultural Revolution." *Journal of World History* 20.2, 187–206.

Decker-Walters, D. S. 1999. "Cucurbits, Sanskrit, and the Indo-Aryas." *Economic Botany* 53, 98–112.

Delmas, A. B., and M. Casanova. 1990. "The Lapis Lazuli Sources in the Ancient East." In M. Taddei (ed.), 493–505. *Proceedings of South Asian Archaeology 1987.* Rome: Istituto italiano per il Medio ed Estremo Oriente.

Desmond, R. 1992. *The European Discovery of the Indian Flora.* Oxford: Clarendon Press.

Desrosiers, S. 2000. "Une culture textile raffinée." In C. Debaine-Francfort and A. Idriss (eds.), *Keriya: mémoires d'un fleuve*, 144–55. Paris: Editions Findakly.

Desset, F. 2012. *Premières écritures iraniennes: les systèmes proto-élamite et élamite linéaire.* Naples: Università degli studi di Napoli "L'Orientale," Dipartimento Asia Africa Mediterraneo.

Dominy, N. J., S. Ikram, G. L. Moritz, J. N. Christensen, P. V. Wheatley, and J. W. Chipman. 2015. "Mummified Baboons Clarify Ancient Red Sea Trade Routes." *American Journal of Physical Anthropology* 156, 122–23.

Dominy, N. J., S. Ikram, G. L. Moritz, P. V. Wheatley, J. N. Christensen, J. W. Chipman, and P. L. Koch. 2020. "Mummified Baboons Reveal the Far Reach of Early Egyptian Mariners." *Elife* 9, 1–28.

Ducène, J. C. 2019. "Le coton, sa culture et son utilisation selon les sources arabes médiévales." *Revue d'ethnoécologie* 15, 1–13.

Dunham, S. 1985. "The Monkey in the Middle." *Zeitschrift für Assyriologie und Vorderasiatische Archäologie* 75.2, 234–64.

Dyson, R. H., Jr. 1962. "The Hasanlu Project." *Science* (New Series) 135, 637–47.

Edde, P. A. 2012. "A Review of the Biology and Control of *Rhyzopertha dominica* (F.) the Lesser Grain Borer." *Journal of Stored Products Research* 48, 1–18.

Edel, E. 1996. "Studien zu den Relieffragmenten aus dem Taltempel des Königs Snofru." In P. Der Manuelian (ed.), *Studies in Honor of William Kelly Simpson, Volume 1*, 198–208. Boston: Museum of Fine Arts.

Edwards, C. J., J. F. Baird, and D. E. MacHugh. 2007. "Taurine and Zebu Admixture in Near Eastern Cattle: A Comparison of Mitochondrial, Autosomal and Y-Chromosomal Data." *Animal Genetics* 38.5, 520–24.

Ehrenberg, E. 2002. "The Rooster in Mesopotamia." In E. Ehrenberg (ed.), *Leaving no Stones Unturned: Essays on the Ancient Near East and Egypt in Honor of Donald P. Hansen*, 53–62. Winona Lake, IN: Eisenbrauns.

Eidem, J., and F. Højlund. 1993. "Trade or Diplomacy? Assyria and Dilmun in the Eighteenth Century BC." *World Archaeology* 24.3, 441–48.

El Awady, T. 2009. *Abusir XVI: Sahure—The Pyramid Causeway: History and Decoration Program in the Old Kingdom.* Prague: Charles University.

El Faïz, M. 1995. *L'agronomie de la Mésopotamie antique: analyse du 'Livre de l'agriculture nabatéenne' de Qûṯâmä.* Leiden: Brill.

El Tahir, I. M., and M. T. Yousif. 2004. "Indigenous Melons (*Cucumis melo* L.) in Sudan: A Review of Their Genetic Resources and Prospects for Use as Sources of Disease and Insect Resistance." *Plant Genetic Resources Newsletter* 138, 36–42.

Elliot, H. M. 1867. *The History of India as Told by Its Own Historians: The Muhammadan Period.* London: Trübner & Co.

Ellison, R., J. Renfrew, D. Brothwell, and N. Seeley. 1978. "Some Food Offerings from Ur, Excavated by Sir Leonard Woolley, and Previously Unpublished." *Journal of Archaeological Science* 5.2, 167–77.

Endl, J., E. G. Achigan-Dako, A. K. Pandey, A. J. Monforte, B. Pico, and H. Schaefer. 2018. "Repeated Domestication of Melon (*Cucumis melo*) in Africa and Asia and a New Close Relative from India." *American Journal of Botany* 105, 1662–71.

Erdosy, G. 1988. *Urbanisation in Early Historic India.* Oxford: British Archaeological Reports.

Ermolli, E. R., and E. Messager. 2013. "The Gardens of Villa A at Oplontis through Pollen and Phytolith Analysis of Soil Samples." In J. R. Clarke and N. K. Muntasser (eds.), *Villa A ("of Poppaea") at Oplontis (Torre Annunziata, Italy): The Ancient Setting and Modern Rediscovery.* New York: ALCS E-book.

Ermolli, E. R., E. Messager, and M. R. B. Lumaga. 2017. "Pollen Morphology Reveals the Presence of *Citrus medica* and *Citrus × limon* in a Garden of Villa Di Poppea in Oplontis, 1st Century BC." In V. Zech-Matterne and G. Fiorentino (eds.), *AGRUMED: Archaeology and History of Citrus Fruit in the Mediterranean.* Naples: Centre Jean Bérard (https://books.openedition.org/pcjb/2190).

Ermolli, E. R., P. Romano, M. R. Ruello, and M. R. B. Lumaga, M. 2014. "The Natural and Cultural Landscape of Naples (Southern Italy) during the Graeco-Roman and Late Antique Periods." *Journal of Archaeological Science* 42, 399–411.

Esquinas-Alcazar, J. T., and P. J. Gulick. 1983. *Genetic Resources of Cucurbitaceae: A Global Report.* Rome: International Board for Plant Genetic Resources.

Essert, S., I. Koncani-Uhač, M. Uhač, and R. Šoštarić. 2018. "Plant Remains and Amphorae from the Roman Harbour under Flacius Street in Pula (Istria, Croatia)." *Archaeological and Anthropological Sciences* 10, 955–71.

Fahmy, A. G. E. D. 2001. "Plant Remains in Gut Contents of Ancient Egyptian Predynastic Mummies (3750–3300 BC)." *Online Journal of Biological Sciences* 1.8, 772–74.

Fales, F. M. 2010. "Production and Consumption at Dūr-Katlimmu: A Survey of the Evidence." In H. Kühne (ed.), *Dūr-katlimmu 2008 and Beyond (Studia Chaburensia Vol. I),* 67–85. Wiesbaden: Harrassowitz.

Fales, F. M., and R. del Fabbro. 2014. "Back to Sennacherib's Aqueduct at Jerwan: A Reassessment of the Textual Evidence." *Iraq* 76.1, 65–98.

Falk, H. 1978. *Quellen des Pañcatantra.* Wiesbaden: Harrassowitz.

Farahani, A. 2021. "Paleoethnobotany and Ancient Agriculture." In D. Hollander and T. Howe (eds.), *A Companion to Ancient Agriculture,* 7–36. Hoboken, NJ: Wiley.

Fauconnier, B. 2015. "Ex Occidente Imperium: Alexander the Great and the Rise of the Maurya Empire." *Histos* 9, 120–73.

Fee, S. 2020. *Cloth that Changed the World: The Art and Fashion of Indian Chintz.* Yale University Press.

Feliks, J. 1963a. *Agriculture in Palestine in the Period of the Mishnah and Talmud.* Jerusalem: Magnes Press (in Hebrew).

———. 1963b. "Rice in Rabbinic Literature." *Bar-Ilan Sefer ha-Shanah* 1, 177–89 (in Hebrew).

———. 2007a. "Etrog." In M. Berenbaun and F. Skolnik (eds.), *Encyclopaedia Judaica VI,* 2nd ed., 540–41. Detroit: Macmillan Reference.

———. 2007b. "Plants." In F. Skolnik (ed.), *Enyclopaedia Judaica XVI,* 219–25. Detroit: Macmillan Reference.

Fernández-Armesto, F. 2001. *Civilizations: Culture, Ambition, and the Transformation of Nature.* New York: Simon and Schuster.

Fernández-Armesto, F., and B. Sacks. 2012. "The Global Exchange of Food and Drugs." In F. Trentmann (ed.), *The Oxford Handbook of the History of Consumption,* 127–44. Oxford University Press.

Filipović, D., et al. 2020. "New AMS 14 C Dates Track the Arrival and Spread of Broomcorn Millet Cultivation and Agricultural Change in Prehistoric Europe." *Scientific Reports* 10.1, 1–18.

Fincke, J. C. 2003/04. "The Babylonian Texts of Nineveh: Report on the British Museum's Ashurbanipal Library Project." *Archiv für Orientforschung* 50, 111–49.

Finkel, I. L., and J. E. Reade. 2002. "On Some Inscribed Babylonian Alabastra." *Journal of the Royal Asiatic Society* (Third Series) 12.1, 31–46.

Fitzpatrick, M. P. 2011. "Provincializing Rome: The Indian Ocean Trade Network and Roman Imperialism." *Journal of World History* 22.1, 27–54.

Fleming, D. 1982. "Achaemenid Sattagydia and the Geography of Vivana's Campaign (DB III, 54–75)." *Journal of the Royal Asiatic Society of Great Britain and Ireland* 114.2, 102–12.

———. 1993. "Where was Achaemenid India?" *Bulletin of the Asia Institute* 7, 67–72.

Floor, W. 2009. "Sugar." *Encyclopaedia Iranica,* online edition (http://www.iranicaonline.org/articles/sugar-cultivation).

Foster, B. R. 1977. "Commercial Activity in Sargonic Mesopotamia." *Iraq* 39, 31–43.

———. 2016. *The Age of Agade: Inventing Empire in Ancient Mesopotamia.* Abingdon: Routledge.

Foster, K. P. 1998. "Gardens of Eden: Exotic Flora and Fauna in the Ancient Near East." In J. Albert, M. Bernhardsson, and R. Kenna (eds), *Transformations of Middle Eastern Environments: Legacies and Lessons*, 320–29. Yale University Press.

Fraenkel, S. 1886. *Die aramäischen Fremdwörter im Arabischen.* Leiden: Brill.

Fragner, B. G. 1984. "Zur Erforschung der kulinarischen Kultur Irans." *Die Welt des Islams* 23/24, 320–60.

Frahm, E. 1997. *Einleitung in die Sanherib-Inschriften.* Vienna: Institut für Orientalistik der Universität.

———. 2011. *Babylonian and Assyrian Text Commentaries: Origins of Interpretation.* Münster: Ugarit.

Frame, G., and A. R. George. 2005. "The Royal Libraries of Nineveh: New Evidence for King Ashurbanipal's Tablet Collecting." *Iraq* 67.1, 265–84.

Frank, K. S., and H.-P. Stika. 1988. *Bearbeitung der makroskopischen Pflanzen- und einiger Tierreste des Römerkastells Sablonetum (Ellingen bei Weissenburg in Bayern).* Kallmünz: Michael Laßleben.

Frankfort, H., T. Jacobsen, S. Lloyd, and G. Martiny. 1940. *The Gimilsin Temple and the Palace of the Rulers at Tell Asmar.* University of Chicago Press.

Frenez, D. 2018. "The Indus Civilization Trade with the Oman Peninsula." In S. Cleuziou and M. Tosi (eds.), *In the Shadow of the Ancestors: The Prehistoric Foundations of the Early Arabian Civilization in Oman,* 2nd ed., 385–96. Muscat: Ministry of Heritage and Culture.

Frenez, D., M. D. Esposti, S. Méry, and J. M. Kenoyer. 2016. "Bronze Age Salūt (ST1) and the Indus Civilization: Recent Discoveries and New Insights on Regional Interaction." *Proceedings of the Seminar for Arabian Studies* 46, 107–24.

Frimmel, F. 1913. "Über einige antike Samen aus dem Orient." In F. Hrozný, *Das getreide im alten Babylonien: Ein beitrag zur kultur- und wirtschaftsgeschichte des alten Orients.* Wien: A. Hölder.

Frisk, H. 1954–72. *Griechisches etymologisches Wörterbuch.* Heidelberg: Carl Winter Universitätsverlag.

Froelicher, Y., et al. 2011. "New Universal Mitochondrial PCR Markers Reveal New Information on Maternal Citrus Phylogeny." *Tree Genetics and Genomes* 7.1, 49–61.

Fujii, H. 1983/4. "Aṭ-Ṭār Caves." *Archiv für Orientforschung* 29/30, 173–82.

Fujino, K., and H. Sekiguchi. 2005. "Mapping of QTLs Conferring Extremely Early Heading in Rice (Oryza sativa L.)." *Theoretical and Applied Genetics* 1112, 393–98.

Fuks, D., O. Amichay, and E. Weiss. 2020. "Innovation or Preservation? Abbasid Aubergines, Archaeobotany, and the Islamic Green Revolution." *Archaeological and Anthropological Sciences* 12:50, 1–16.

Fuller, D. Q. 2002. "Fifty Years of Archaeobotanical Studies in India: Laying a Solid Foundation." In S. Settar and R. Korisettar (eds.), *Indian Archaeology in Retrospect*, Vol. III, 247–364. New Delhi: Manohar.

———. 2003a. "African Crops in Prehistoric South Asia: A Critical Review." In K. Neumann, A. Butler, and S. Kahlheber (eds.), *Food, Fuel and Fields: Progress in African Archaeobotany*, 239–71. Cologne: Heinrich Barth Institut.

———. 2003b. "An Agricultural Perspective on Dravidian Historical Linguistics." In P. Bellwood and C. Renfrew (eds.), *Examining the Farming/Language Dispersal Hypothesis*, 191–213. McDonald Institute for Archaeological Research, University of Cambridge.

———. 2003c. "Further Evidence on the Prehistory of Sesame." *Asian Agri-History* 7.2, 127–37.

———. 2004. "Early Kushite Agriculture: Archaeobotanical Evidence from Kawa." *Sudan and Nubia: The Sudan Archaeological Research Society Bulletin* 8, 70–74.

———. 2006a. "Agricultural Origins and Frontiers in South Asia: A Working Synthesis." *Journal of World Prehistory* 20.1, 1–86.

———. 2006b. "Silence before Sedentism and the Advent of Cash-Crops: A Status Report on Early Agriculture in South Asia from Plant Domestication to the Development of Political Economies (with an Excursus on the Problem of Semantic Shift among Millets and Rice)." In T. Osada (ed.), *Proceedings of the Pre-symposium of RIHN and 7th ESCA Harvard-Kyoto Roundtable*, 175–211. Kyoto: Research Institute for Humanity and Nature.

———. 2007. "Non-Human Genetics, Agricultural Origins and Historical Linguistics in South Asia." In M. D. Petraglia and B. Allchin (eds.), *The Evolution and History of Human Populations in South Asia*, 393–443. Dordrecht: Springer.

———. 2008. "The Spread of Textile Production and Textile Crops in India Beyond the Harappan Zone: An Aspect of the Emergence of Craft Specialization and Systematic Trade." In T. Osada and A. Uesugi (eds.), *Linguistics, Archaeology and the Human Past*, 1–26. Kyoto: Research Institute for Humanity and Nature.

———. 2009. "Silence before Sedentism and the Advent of Cash-Crops: A Revised Summary of Early Agriculture in South Asia from Plant Domestication to the Development of Political Economies (with an Excursus on the Problem of Semantic Shift among Millets and Rice)." In T. Osada (ed.), *Linguistics, Archaeology and the Human Past*, 147–87. New Delhi: Manohar.

———. 2011. "Pathways to Asian Civilizations: Tracing the Origins and Spread of Rice and Rice Cultures." *Rice* 4, 78–92.

———. 2013. "Agricultural Innovation and State Collapse in Meroitic Nubia." In C. J. Stevens et al. (eds.), *Archaeology of African Plant Use*, 165–77. Walnut Creek, CA: Left Coast Press.

Fuller, D. Q., and N. Boivin. 2009. "Crops, Cattle and Commensals across the Indian Ocean: Current and Potential Archaeobiological Evidence." *Études Océan Indien* 42–43, 13–46.

Fuller, D. Q., N. Boivin, C. Castillo, T. Hoogervorst, and R. G. Allaby. 2015. "The Archaeobiology of Indian Ocean Translocation: Current Outlines of Cultural Exchanges by Proto-Historic Seafarers." In S. Tripati (ed.), *Maritime Contacts of the Past: Deciphering Connection amongst Communities*, 1–23. New Delhi: Kaveri Book Service.

Fuller, D. Q., N. Boivin, T. Hoogervorst, and R. G. Allaby. 2011. "Across the Indian Ocean: The Prehistoric Movement of Plants and Animals." *Antiquity* 85, 544–58.

Fuller, D. Q., C. Castillo, and C. Murphy. 2017. "How Rice Failed to Unify Asia: Globalization and Regionalism of Early Farming Traditions in the Monsoon World." In T. Hodos (ed.), *The Routledge Handbook of Globalization and Archaeology*, 711–29. Abingdon: Routledge.

Fuller, D. Q., and L. Lucas. 2017. "Adapting Crops, Landscapes, and Food Choices: Patterns in the Dispersal of Domesticated Plants across Eurasia." In N. Boivin, R. Crassard, and M. Petraglia (eds.), *Human Dispersal and Species Movement: From Prehistory to the Present*, 304–11. Cambridge University Press.

Fuller, D. Q., and M. Madella. 2001. "Issues in Harappan Archaeobotany: Retrospect and Prospect." In S. Settar and R. Korisettar (eds.), *Indian Archaeology in Retrospect, Vol. II: Protohistory*, 317–90. New Delhi: Manohar.

Fuller, D. Q., G. Willcox, and R. G. Allaby. 2011. "Cultivation and Domestication Had Multiple Origins: Arguments Against the Core Area Hypothesis for the Origins of Agriculture in the Near East." *World Archaeology* 43.4, 628–52.

Fuller, D. Q., et al. 2010. "Consilience of Genetics and Archaeobotany in the Entangled History of Rice." *Archaeological and Anthropological Science* 2, 115–31.

Furger, A. R. 1995. "Vom Essen und Trinken im römischen Augst: Kochen, Essen und Trinken im Spiegel einiger Funde." *Archäologie der Schweiz* 8, 168–84.

Garner, J. 2015. "Bronze Age Tin Mines in Central Asia." In A. Hauptmann and D. Modarressi-Tehrani (eds.), *Archaeometallurgy in Europe III*, 135–43. Bergbau-Museum Bochum.

Garzilli, E. 2003. "The Flowers of Rgveda Hymns: Lotus in V. 78.7, X. 184.2, X. 107.10, VI. 16.13, and VII. 33.11, VI. 61.2, VIII. 1.33, X. 142.8." *Indo-Iranian Journal* 46.4, 293–314.

Gaspa, S. 2012. *Alimenti e pratiche alimentari in Assiria: le materie alimentari nel culto ufficiale dell'Assiria del primo millennio A.C.* Padua: S.A.R.G.O.N.

———. 2013. "Textile Production and Consumption in the Neo-Assyrian Empire." In M-L. Nosch, H. Koefoed, and E. A. Strand (eds.), *Textile Production and Consumption in the Ancient Near East: Archaeology, Epigraphy, Iconography*, 224–47. Oxford: Oxbow Books.

Gaud, W. S. 1968. "The Green Revolution: Accomplishments and Apprehensions." Address by the Honourable William S. Gaud, Administrator, U.S. Agency for International Development, Department of State, before the Society for International Development, Shorehan Hotel, Washington, D.C., March 8 (http://www.agbioworld.org/biotech-info/topics/borlaug/borlaug-green.html).

Gelb, I. J. 1982. "Sumerian and Akkadian Words for 'String of Fruit.'" In G. van Driel (ed.), *Zikir Šumim: Assyriological studies Presented to F.R. Kraus on the Occasion of His Seventieth Birthday*, 67–82, 484 (addenda). Leiden: Nederlands Instituut voor het Nabije Oosten.

Genaust, H. 1996. *Etymologisches Wörterbuch der botanischen Pflanzennamen*, 3rd ed. Basel: Birkhäuser.

Gentilcore, D. 2010. *Pomodoro!: A History of the Tomato in Italy*. New York: Columbia University Press.

George, A.R. 1988. “Babylonian Texts from the Folios of Sidney Smith, Part 1.” *Revue d'Assyriologie et d'archéologie orientale* 82, 139–62.

———. 2005/06. “The Tower of Babel: Archaeology, History and Cuneiform Texts.” *Archiv für Orientforschung* 51, 75–95.

Gerisch, R. 2013. “Die Holzkohlefunde des Roten Hauses.” In F.J. Kreppner and J. Schmid, *Stratigraphie und Architektur des “Roten Hauses” von Tall Šēḫ Ḥamad/Dūr-Katlimmu*, 443–67. Wiesbaden: Harrassowitz.

Germer, R. 1985. *Flora des pharaonischen Ägypten*. Mainz: Philipp von Zabern.

———. 1987. “Ancient Egyptian Plant-Remains in the Manchester Museum.” *Journal of Egyptian Archaeology* 73, 245–46.

———. 1988. *Katalog der altägyptischen Pflanzenreste der Berliner Museen*. Wiesbaden: Harrassowitz.

Gershevitch, I. 1957. “Sissoo at Susa (OPers. *yakā* = *Dalbergia sissoo* Roxb.).” *Bulletin of the School of Oriental and African Studies* 19.2, 317–20.

Gharmarzadeh, M., and A.J. Ghasemov. 2015. “The Investigation of a Few Lexicons of Comestibles in Ferdowsi's Shahnama (The Book of Kings).” *Journal of Scientific Research and Development* 2.1, 28–31.

Ghosh, A. 2000. *Non-Aryan Linguistic Elements in the Atharvaveda: A Study of Some Words of Austric Origin*. Kolkata: Sanskrit Pustak Bhandar.

Gibbon, E. 1776–88. *The History of the Decline and Fall of the Roman Empire*. 6 vols. London: Strahan and Cadell.

Gilboa, A., and D. Namdar. 2015. “On the Beginnings of South Asian Spice Trade with the Mediterranean Region: A Review.” *Radiocarbon* 57.2, 265–83.

Giovinazzo, G. 2000–01. “Les Indiens à Suse.” *Annali dell'Università degli Studi di Napoli, Sezione orientale* 60–61, 70–76.

Gleba, M. 2008. *Textile Production in Pre-Roman Italy*. Oxford: Oxbow.

Glidden, H.W. 1937. “The Lemon in Asia and Europe.” *Journal of the American Oriental Society* 57.4, 381–96.

Glindemann, D., A. Dietrich, H.J. Staerk, and P. Kuschk. 2006. “The Two Odors of Iron When Touched or Pickled: (Skin) Carbonyl Compounds and Organophosphines.” *Angewandte Chemie International Edition* 45, 7006–09.

Gmitter, F.G., and X. Hu. 1990. “The Possible Role of Yunnan, China, in the Origin of Contemporary *Citrus* Species (Rutaceae).” *Economic Botany* 44.2, 267–77.

Goede, B. 2005. “Lotos und Seerosen in Ägypten: Ein Nachweis des Roten Lotos in ägyptischer Glaskunst ptolemäisch-römischer Zeit.” *Göttinger Miszellen* 207, 49–73.

Goddeeris, A. 2007. “The Old Babylonian Economy.” In G. Leick (ed.), *The Babylonian World*, 198–209. Abingdon: Routledge.

Goitein, S. 1983. *A Mediterranean Society: The Jewish Communities of the Arab World as Portrayed in the Documents of the Cairo Geniza. Vol. IV: Daily Life*. University of California Press.

Gómez-Ariza, J., et al. 2015. “Loss of Floral Repressor Function Adapts Rice to Higher Latitudes in Europe.” *Journal of Experimental Botany* 66(7), 2027–39.

Gonda, J. 1977. *The Ritual Sūtras.* Wiesbaden: Harrassowitz.
Gonzalo, M. J., A. Díaz, N. P. Dhillon, U. K. Reddy, B. Picó, B., and A. J. Monforte. 2019. "Re-evaluation of the Role of Indian Germplasm as Center of Melon Diversification Based on Genotyping-by-Sequencing Analysis." *BMC Genomics*, 20:448, 1–13.
Good, I. 2011. "Up from the Ice: A Look at Dress in the Iron Age Altai. Review of L. Barkova and N. V. Polosmak, *Kostium i tekstil' pazyryktsev Altaia.*" *Silk Road* 9, 146–53.
Government of Iraq. 1962. *The Iraqi Revolution in Its Fourth Year.* Baghdad: Times Press.
Granger-Taylor, H. 2000. "The Textiles from Khirbet Qazone (Jordan)." In D. Cardon and M. Feugère (eds.), *Archéologie des textiles des origines au Ve siècle,* 149–162. Montagnac: Editions Monique Mergoil.
Grant, M. 1997. *Dieting for an Emperor: a translation of Books 1 and 4 of Oribasius' Medical Compilations with an Introduction and Commentary.* Leiden: Brill.
Graslin-Thomé, L. 2009. *Les échanges à longue distance en Mésopotamie au 1er millénaire: Une approche économique.* Paris: De Boccard.
Green, M. H. 2017. "The Globalisations of Disease." In N. Boivin, R. Crassard, and M. Petraglia (eds.), *Human Dispersal and Species Movement: From Prehistory to the Present,* 494–520. Cambridge University Press.
Greppin, J. A. C. 1999. "Gk. *Kostos*: A Fragrant Plant and Its Eastern Origin." *Journal of Indo-European Studies* 29, 395–408.
Grimaldi, I. M. 2014. *Food for Thought: Genetic, Historical and Ethnobotanical Studies of Taro (Colocasia esculenta (L.) Schott) in Africa.* Doctoral dissertation, University of Oxford.
Groneberg, B. 1992. "Le Golfe Arabo-Persique vu depuis Mari." In J-M. Durand (ed.), *Florilegium Marianum: Recueil d'études en l'honneur de Michel Fleury,* 69–80. Paris: Société pour l'étude du Proche-Orient ancien.
Groom, Q. J., J. Van der Straeten, and I. Hoste. 2019. "The Origin of *Oxalis corniculata* L." *PeerJ* 7, 1–19.
Grüger, E., and B. Thulin. 1998. "First Results of Biostratigraphical Investigations of Lago d'Averno near Naples Relating to the Period 800 BC–800 AD." *Quaternary International* 47, 35–40.
Grüger, E., et al. 2002. "Environmental Changes in and around Lake Avernus in Greek and Roman Times: A Study of the Plant and Animal Remains Preserved in the Lake's Sediments." In W. F. Jashemski and F. G. Meyer (eds.), *The Natural History of Pompeii,* 240–73. Cambridge University Press.
Gulsen, O., and M. L. Roose. 2001. "Lemons: Diversity and Relationships with Selected Citrus Genotypes as Measured with Nuclear Genome Markers." *Journal of the American Society for Horticultural Science* 126.3, 309–17.
Gumerman, G. 1997. "Food and Complex Societies." *Journal of Archaeological Method and Theory* 4.2, 105–39.
Gurjazkaite, K., J. Routh, M. Djamali, A. Vaezi, Y. Poher, A. N. Beni, and H. Kylin. 2018. "Vegetation History and Human-Environment Interactions through the Late Holocene in Konar Sandal, SE Iran." *Quaternary Science Reviews* 194, 143–55.
Guzmán-Solís, A. A., et al. 2021. "Ancient Viral Genomes Reveal Introduction of Human Pathogenic Viruses into Mexico during the Transatlantic Slave Trade." *Elife* 10, e68612.
Halevi, L. 2007. *Muhammad's Grave: Death Rites and the Making of Islamic Society.* Columbia University Press.

Hallo, W. W. 1996. *Origins: The Ancient Near Eastern Background of Some Modern Western Institutions.* Leiden: Brill.

Hämeen-Anttila, J. 2006. *The Last Pagans of Iraq: Ibn Waḥshiyya and his Nabatean Agriculture.* Leiden: Brill.

Hanafi-Bojd, A. A., M. M. Sedaghat, H. Vatandoost, S. Azari-Hamidian, and K. Pakdad, K. 2018. "Predicting Environmentally Suitable Areas for *Anopheles superpictus* Grassi (s.l.), *Anopheles maculipennis* Meigen (s.l.) and *Anopheles sacharovi* Favre (Diptera: Culicidae) in Iran." *Parasites and Vectors* 11:382.

Hansman, J. 1973. "A Periplus of Magan and Meluḫḫa." *Bulletin of the School of Oriental and African Studies* 36.3, 553–87.

Hardy, G., and L. Totelin. 2016. *Ancient Botany.* Abingdon: Routledge.

Harris, D. R. 1998. "The Multi-disciplinary Study of Cross-Cultural Plant Exchange." In H. D. V. Prendergast et al. (eds.), *Plants for Food and Medicine. Proceedings of the Joint Conference of the Society for Economic Botany and the International Society for Ethnopharmacology, London, 1–6 July 1996*, 85–91. London: Royal Botanic Gardens, Kew.

Harvey, E. L., D. Q. Fuller, K. K. Basa, R. Mohany, and B. Mohanta. 2006. "Early Agriculture in Orissa: Some Archaeobotanical Results and Field Observations on the Neolithic." *Man and Environment* 31.2, 21–32.

Hauenschild, I. 2006. *Botanica und Zoologica im Babur-name: eine lexikologische und kulturhistorische Untersuchung*, Wiesbaden: Harrassowitz.

Haw, S. G. 2017. "Cinnamon, Cassia and Ancient Trade." *Journal of Ancient History and Archaeology* 4.1, 5–18.

Hawkes, J. G. 1998. "The Introduction of New World Crops into Europe after 1492." In H. D. V. Prendergast et al. (eds.), *Plants for Food and Medicine: Proceedings of the Joint Conference of the Society for Economic Botany and the International Society for Ethnopharmacology, London, 1–6 July 1996*, 147–59, London: Royal Botanic Gardens, Kew.

He, X., R. Shen, and J. Jin. 2010. "A New Species of *Nelumbo* from South China and Its Palae oecological Implications." *Review of Palaeobotany and Palynology* 162.2, 159–67.

Heeßel, N. P. 2009. "The Babylonian Physician Rabâ-Ša-Marduk: Another Look at Physicians and Exorcists in the Ancient Near East." In A. Attia and G. Buisoon (eds.), *Advances in Mesopotamian Medicine from Hammurabi to Hippocrates*, 13–28. Leiden: Brill.

Heffron, Y. 2017. "Testing the Middle Ground in Assyro-Anatolian Marriages of the Karum Period." *Iraq* 79, 71–83.

Hehn, V. 1887. *Kulturpflanzen und Hausthiere in ihrem Übergang aus Asien nach Griechenland und Italien sowie in das übrige Europa.* Berlin: Gebrüder Borntræger.

Heide, M. 2010. "The Domestication of the Camel: Biological, Archaeological and Inscriptional Evidence from Mesopotamia, Egypt, Israel and Arabia, and Traditional Evidence from the Hebrew Bible." *Ugarit-Forschungen* 42, 331–82.

Heimpel, W. 1987–90. "Magan." In D. O. Edzard et al. (eds.), *Reallexikon der Assyriologie und Vorderasiatischen Archäologie* 7, 195–99. Berlin: de Gruyter.

———. 2003. *Letters to the King of Mari: A New Translation, with Historical Introduction, Notes, and Commentary.* Winona Lake, IN: Eisenbrauns.

Helbaek, H. 1966. "The Plant Remains from Nimrud." In M. E. L. Mallowan, *Nimrud and Its Remains*, 613–20. London: Collins.

Helwing, B. 2009. "Rethinking the Tin Mountains: Patterns of Usage and Circulation of Tin in Greater Iran from the 4th to the 1st Millennium BC." *Türkiye Bilimler Akademisi Arkeoloji Dergisi (Turkish Academy of Sciences–Journal of Archaeology)* 12, 209–21.

Hemmerdinger, B. 1970. "De la méconnaissance de quelques etymologies grecques." *Glotta* 48, 40–66.

Hendy, J., et al. 2018. "A Guide to Ancient Protein Studies." *Nature Ecology and Evolution* 2(5), 791–99.

Henkelman, W. 2008. *The Other Gods Who Are: Studies in Elamite-Iranian Acculturation based on the Persepolis Fortification Texts.* Leiden: Nederlands Instituut voor het Nabije Oosten.

———. 2010. "'Consumed before the King': The Table of Darius, That of Irdabama and Irtaštuna, and That of his Satrap, Karkis." In B. Jacobs and R. Rollinger (eds.), *Der Achämenidenhof*, 667–775. Wiesbaden: Harrassowitz.

———. 2017. "Imperial Signature and Imperial Paradigm: Achaemenid Administrative Structure and System across and beyond the Iranian Plateau." In B. Jacobs, W. Henkelman, and M. Stolper (eds.), *Administration in the Achaemenid Empire.* Wiesbaden: Harrassowitz.

Henning, W. B. 1977. "Sogdica." In W. B. Henning, *Selected Papers (Acta Iranica VI)*, 1–68. Tehran: Bibliothèque Pahlavi.

Hepper, F. N. 1981. "Plant Material." In G. T. Martin, *The Sacred Animal Necropolis at North Saqqara: The Southern Dependencies of the Main Temple Complex*, 146–51. London: Egypt Exploration Society.

Herrmann, G. 1968. "Lapis Lazuli: The Early Phases of Its Trade." *Iraq* 30.1, 21–57.

Herrscher, E., et al. 2018. "The Origins of Millet Cultivation in the Caucasus: Archaeological and Archaeometric Approaches." *Préhistoires Méditerranéennes* 6, 1–27.

Heywood, V. H. 2012. "The Role of New World Biodiversity in the Transformation of Mediterranean Landscapes and Culture." *Bocconea* 24, 69–93.

———. 2013. "Overview of Agricultural Biodiversity and Its Contribution to Nutrition and Health." In J. Fanzo et al. (eds.), *Diversifying Food and Diets: Using Agricultural Biodiversity to Improve Nutrition and Health*, 35–67. Abingdon: Routledge.

Hiltebeitel, A. 1991. *The Cult of Draupadi, Vol. 2: On Hindu Ritual and the Goddess.* University of Chicago Press.

Hinz, W. 1975. *Altiranisches Sprachgut der Nebenüberlieferungen.* Wiesbaden: Harrassowitz.

Hirschfeld, Y. 2007. "The Rose and the Balsam: The Garden as a Source of Perfume and Medicine." In M. Conan (ed.), *Middle East Garden Traditions: Unity and Diversity*, 21–39. Washington, DC: Dumbarton Oaks.

Hitch, S. 2018. "Tastes of Greek Poetry: From Homer to Aristophanes." In K. C. Rudolph (ed.), *Taste and the Ancient Senses*, 23–44. Abingdon: Routledge.

Hjelmqvist, H. 1973. "Some Economic Plants from Ancient Cyprus." In V. Karageorghis, *Excavations in the Necropolis of Salamis III*, 243–55. Nicosia: Department of Antiquities.

———. 1979. "Some Economic Plants and Weeds from the Bronze Age of Cyprus." In U. Öbrink, *Hala Sultan Tekke 5: Excavations in Area 22 1971–1973 and 1975–1978*, 110–33. Göteborg: Paul Åströms.

Hoch, J. E. 1994. *Semitic Words in Egyptian Texts of the New Kingdom and Third Intermediate Period.* Princeton University Press.

Højlund, F., and H. H. Andersen. 1994. *Qala'at al-Bahrain, Vol. 1. The Northern City Wall and the Islamic Fortress.* Aarhus University Press.

Holtz, S. E. 2014. *Neo-Babylonian Trial Records.* Atlanta: Society of Biblical Literature.

Hoogervorst, T. 2013. *Southeast Asia in the Ancient Indian Ocean World.* Oxford: Archaeopress.

Horden, P., and N. Purcell. 2000. *The Corrupting Sea: A Study of Mediterranean History.* Oxford: Wiley-Blackwell.

Horowitz, W. 2008. "'The Ship of the Desert, the Donkey of the Sea': The Camel in Early Mesopotamia Revisited." In C. Cohen et al. (eds.), *Birkat Shalom: Studies in the Bible, Ancient Near Eastern Literature and Postbiblical Judaism Presented to Shalom M. Paul on the Occasion of His Seventieth Birthday, Vol. 2,* 597–611. Winona Lake, IN: Eisenbrauns.

Howard, J. K., and W. A. Starck. 1975. *Distribution and Relative Abundance of Billfishes (Istiophoridae) of the Indian Ocean.* University of Miami.

Hoyland, R. G. 2001. *Arabia and the Arabs: From the Bronze Age to the Coming of Islam.* London: Routledge.

Høyrup, J. 1992. "Sumerian: The Descendant of a Proto-Historical Creole? An Alternative Approach to the Sumerian Problem." *AION: Annali del Seminario di studi del mundo classico, Sezione linguistica* 14, 21–72.

Hunt, H. V., M. V. Linden, X. Liu, G. Motuzaite-Matuzeviciute, S. Colledge, and M. K. Jones. 2008. "Millets across Eurasia: Chronology and Context of Early Records of the Genera *Panicum* and *Setaria* from Archaeological Sites in the Old World." *Vegetation History and Archaeobotany* 17.1, 5–18.

Hutchinson, J. B., and R. L. M. Ghose. 1937. "The Classification of Cottons of Asia and Africa." *Indian Journal of Agricultural Science* 7, 233–57.

Iorizzo, M., et al. 2013. "Genetic Structure and Domestication of Carrot (*Daucus carota* subsp. *sativus*) (Apiaceae)." *American Journal of Botany* 100.5, 930–38.

Isaac, E. 1959. "Influence of Religion on the Spread of Citrus." *Science* 129, 179–86.

Ishikawa, R., C. C. Castillo, and D. Q. Fuller. 2020. "Genetic Evaluation of Domestication-Related Traits in Rice: Implications for the Archaeobotany of Rice Origins." *Archaeological and Anthropological Sciences* 12(8), 1–14.

Jain, D. C. 1980. *Economic Life in Ancient India as Depicted in Jain Canonical Literature.* Vaishali: Research Institute of Prakrit, Jainology and Ahimsa.

Jaksch, G. 2012. *Ausgewählte Pflanzendarstellungen auf Textilien des spätantiken Ägypten.* Doctoral dissertation, University of Vienna.

Janick, J., H. S. Paris, and D. C. Parrish. 2007. "The Cucurbits of Mediterranean Antiquity: Identification of Taxa from Ancient Images and Descriptions." *Annals of Botany* 100, 1441–57.

Jashemski, W. F. 1979. *The Gardens of Pompeii, Herculaneum and the Villas Destroyed by Vesuvius.* New Rochelle: Caratzas Brothers.

Jashemski, W. F., G. M. Meyer, and M. Ricciardi. 2002. "Plants: Evidence from Wall Paintings, Mosaics, Sculpture, Plant Remains, Graffiti, Inscriptions and Ancient Authors." In W. F. Jashemski and F. G. Meyer (eds.), *The Natural History of Pompeii,* 80–180. Cambridge University Press.

Joannès, F. 1989. *Archives de Borsippa: la famille Ea-Ilûta-Bâni: étude d'un lot d'archives familiales en Babylonie du VIIIe au Ve siècle av. J.-C.* Geneva: Librairie Droz.

———. 1992. "Les temples de Sippar et leurs trésors à l'époque néo-babylonienne." *Revue d'assyriologie et d'archéologie orientale* 86, 159–84.

———. 2014. "Fabrics and Clothes from Mesopotamia during the Achaemenid and Seleucid Periods: The Textual References." In C. Breniquet and C. Michel (eds.), *Wool Economy in the Ancient Near East and the Aegean*, 453–64. Oxford: Oxbow.

Johannessen, C. L., and A. Z. Parker. 1989. "Maize Ears Sculptured in 12th and 13th Century a.D. India as Indicators of Pre-Columbian Diffusion." *Economic Botany* 43.2, 164–80.

Johns, J. 1984. "A Green Revolution? Review of Agricultural Innovation in the Early Islamic World by Andrew Watson." *Journal of African History* 25, 343–44.

Johnson, A. W. 1972. "Individuality and Experimentation in Traditional Agriculture." *Human Ecology* 12, 149–59.

Jones, J., and R. Oldfield. 2006. "'A Kind of Wool Is Made by the Egyptians from a Tree . . .'" *Archaeological Textiles Newsletter* 43, 27–32.

Jones, M., H. Hunt, E. Lightfoot, D. Lister, X. Liu, and G. Motuzaite-Matuzeviciute. 2011. "Food Globalization in Prehistory." *World Archaeology* 43.4, 665–75.

Jursa, M. 1995. *Die Landwirtschaft in Sippar in Neubabylonischer Zeit.* Vienna: Selbstverlag des Instituts fur Orientalistik der Universitat.

———. 1998. *Der Tempelzehnt in Babylonien vom siebenten bis zum dritten Jahrhundert v. Chr.* Münster: Ugarit.

———. 1999/2000. "Review of D. T. Potts, Mesopotamian Civilization: The Material Foundations." *Archiv für Orientforschung* 46/47, 290–95.

———. 2003. "Spätachämenidische texte aus Kutha." *Revue d'assyriologie et d'archéologie orientale* 97.1, 43–140.

———. 2009. "Die Kralle des Meeres und andere Aromata." In W. Arnold, M. Jursa, W. W. Müller, and S. Procházka (eds.), *Philologisches und Historisches zwischen Anatolien und Soqotra: Analecta Semitica In Memoriam Alexander Sima*, 147–80. Wiesbaden: Harrassowitz.

———. 2010. *Aspects of the Economic History of Babylonia in the First Millennium BC.* Münster: Ugarit.

Kamash, Z. 2012. "Irrigation Technology, Society and Environment in the Roman Near East." *Journal of Arid Environments* 86, 65–74.

Kaniewski, D., E. Paulissen, E. Van Campo, M. Al-Maqdissi, J Bretschneider, and K. Van Lerberghe. 2008. "Middle East Coastal Ecosystem Response to Middle-to-Late Holocene Abrupt Climate Changes." *Proceedings of the National Academy of Sciences* 105(37), 13941–46.

Karttunen, K. 1989. *India in Early Greek literature.* Helsinki: Finnish Oriental Society.

———. 1997. *India and the Hellenistic World.* Helsinki: Finnish Oriental Society.

———. 2014. "Aśoka, the Buddhist Saṁgha and the Graeco-Roman World." *Studia Orientalia Electronica* 112, 35–40.

Kashyap, A., and S. Weber. 2013. "Starch Grain Analysis and Experiments Provide Insights into Harappan Cooking Practices." In S. Abraham et al. (eds.), *Connections and Complexity: New Approaches to the Archaeology of South Asia*, 177–94. Walnut Creek, CA: Left Coast Press.

Kawami, T. S. 1992. "Archaeological Evidence for Textiles in Pre-Islamic Iran." *Iranian Studies* 25, 7–18.

Kelley, A. 2019. "By Land or by Sea: Tracing the Adoption of Cotton in the Economies of the Mediterranean." In M. Ivanova and H. Jeffery (eds.), *Transmitting and Circulating the Late Antique and Byzantine Worlds*, 274–97. Leiden: Brill.

King, A. 2015. "The New *Materia Medica* of the Islamicate Tradition: The Pre-Islamic Context." *Journal of the American Oriental Society* 135.3, 499–528.

Kingwell-Banham, E.J. 2015. *Early Rice Agriculture in South Asia: Identifying Cultivation Systems Using Archaeobotany.* Doctoral dissertation, Institute of Archaeology, University College London.

Kingwell-Banham, E., C. Petrie, and D. Fuller. 2015. "Early Agriculture in South Asia." In G. Barker and C. Goucher (eds.), *The Cambridge World History, Vol. 2: A World with Agriculture, 12,000 BCE–500 CE*, 261–88. Cambridge University Press.

Kingwell-Banham, E., E. Harvey, R. Mohanty, and D. Fuller. 2018. "Archaeobotanical Investigations into Golbai Sasan and Gopalpur, Two Neolithic-Chalcolithic Settlements of Odisha." *Ancient Asia* 9, 1–14.

Kenoyer, J.M. 1997. "Trade and Technology of the Indus Valley: New Insights from Harappa, Pakistan." *World Archaeology* 29.2, 262–80.

———. 2008. "Indus and Mesopotamian Trade Networks: New Insights from Shell and Carnelian Artifacts." In E. Olijdam and R.H. Spoor (eds.), *Intercultural Relations between South and Southwest Asia: Studies in commemoration of E.C.L. During Caspers (1934–1996)*, 19–28. Oxford: Archaeopress.

Kenoyer, J.M., T.D. Price, and J.H. Burton. 2013. "A New Approach to Tracking Connections between the Indus Valley and Mesopotamia: Initial Results of Strontium Isotope Analyses from Harappa and Ur." *Journal of Archaeological Science* 40.5, 2286–97.

Kessler, K. 2002. "Sittake, Sittakene, Sattagū." *Altorientalische Forschungen* 29, 238–48.

Khan, M.M.H., M.Z. Alam, M.M. Rahman, M.I.H. Miah, and M.M. Hossain. 2012. "Influence of Weather Factors on the Incidence and Distribution of Pumpkin Beetle Infesting Cucurbits." *Bangladesh Journal of Agricultural Research* 37.2, 361–67.

Khoury, C.K., et al. 2016. "Origins of Food Crops Connect Countries Worldwide." *Proceedings of the Royal Society: Biological Sciences* 283, 1–9.

Killermann, S. 1916. "Die Zitronen und Orangen in Geschichte und Kunst." *Naturwissenschaftliche Wochenschrift* 31, 201–08.

Kintaert, T. 2010. "On the Cultural Significance of the Leaf of the Indian Lotus: Introduction and Uses." In E. Franco and M. Zin (eds.), *From Turfan to Ajanta: Festschrift for Dieter Schlingloff on the Occasion of his Eightieth Birthday*, 481–512. Lumbini International Research Institute.

Kintaert, T. 2011–12. "On the Role of the Lotus Leaf in South Asian Cosmography." *Wiener Zeitschrift für die Kunde Südasiens* 54, 85–120.

Kislev, M.E. 2015. "Infested Stored Crops in the Iron Age I Granary at Tel Hadar." *Israel Journal of Plant Sciences* 62.1–2, 86–97.

Kleber, K. 2010. "Eanna's Trade in Wool." In M. Jursa et al. (eds.), *Aspects of the Economic History of Babylonia in the First Millennium B.C.: Economic Geography, Economic Mentalities, Agriculture, the Use of Money and the Problem of Economic Growth*, 595–616. Münster: Ugarit Verlag.

———. 2011. "Review of S. Zawadzki, Garments of the Gods: Studies on the Textile Industry and the Pantheon of Sippar according to the Texts from the Ebabbar Archive." *Orientalistische Literaturzeitung* 106, 86–90.

Knörzer, K-H. 1966. "Über Funde römischer Importfrüchte in Novaesium (Neuss/Rh)." *Bonner Jahrbücher* 166, 433–43.

———. 1970. *Römerzeitliche Pflanzenfunde in Neuss (Novaesium 4).* Berlin: Gebr. Mann.
Konen, H. 1995. "Die Kürbisgewächse (Cucurbitaceen) als Kulturpflanzen im römischen Ägypten, 1.-3. Jh. n. Chr." *Münstersche Beiträge zur Antiken Handelsgeschichte* 14.1, 43–81.
———. 1999. "Reis im Imperium Romanum: Bemerkungen zu seinem Anbau und seiner Stellung ah Bedarfs- und Handelsartikel in der ròmischen Kaiserzeit." *Münstersche Beiträge zur Antiken Handelsgeschichte* 18, 23–47.
König, M. 2001. "Die Grundlagen der Ernährung im römischen Trier." In H. P. Kuhnen (ed.), *Das römische Trier.* Stuttgart: Theiss.
Kopp, G. H., C. Roos, T. M. Butynski, D. E. Wildman, A. N. Alagaili, L. F. Groeneveld, and D. Zinner. 2014. "Out of Africa, but How and When? The Case of Hamadryas Baboons (*Papio hamadryas*)." *Journal of Human Evolution* 76, 154–64.
Koren, O. G., M. S. Yatsunskaya, and O. V. Nakonechnaya. 2012. "Low Level of Allozyme Polymorphism in Relict Aquatic Plants of the Far East *Nelumbo komarovii* Grossh. and *Euryale ferox* Salisb." *Russian Journal of Genetics* 48, 912–19.
Kosmin, P. 2014. *The Land of the Elephant Kings: Space, Territory and Ideology in the Seleucid Empire.* Harvard University Press.
Kothari, A. S. 2007. *A Celebration of Indian Trees.* Mumbai: Marg.
Krauss, B. H. 1993. *Plants in Hawaiian Culture.* University of Hawaii Press.
Krishnamurti, B. 2003. *The Dravidian Languages.* Cambridge University Press.
Kroll, H. 1982. "Kulturpflanzen von Tiryns." *Archäologischer anzeiger* 3, 467–85.
Kronasser, H. 1960. "κολο- 'groß'." *Die Sprache* 6, 172–78.
Kulke, H., and D. Rothermund. 2004. *A History of India,* 4th ed. Abingdon: Routledge.
Kumar, S., D. C. Rai, K. Niranjan, and Z. F. Bhat. 2014. "Paneer: An Indian Soft Cheese Variant: A Review." *Journal of Food Science and Technology* 51.5, 821–31.
Kučan, D. 1992. "Die Pflanzenreste aus dem römischen Militärlager Oberaden." In J. S. Kühlborn (ed.), *Das Römerlager in Oberaden III. Die Ausgrabungen im Nordwestlichen Lagerbereich und weitere Baustellenuntersuchungen der Jahre 1962–1988*, 237–65. Münster: Aschendorff.
———. 1995. "Zur Ernährung und dem Gebrauch von Pflanzen im Heraion von Samos im 7. Jahrhundert v. Chr." *Jahrbuch des Deutschen Archäologischen Instituts* 110, 1–64.
Kuhrt, A. 1995. *The Ancient Near East,* 2 vols. Abingdon: Routledge.
Kumbaric, A., and G. Caneva. 2014. "Updated Outline of Floristic Richness in Roman Iconography." *Rendiconti Lincei* 25.2, 181–93.
Kupper, J. R. 1992. "Le bois à Mari." *Bulletin on Sumerian Agriculture* 6, 163–70.
Kuttner, A. 2005. "Cabinet Fit for a Queen: The Λιθικά as Posidippus' Gem Museum." In K. Gutzwiller (ed.), *The New Posidippus: A Hellenistic Poetry Book*, 141–63. Oxford University Press.
Kuzucuoğlu, C., W. Dörfler, S. Kunesch, and F. Goupille. 2011. "Mid-to Late-Holocene Climate Change in Central Turkey: The Tecer Lake Record." *Holocene* 21.1, 173–88.
Lafont, B. 2020. "Note sur les chameaux bactriens attestés à Sumer." In D. Agut-Labordère and B. Redon (eds.), *Les vaisseaux du désert et des steppes: Les camélidés dans l'Antiquité*, 59–64. Lyon: MOM.
Lagardère, V. 1996. "La riziculture en al-Andalus (VIIIe-XVe siècles)." *Studia Islamica* 83, 71–87.
Lamberg-Karlovsky, C. C. 2013. "Iran and Its Neighbours." In H. Crawford (ed.), *The Sumerian World*, 559–78. Abingdon: Routledge.

Lambert, W. G. 1987. "Devotion: The Languages of Religion and Love." In M. Mindlin et al. (eds.), *Figurative Language in the Ancient Near East*, 25–40. London: School of Oriental and African Studies.

Lamm, C. J. 1937. *Cotton in Mediaeval Textiles of the Near East.* Paris: Paul Geuthner.

Landsberger, B. 1934. *Die Fauna des alten Mesopotamien nach der 14. Tafel der Serie ḪAR-ra = ḫubullu.* Leipzig: Hirzel.

Laneri, N. 2014. "The Lifestyle of Ancient Entrepreneurs: Trade and Urbanisation in the Ancient Near East during the Early 2nd Millennium BC." In C. C. Lamberg-Karlovsky and B. Genito (eds.), *"My Life Is Like the Summer Rose": Maurizio Tosi e l'Archeologia come modo di vivere: Papers in Honour of Maurizio Tosi for his 70th birthday*, 401–09. Oxford: Archaeopress.

Lang, P. 2012. "Hekataios (264)." *Brill's New Jacoby*, online edition.

Langer, R. H. M., and G. D. Hill. 1982. *Agricultural Plants.* Cambridge University Press.

Langgut, D. 2014. "Prestigious Fruit Trees in Ancient Israel: First Palynological Evidence for Growing *Juglans regia* and *Citrus medica*." *Israel Journal of Plant Sciences* 62, 98–110.

———. 2017. "The History of *Citrus medica* (Citron) in the Near East: Botanical Remains and Ancient Art and Texts." In V. Zech-Matterne and G. Fiorentino (eds.), *AGRUMED: Archaeology and History of Citrus Fruit in the Mediterranean.* Naples: Centre Jean Bérard (https://books.openedition.org/pcjb/2184).

Langgut, D., Y. Gadot, N. Porat, and O. Lipschits. 2013. "Fossil Pollen Reveals the Secrets of the Royal Persian Garden at Ramat Rahel, Jerusalem." *Palynology* 37.1, 115–29.

Langgut, D., I. Finkelstein, T. Litt, F. H. Neumann, and M. Stein. 2015. "Vegetation and Climate Changes during the Bronze and Iron Ages (~ 3600–600 BCE) in the Southern Levant based on Palynological Records." *Radiocarbon* 572, 217–35.

Larsen, M. T. 2015. *Ancient Kanesh: A Merchant Colony in Bronze Age Anatolia.* Cambridge University Press.

Laufer, B. 1919. *Sino-Iranica: Chinese Contributions to the History of Civilization in Ancient Iran: With Special Reference to the History of Cultivated Plants and Products.* Chicago: Field Museum of Natural History.

Laursen, S., and P. Steinkeller. 2017. *Babylonia, the Gulf Region and the Indus: Archaeological and Textual Evidence for Contact in the Third and Early Second Millennia B.C.* Winona Lake, IN: Eisenbrauns.

Law, R. 2014. "Evaluating Potential Lapis Lazuli Sources for Ancient South Asia using Sulphur Isotope Analysis." In C. C. Lamberg-Karlovsky and B. Genito (eds.), *"My Life is Like the Summer Rose": Maurizio Tosi e l'Archeologia come modo di vivere: Papers in Honour of Maurizio Tosi for his 70th Birthday*, 419–29. Oxford: Archaeopress.

Lawal, R. A., and O. Hanotte. 2021. "Domestic Chicken Diversity: Origin, Distribution, and Adaptation." *Animal Genetics* 52(4), 385–94.

Lawrence, D., G. Philip, H. Hunt, L. Snape-Kennedy, and T. J. Wilkinson. 2016. "Long Term Population, City Size and Climate Trends in the Fertile Crescent: A First Approximation." *PloS One* 11(3), 1–16.

Lazaridis, I., et al. 2016. "Genomic Insights into the Origin of Farming in the Ancient Near East." *Nature* 536, 419–24.

Lecce, M. 1958. *La coltura del riso in territorio veronese (secoli XVI–XVIII).* Verona: Vittore Gualandi.

Lechevallier, M., and G. Quivron. 1981. "The Neolithic in Baluchistan: New Evidence from Mehrgarh." In H. Hartel (ed.), *South Asian Archaeology 1979*, 93–114. Berlin: Dietrich Reimer.

Leemans, W. F. 1960. *Foreign Trade in the Old Babylonian Period as Revealed by Texts from Southern Mesopotamia*. Leiden: Brill.

Leroi-Gourhan, A. 1985. "Les pollens et l'embaumement." In L. Balout and C. Roubet (eds.), *La momie de Ramsès II: Contribution scientifique à l'égyptologie*, 162–65. Paris: Editions Recherche sur les Civilisations.

Letellier-Willemin, F. 2019. "Le coton à El Deir: Premières observations sur l'existence d'une nouvelle fibre textile dans l'oasis de Kharga." *Revue d'ethnoécologie* 15, 1–19.

Letellier-Willemin, F., and C. Moulherat. 2006. "Le découverte de coton dans une nécropole du site d'El Deir, Oasis de Kharga, désert occidental egyptien." *Archaeological Textiles Newsletter* 43, 20–27.

Lévi-Provençal, E. 1932. *L'Espagne musulmane au Xème siècle: institutions et vie sociale*. Paris: Larose.

Lewicka, P. 2011. *Food and Foodways of Medieval Cairenes: Aspects of Life in an Islamic Metropolis of the Eastern Mediterranean*. Leiden: Brill.

Lewis, B. 1976. *Islam from the Prophet Muhammad to the Capture of Constantinople*, Vol. II. New York: Harper and Row.

Li, S., M.H. Pan, C.Y. Lo, D. Tan, Y. Wang, F. Shahidi, and C.T. Ho. 2009. "Chemistry and Health Effects of Polymethoxyflavones and Hydroxylated Polymethoxyflavones." *Journal of Functional Foods* 1.1, 2–12.

Li, Y., T. Smith, J. Yang, J.H. Jin, and C.S. Li. 2014a. "Paleobiogeography of the Lotus Plant (Nelumbonaceae: *Nelumbo*) and Its Bearing on the Paleoclimatic Changes." *Palaeogeography, Palaeoclimatology, Palaeoecology* 399, 284–93.

Li, Y., J. Yao, and C. Li. 2014b. "A Review on the Taxonomic, Evolutionary and Phytogeographic Studies of the Lotus Plant (Nelumbonaceae: *Nelumbo*)." *Acta Geologica Sinica* (English Edition) 88.4, 1252–61.

Librado, P., et al. 2021. "The Origins and Spread of Domestic Horses from the Western Eurasian Steppes." *Nature* 598, 634–40.

Lichtenberg, R.J., and A.C. Thuilliez. 1981. "Sur quelques aspects insolites de la radiologie de Ramsès II." *Bulletins et mémoires de la société d'anthropologie de Paris* 8.3, 323–30.

Lienhard, S. 2000. "On a Number of Names for Lotos in Particular Aravinda." *Indo-Iranian Journal* 43.4, 397–402.

———. 2007. "'Blood' and 'Lac': On Skt. *rakta*/ *lakta* and Related Lexemes Denoting Redness." In O. von Hinüber (ed.), *Siegfried Lienhard: Kleine Schriften*, 419–29. Wiesbaden: Harrassowitz.

Lindtner, C. 1988. "Buddhist References to Old Iranian Religion." In W. Sundermann et al. (eds.), *A Green Leaf: Papers in Honour of Professor J.P. Asmussen*, 433–44. Leiden: Brill.

Lipiński, E. 2000. *The Aramaeans: Their Ancient History, Culture, Religion*. Leuven: Peeters.

Lippi, M. 2000. "The Garden of the 'Casa delle Nozze di Ercole ed Ebe' in Pompeii (Italy): Palynological Investigations." *Plant Biosystems* 134.2, 205–11.

———. 2012. "Ancient Floras, Vegetational Reconstruction and Man-Plant Relationships: Case Studies from Archaeological Sites." *Bocconea* 24, 105–13.

Lipschits, O., Y. Gadot and D. Langgut. 2012. "The Riddle of Ramat Raḥel: The Archaeology of a Royal Persian Period Edifice." *Transeuphraten* 41, 57–79.

Liu, X., and M. K. Jones. 2014. "Food Globalisation in Prehistory: Top Down or Bottom Up?" *Antiquity* 88, 956–63.

Livarda, A. 2008. *Introduction and Dispersal of Exotic Food Plants into Europe during the Roman and Medieval Periods.* Doctoral dissertation, School of Archaeology and Ancient History, University of Leicester.

———. 2011. "Spicing Up Life in Northwestern Europe: Exotic Food Plant Imports in the Roman and Medieval World." *Vegetation History and Archaeobotany* 20, 143–64.

Llamas, B., et al. 2016. "Ancient Mitochondrial DNA Provides High-Resolution Time Scale of the Peopling of the Americas." *Science Advances* 2.4, e1501385.

Lodwick, L., and E. Rowan. 2022. "Archaeobotanical Research in Classical Archaeology." *American Journal of Archaeology* 126(4), 593–623.

Lombard, P. 1999. "Bahrein à l'âge du Fer: les derniers siècles de Dilmoun, 1000–500 av. J.–C." In P. Lombard, *Bahreïn, la civilisation des deux mers, de Dilmoun à Tylos*, 130–44. Paris: Institut du Monde Arabe.

Londo, J. P., et al. 2006. "Phylogeography of Asian Wild Rice, *Oryza rufipogon*, Reveals Multiple Independent Domestications of Cultivated Rice, *Oryza sativa*." *Proceedings of the National Academy of Sciences* 103, 9578–83.

Löw, I. 1881. *Aramäische Pflanzennamen.* Leipzig: W. Engelmann.

———. 1924–34. *Die Flora der Juden.* 4 vols. Wien: Löwit.

Lückge, A., H. Doose-Rolinski, A. A. Khan, H. Schulz, and U. Von Rad. 2001. "Monsoonal Variability in the Northeastern Arabian Sea during the Past 5000 Years: Geochemical Evidence from Laminated Sediments." *Palaeogeography, Palaeoclimatology, Palaeoecology* 167(3–4), 273–86.

Lumaga, M. R. B., E. R. Ermolli, B. Menale, and S. Vitale. 2020. "Exine Morphometric Analysis as a New Tool for Citrus Species Identification: A Case Study from Oplontis (Vesuvius Area, Italy)." *Vegetation History and Archaeobotany* 29, 671–80.

Maaijer, R., and B. Jagersma. 2003/2004. "Review of The Sumerian Dictionary of the University of Pennsylvania Museum, A, Part III." *Archiv für Orientforschung* 50, 351–55.

Mabberley, D. J. 1997. "A Classification for Edible *Citrus*." *Telopea* 7, 167–72.

———. 2004. "*Citrus* (Rutaceae): A Review of Recent Advances in Etymology, Systematics and Medical Applications." *Blumea* 49, 481–98.

Macdonell, A. A., and A. B. Keith. 1912. *Vedic Index of Names and Subjects*, 2 vols. London: John Murray.

Magee, P. 2014. *The Archaeology of Prehistoric Arabia: Adaptation and Social Formation from the Neolithic to the Iron Age.* Cambridge University Press.

Magee, P., and C. A. Petrie. 2010. "West of the Indus—East of the Empire: The Archaeology of the Pre-Achaemenid and Achaemenid Periods in Baluchistan and the North-West Frontier Province, Pakistan." In J. Curtis and S. J. Simpson (eds.), *The World of Achaemenid Persia: History, Art and Society in Iran and the Ancient Near East*, 503–22. London: I. B. Tauris.

Mahadevan, I. 2003. *Early Tamil Epigraphy: From the Earliest Times to the Sixth Century AD.* Harvard University Press.

Mai, B. T., and M. Girard. 2014. "Citrus (Rutaceae) was Present in the Western Mediterranean in Antiquity." In A. Chevalier, E. Marinova, and L. Peña-Chocarro (eds.), *Plants and People: Choices and Diversity through Time*, 170–74. Oxford: Oxbow.

Mairs, R. 2014. *The Hellenistic Far East: Archaeology, Language, and Identity in Greek Central Asia*. University of California Press.

Maloney, C. 1974. *Peoples of South Asia*. Holt, Rinehart and Winston.

Manniche, L. 2006. *An Ancient Egyptian Herbal*. London: British Museum.

Marcus, A. 1989. *The Middle East on the Eve of Modernity: Aleppo in the Eighteenth Century*. Columbia University Press.

Marcus, D. 2007. "Books of Ezra and Nehemiah." In M. Berenbaum and F. Skolnik (ed.), *Encyclopaedia Judaica*, Vol. 6, 656–62. Detroit: Macmillan Reference USA.

Margariti, C., S. Protopapas, and V. Orphanou. 2010. "Recent Analyses of the Excavated Textile Find from Grave 35 HTR73, Kerameikos Cemetery, Athens, Greece." *Journal of Archaeological Science* 38, 522–27.

Marinval, P. 2000. "Agriculture et structuration du paysage agricole à Marseille grec et dans les sociétés indigènes aux premiers et second âges du Fer." In J. M. Luce (ed.), *Paysage et alimentation dans le monde grec*, 183–94. Toulouse: Presses Universitaires du Mirail.

———. 2003. "Les cucurbitacées antiques." *Archéologia* 405, 22–29.

Marinval, P., D. Maréchal, and D. Labadie. 2002. "Arbres fruitiers et cultures jardinées gallo-romains à Longueil-Sainte-Marie (Oise)." *Gallia* 59.1, 253–71.

Marr, K. L., Y. M. Xia, and N. K. Bhattarai. 2005. "Allozymic, Morphological, Phenological, Linguistic, Plant Use, and Nutritional Data on Wild and Cultivated Collections of *Luffa aegyptiaca* Mill. (Cucurbitaceae) from Nepal, Southern China, and Northern Laos." *Economic Botany* 59.2, 137–53.

Marx, K. 1853. "The East India Company: Its History and Results." *New York Daily Tribune*, July 11.

Mashkour, M., M. Tengberg, Z. Shirazi and Y. Madjidzadeh. 2013. "Bioarchaeological Studies at Konar Sandal, Halil Rud Basin, Southeastern Iran." *Environmental Archaeology* 18.3, 222–46.

Masson, V. M., and V. I. Sarianidi. 1972. *Central Asia*. London: Thames and Hudson.

Matthews, P. 1991. "A Possible Tropical Wildtype Taro: *Colocasia esculenta* var. *aquatilis*." *Bulletin of the Indo-Pacific Prehistory Association* 11, 69–81.

Matthews, P. J., and M. E. Ghanem. 2021. "Perception Gaps That May Explain the Status of Taro (*Colocasia esculenta*) as an 'Orphan Crop'." *Plants, People, Planet* 3(2), 99–112.

Matthews, R. 2002. "Zebu: Harbingers of Doom in Bronze Age Western Asia?" *Antiquity* 76, 438–46.

Mattioli, P. A. 1565. *Commentarii in sex libros Pedacii Dioscoridis Anazarbei de Medica material*. Venice: Officina Valgrisiana.

Maxwell-Hyslop, K. R. 1983. "*Dalbergia sissoo* Roxburgh." *Anatolian Studies* 33, 67–72.

Mayrhofer, M. 1971. "Neuere Forschungen zum Altpersischen." In R. Schmitt-Brandt (ed.), *Donum Indogermanicum: Festgabe für Anton Scherer*, 41–66. Heidelberg: Carl Winter Universitätsverlag.

———. 1956–80. *Kurzgefasstes etymologisches Wörterbuch des Altindischen I-IV*. Heidelberg: Carl Winter Universitätsverlag.

McCorriston, J. 1997. "The Fibre Revolution: Textile Extensification, Alientation, and Social Stratification in Ancient Mesopotamia." *Current Anthropology* 38.4, 517–49.

McCreight, J.D., J.E. Staub, T.C. Wehner, and N.P.S. Dhillon. 2013. "Gone Global: Familiar and Exotic Cucurbits Have Asian Origins." *Horticultural Science* 48.9, 1078–89.

McGill, S. 2012. *Plagiarism in Latin Literature.* Cambridge University Press.

McGrail, S., L. Blue, E. Kentley, and C. Palmer. 2003. *Boats of South Asia.* Abingdon: Routledge.

McHugh, J. 2012. *Sandalwood and Carrion: Smell in Indian Religion and Culture.* Oxford University Press.

McNeill, J.R. 2001. "Biological Exchange and Biological Invasion in World History." In S. Sogner (ed.), *Making Sense of Global History*, 106–19. Oslo: Universitetsforlaget.

———. 2014. "Biological Exchange in Global Environmental History." In J.R. McNeill and E.S. Mauldin (eds.), *A Companion to Global Environmental History*, 433–51. Oxford: Wiley Blackwell.

Mears, J.A. 2011. "Agriculture." In J.H. Bentley (ed.), *The Oxford Handbook of World History*, 143–57. Oxford University Press.

Meissner, B. 1913. "Haben die Assyrer den Pfau gekannt?" *Orientalistische Literaturzeitung* 16, 292–93.

Messedaglia, L. 1938. "Per la storia delle nostre piante alimentari: Il riso." *Rivista di storia delle scienze mediche e naturali* 29, 2–15, 50–64.

Meyboom, P.G. 1995. *The Nile Mosaic of Palestrina: Early Evidence of Egyptian Religion in Italy.* Leiden: Brill.

Meyer, C., et al. 1991. "From Zanzibar to Zagros: A Copal Pendant from Eshnuna." *Journal of Near Eastern Studies* 50.4, 289–98.

Meyer, R.S., A.E. Duval, and H.R. Jensen. 2012. "Patterns and Processes in Crop Domestication: An Historical Review and Quantitative Analysis of 203 Global Food Crops." *New Phytologist* 196.1, 29–48.

Meyerhof, M. 1943–44. "Sur un traité d'agriculture composé par un sultan yéménite du XIVe siècle." *Bulletin de l'Institut d'Egypte* 26, 51–65.

Michalowski, P. 1988. "Magan and Meluḫḫa Once Again." *Journal of Cuneiform Studies* 40.2, 156–64.

Michel, C. 2010. "Women of Aššur and Kaniš." In F. Kulakoğlu and S. Kangal (eds.), *Anatolia's Prologue Kültepe Kanesh Karum: Assyrians in Istanbul*, 124–33. Istanbul: Kayseri Metropolitan Municipality.

———. 2014. "Akkadian Texts: Women in Letters—Old Assyrian Kaniš." In M.W. Chavalas (ed.), *Women in the Ancient Near East*, 205–12. Abingdon: Routledge.

Mikhail, A. 2011. "Global Implications of the Middle Eastern Environment." *History Compass* 9.12, 952–70.

Milano, L. 2004. "Food and Identity in Mesopotamia: A New Look at the Aluzinnu's Recipes." In C. Grottanelli and L. Milano (eds.), *Food and Identity in the Ancient World*, 243–56. Padua: S.A.R.G.O.N.

Miller, J.I. 1969. *The Spice Trade of the Roman Empire, 29 BC to AD 641.* Oxford: Clarendon Press.

Miller, M.C. 2004. *Athens and Persia in the Fifth Century BC: A Study in Cultural Receptivity.* Cambridge University Press.

Miller, N.F. 1981. "Plant Remains from Ville Royale II Susa." *Cahiers de la Délégation Archéologique Française en Iran* 12, 137–42.

Miller, N. F. 2000. "Plant Forms in Jewellery from the Royal Cemetery at Ur." *Iraq* 62, 149–55.

———. 2013. "Symbols of Fertility and Abundance in the Royal Cemetery at Ur, Iraq." *American Journal of Archaeology* 117, 127–33.

Miller, N. F., R. N. Spengler, and M. Frachetti. 2016. "Millet Cultivation across Eurasia: Origins, Spread, and the Influence of Seasonal Climate." *Holocene* 26, 1–10.

Minniti, C., and S. M. S. Sajjadi. 2019. "New Data on Non-Human Primates from the Ancient Near East: The Recent Discovery of a Rhesus Macaque Burial at Shahr-i Sokhta (Iran)." *International Journal of Osteoarchaeology* 29(4), 538–48.

Mitchell, P. 2018. *The Donkey in Human History: An Archaeological Perspective.* Oxford University Press.

Mittermayer, C. 2009. *Enmerkara und der Herr von Arata.* Fribourg: Academic Press.

Molina, J., et al. 2011. "Molecular Evidence for a Single Evolutionary Origin of Domesticated Rice." *Proceedings of the National Academy of Sciences* 108, 8351–56.

Montanari, M. 1994. *The Culture of Food.* Oxford: Blackwell.

Moorey, P. R. S. 1999. *Ancient Mesopotamian Materials and Industries: The Archaeological Evidence.* Winona Lake, IN: Eisenbrauns.

Moraitou, G. 2007. "The Funeral Pyre Textile from Royal Tomb II in Vergina. Report on the 1997 Documentation, Treatment and Display." *Archaeological Textiles Newsletter* 44, 5–10.

Morello, M. 2014. "L'Avorio nella Civiltà dell'Indo: Origini dell'uso e dell'ammaestramento di *Elephas maximus indicus.*" In C. C. Lamberg-Karlovsky and B. Genito (eds.), *"My Life is Like the Summer Rose": Maurizio Tosi e l'Archeologia come modo di vivere: Papers in Honour of Maurizio Tosi for His 70th Birthday*, 531–47. Oxford: Archaeopress.

Motaghed, S. 1990. "Pārche-ye makshuf-e az tābut-e bronzi-ye Kidin Hutran dar Arjan, Behbahān" [Textiles discovered in the bronze coffin of Kidin Hutran in Arjan, Behbahān]. *Asar* 17, 64–147.

Motta, E. 1905. "Per la storia della coltura del riso in Lombardia." *Archivo storico Lombardo ser. IV* 32, 392–400.

Motuzaite-Matuzeviciute, G., R. A. Staff, H. V. Hunt, X. Liu, and M. K. Jones. 2013. "The Early Chronology of Broomcorn Millet (*Panicum miliaceum*) in Europe." *Antiquity* 87, 1073–85.

Moulherat, Ch., M. Tengberg, J.-F. Haquet, and B. Mille. 2002. "First Evidence of Cotton at Neolithic Mehrgarh, Pakistan: Analysis of Mineralized Fibres from a Copper Bead." *Journal of Archaeological Science* 29, 1393–1401.

Mukhopadhyay, B. A. 2021. "Ancestral Dravidian Languages in Indus Civilization: Ultraconserved Dravidian Tooth-Word Reveals Deep Linguistic Ancestry and Supports Genetics." *Humanities and Social Sciences Communications* (https://doi.org/10.1057/s41599-021-00868-w).

Müller, M., et al. 2004. "Identification of Ancient Textile Fibres from Khirbet Qumran Caves Using Synchrotron Radiation Microbeam Diffraction." *Spectrochimica Acta Part B—Atomic Spectroscopy* 59, 1669–74.

Muntz, C. E. 2012. "Diodorus Siculus and Megasthenes: A Reappraisal." *Classical Philology* 107.1, 21–37.

Murphy, C. A., and D. Q. Fuller. 2016. "The Transition to Agricultural Production in India: South Asian Entanglements of Domestication." In G. R. Schug and S. R. Walimbe (eds.), *A Companion to South Asia in the Past*, 344–57. Chichester: Wiley.

Muthukumaran, S. 2012. *Between the Country of the Yavanas and the Continent of the Black Plum: Contact and Commerce between South Asia and the Hellenistic World.* Master's dissertation, University of Oxford.

———. 2016. "Tree Cotton (*G. arboreum*) in Babylonia." In E. Foietta et al. (eds.), *Cultural and Material Contacts in the Ancient Near East*, 98–105. Florence: Apice Libri.

———. 2019. "The Tamil Diaspora in Pre-modern Southeast Asia." In A. Mahizhnan and N. Gopal (eds.), *Sojourners to Settlers: Tamils in Southeast Asia and Singapore*, 26–44. Singapore: Institute of Policy Studies and Indian Heritage Centre.

———. 2021. "The Language(s) behind the Indus Script: The Evidence from Mesopotamia, Genetics and Vedic Substrates." In N. Gopal (ed.), *The Word and the Image*, 12–21. Singapore: Indian Heritage Centre.

Nadeau, R. 2015. "Cookery Books." In J. Wilkins and R. Nadeau (eds.), *A Companion to Food in the Ancient World*, 53–58. Chichester: Wiley-Blackwell.

Naik, K. C. 1963. *South Indian Fruits and Their Culture.* Madras: Varadachary.

Nagarajan, M., K. Nimisha, and S. Kumar. 2015. "Mitochondrial DNA Variability of Domestic River Buffalo (*Bubalus bubalis*) Populations: Genetic Evidence for Domestication of River Buffalo in Indian Subcontinent." *Genome Biology and Evolution* 7.5, 1252–59.

Namdar, D., A. Gilboa, R. Neumann, I. Finkelstein, and S. Weiner. 2013. "Cinnamaldehyde in Early Iron Age Phoenician Flasks Raises the Possibility of Levantine Trade with South East Asia." *Mediterranean Archaeology and Archaeometry* 13.2, 1–19.

Nasrallah, N. 2013. *Delights from the Garden of Eden: A Cookbook and History of the Iraqi Cuisine.* Sheffield: Equinox.

Nazarova, G. 2005. "Indo-European Elements in Turkic Fitonimy (on the Material of Uzbek, Uigur, Kazakh and Turkic Languages)." In H. Bismark (ed.), *Usbekisch-deutsche Studien: Indogermanische und ausserindogermanische Kontakte in Sprache, Literatur und Kultur*, 77–88. Münster: Lit.

N'dri, A. N. A., B. I. A. Zoro, L. P. Kouamé, D. Dumet, and I. Vroh-Bi. 2016. "On the Dispersal of Bottle Gourd [*Lagenaria siceraria* (Mol.) Standl.] Out of Africa: A Contribution from the Analysis of Nuclear Ribosomal DNA Haplotypes, Divergent Paralogs and Variants of 5.8S Protein Sequences." *Plant Molecular Biology Reporter* 34.2, 454–66.

Nesbitt, M., St. J. Simpson, and I. Svanberg. 2010. "History of Rice in Western and Central Asia." In S. D. Sharma (ed.), *Rice: Origin, Antiquity and History*, 308–40. Enfield, NH: Science.

Nesbitt, M., J. Bates, S. Mitchell, and G. Hillman 2017. *The Archaeobotany of Aşvan: Environment and Cultivation in Eastern Anatolia from the Chalcolithic to the Medieval Period.* London: British Institute at Ankara.

Nezafati, N., M. Momenzadeh, and E. Pernicka, E. 2008. "New Insights into the Ancient Mining and Metallurgical Researches in Iran." In Ü. Yalçin, H. Özbal, and A. G. Paşamehmetoğlu (eds.), *Ancient Mining in Turkey and the Eastern Mediterranean*, 307–28. Ankara: Atılım University.

Newberry, P. E. 1889. "On the Vegetable Remains Discovered in the Cemetery of Hawara." In W. M. F. Petrie, *Hawara, Biahmu and Arsinoe*, 46–53. London: Leadenhall Press.

———. 1890. "The Ancient Botany." In W. M. F. Petrie, *Kahun, Gurob and Hawara*, 46–50. London: Kegan Paul.

Newton, C. 2007. "Growing, Gathering and Offering: Predynastic Plant Economy at Adaïma (Upper Egypt)." In R. T. Cappers, *Fields of Change: Progress in African Archaeobotany*, Vol. 5, 139–55. Groningen: Barkhuis.

Nicolson, D. H. 1987. "Derivation of Aroid Generic Names." *Aroideana* 10.3, 15–25.

Niese, B. 1922. "Excerpta ex Eudemi codice Parisino n. 2635." *Philologus Supplement* 15, 145–60.

Norman, K. R. 1983. *Pāli Literature.* Wiesbaden: Harrassowitz.

Oates, D. 1968. *Studies in the Ancient History of Northern Iraq.* London: British Academy.

Olijdam, E. 1997. "Babylonian Quest for Lapis Lazuli and Dilmun during the City III Period." In R. Allchin and B. Allchin (eds.), *South Asian Archaeology 1995*, 119–26. New Delhi: Oxford and IBH.

———. 2000. "Additional Evidence of a Late Second Millennium Lapis Lazuli Route: The Fullol Hoard." In M. Taddei and G. de Marco (eds.), *South Asian Archaeology 1997*, 397–407. Rome: Istituto italiano per l'Africa e l'Oriente.

Oppenheim, A. L. 1967. "Essay on Overland Trade in the First Millennium BC." *Journal of Cuneiform Studies* 21, 236–54.

Ottoni, C., et al. 2017. "The Palaeogenetics of Cat Dispersal in the Ancient World." *Nature Ecology and Evolution* 1(7), 1–7.

Overton, M. 1996. *Agricultural Revolution in England: The Transformation of the Agrarian Economy 1500–1850.* Cambridge University Press.

Pagnoux, C., A. Celant, S. Coubray, G. Fiorentino, and V. Zech-Matterne. 2013. "The Introduction of Citrus to Italy, with Reference to the Identification Problems of Seed Remains." *Vegetation History and Archaeobotany* 22.5, 421–38.

Palmer, S. A., et al. 2012. "Archaeogenomic Evidence of Punctuated Genome Evolution in Gossypium." *Molecular Biology and Evolution* 29(8), 2031–38.

Panagiotakopulu, E. 1998. "An Insect Study from Egyptian Stored Products in the Liverpool Museum." *Journal of Egyptian Archaeology* 84, 231–34.

———. 2001. "New Records for Ancient Pests: Archaeoentomology in Egypt." *Journal of Archaeological Science* 28.11, 1235–46.

———. 2003. "Insect Remains from the Collections in the Egyptian Museum of Turin." *Archaeometry* 452, 355–62.

Panagiotakopulu, E., and M. van der Veen. 1997. "Synanthropic Insect Faunas from Mons Claudianus, a Roman Quarry Site in the Eastern Desert, Egypt." *Quaternary Proceedings* 5, 199–206.

Pang, X. M., C. G. Hu, and X. X. Deng. 2007. "Phylogenetic Relationships within Citrus and Its Related Genera as Inferred from AFLP Markers." *Genetic Resources and Crop Evolution* 54.2 429–36.

Pareja, M. N., T. McKinney, J. A. Mayhew, J. M. Setchell, S. D. Nash, and R. Heaton. 2020a. "A New Identification of the Monkeys Depicted in a Bronze Age Wall Painting from Akrotiri, Thera." *Primates* 61, 159–68.

Pareja, M. N., T. McKinney, and J. Setchell. M. 2020b. "Aegean Monkeys and the Importance of Cross-Disciplinary Collaboration in Archaeoprimatology: A Reply to Urbani and Youlatos." *Primates* 61, 767–74.

Paris, H. S. 2012. "Semitic-Language Records of Snake Melons (*Cucumis melo,* Cucurbitaceae) in the Medieval Period and the 'Piqqus' of the 'Faqqous.'" *Genetic Resources and Crop Evolution* 59, 31–38.

———. 2016. "Overview of the Origins and History of the Five Major Cucurbit Crops: Issues for Ancient DNA Analysis of Archaeological Specimens." *Vegetation History and Archaeobotany* 25, 405–14.

Paris, H. S., Z. Amar, and E. Lev. 2012a. "Medieval History of the Duda'im Melon (*Cucumis melo,* Cucurbitaceae)." *Economic Botany* 66.3, 276–84.

———. 2012b. "Medieval Emergence of Sweet Melons, *Cucumis melo* (Cucurbitaceae)." *Annals of Botany* 110, 23–33.

Paris, H. S., M. C. Daunay, and J. Janick. 2012. "Occidental Diffusion of Cucumber (*Cucumis sativus*) 500–1300 CE: Two Routes to Europe." *Annals of Botany* 109, 117–26.

Paris, H. S., and J. Janick. 2008a. "Reflections on Linguistics as an Aid to Taxonomical Identification of Ancient Mediterranean Cucurbits: The Piqqus of the Faqqous." In M. Pitrat (ed.), *Cucurbitaceae 2008: Proceedings of the IXth EUCARPIA Meeting on Genetics and Breeding of Cucurbitaceae*, 43–51. Avignon: INRA.

———. 2008b. "What the Roman Emperor Tiberius Grew in His Greenhouses." In M. Pitrat (ed.), *Cucurbitaceae 2008: Proceedings of the IXth EUCARPIA Meeting on Genetics and Breeding of Cucurbitaceae*, 33–42. Avignon: INRA.

———. 2010–11. "The *Cucumis* of Antiquity: A Case of Mistaken Identity." *Cucurbit Genetics Cooperative Report* 33-34, 1–2.

Paris, H. S., J. Janick, and M-C. Daunay. 2011. "Medieval Herbal Iconography and Lexicography of *Cucumis* (Cucumber and Melon, Cucurbitaceae) in the Occident, 1300–1458." *Annals of Botany* 108, 471–84.

Parker, G. 2002. "Ex Oriente Luxuria: Indian Commodities and Roman Experience." *Journal of the Economic and Social History of the Orient* 45.1, 40–95.

———. 2008. *The Making of Roman India*. Cambridge University Press.

Parpola, A. 1975. "India's Name in Early Foreign Sources." *Sri Venkateswara University Oriental Journal* 18.1–2, 9–19.

———. 2015. *The Roots of Hinduism: The Early Aryans and the Indus Civilization*. Oxford University Press.

———. 2020. "Iconographic Evidence of Mesopotamian Influence on Harappan Ideology and Its Survival in the Royal Rites of the Veda and Hinduism." In I. Finkel and St. J. Simpson (eds.), *In Context: The Reade Festschrift*, 183–90. Oxford: Archaeopress.

Parpola, S., A. Parpola, and R. H. Brunswig. 1977. "The Meluḫḫa Village: Evidence of Acculturation of Harappan Traders in Late Third Millennium Mesopotamia?" *Journal of the Economic and Social History of the Orient/Journal de l'histoire economique et sociale de l'Orient* 20, 129–65.

Parthasarathi, P., and G. Riello. 2012. "From India to the World: Cotton and Fashionability." In F. Trentmann (ed.), *The Oxford Handbook of the History of Consumption*, 145–66. Oxford University Press.

Parzinger, H., and N. Boroffka. 2003. *Das Zinn der Bronzezeit in Mittelasien. 1: Die siedlungsarchäologischen Forschungen im Umfeldt der Zinnlagerstätten*. Mainz: Philipp von Zabern.

Paspalas, S. 2006. "The Achaemenid Empire and the North-Western Aegean." *Ancient West and East* 5, 90–120.

Paulus, S. 2018. "Fraud, Forgery, and Fiction: Is There Still Hope for Agum-Kakrime?" *Journal of Cuneiform Studies* 70, 115–66.

Payne, E. E. 2011. "Review of S. Zawadzki, Garments of the Gods: Studies on the Textile Industry and the Pantheon of Sippar According to the Texts from the Ebabbar Archive." *Archiv für Orientforschung* 52, 250–52.

Pedersen, R. K. 2003. *The Boatbuilding Sequence in the Gilgamesh Epic and the Sewn Boat Relation*. Doctoral dissertation, Texas A&M University.

———. 2004. "Traditional Arabian Watercraft and the Ark of the Gilgamesh Epic: Interpretations and Realizations." *Proceedings of the Seminar for Arabian Studies* 34, 231–38.

Pelling, R. 2005. "Garamantian Agriculture and Its Significance in a Wider North African Context: The Evidence of Plant Remains from the Fazzan Project." *Journal of North African Studies* 10, 397–411.

Pelling, R. 2007. *Agriculture and Trade amongst the Garamantes: 3000 years of Archaeobotanical Data from the Sahara and Its Margins*. Doctoral dissertation, Institute of Archaeology, University College London.

———. 2008. "Garamantian Agriculture: The Plant Remains from Jarma, Fazzan." *Libyan Studies* 39, 41–71.

———. 2013. "The Archaeobotanical Remains." In D. J. Mattingly et al. (eds.), *The Archaeology of Fazzan, Vol. 4: Survey and Excavations at Old Jarma (Ancient Garama) carried out by C. M. Daniels (1962–69) and the Fazzān Project, 1997–2001*, 473–94. London: Society for Libyan Studies.

Pelliot, P. 1959. *Notes on Marco Polo*, Vol. I. Paris: Imprimerie Nationale.

Perrot, J., and Y. Madjidzadeh. 2003. "Découvertes récentes à Jiroft (Sud du plateau iranien)." *Comptes rendus des séances de l'Académie des Inscriptions et Belles-Lettres* 147e année, 3, 1087–1102.

Perry, C. 1994. "Tenth-Century Arab Food in Poetry." In H. Walker (ed.), *Look and Feel: Studies in Texture, Appearance and Incidental Characteristics of Food*, 141–43. Totnes: Prospect Books.

Perry, J. 2005. "Šāh-Nāma v. Arabic Words." *Encyclopaedia Iranica*, online edition (https://www.iranicaonline.org/articles/sah-nama-v-arabic-words).

Perry-Gal, L., A. Erlich, A. Gilboa, and G. Bar-Oz. 2015. "Earliest Economic Exploitation of Chicken outside East Asia: Evidence from the Hellenistic Southern Levant." *Proceedings of the National Academy of Sciences* 112, 9849–54.

Peters, J. P. 1913. "The Cock." *Journal of the American Oriental Society* 33, 363–396.

Peters, J., et al. 2022. "The Biocultural Origins and Dispersal of Domestic Chickens." *Proceedings of the National Academy of Sciences* 119(24), e2121978119.

Petrie, C. A. 2013. "Ancient Iran and its Neighbours: The State of Play." In C. A. Petrie (ed.), *Ancient Iran and its Neighbours: Local Developments and Long-Range Interactions in the 4th Millennium BC*, 1–24. Oxford: Oxbow Books.

Petrie, C. A., J. Bates, T. Higham, and R. N. Singh. 2016. "Feeding Ancient Cities in South Asia: Dating the Adoption of Rice, Millet and Tropical Pulses in the Indus Civilisation." *Antiquity* 90, 1489–1504.

Pfister, R., and L. Bellinger, L. 1945. *The Excavations at Dura Europos IV. Part 2: The Textiles*. Yale University Press.

Pingree, D. 1976. "The Indian and Pseudo-Indian Passages in Greek and Latin Astronomical and Astrological Texts." *Viator* 7, 141–95.

Pinnock, F. 1995. "Il commercio e i livelli di scambio nel Periodo Protosiriano." In P. Matthiae, F. Pinnock, and G.S. Matthiae (eds.), *Ebla: Alle origini della civiltà urbana*, 148–55. Milan: Electa.

Pirngruber, R. 2017. *The Economy of Late Achaemenid and Seleucid Babylonia.* Cambridge University Press.

Pisani, V. 1940. "Aethiopes d'Asia e Paricani." *Rivista degli Studi Orientali* 18, 97.

Pitrat, M. 2008. "Melon." In J. Prohens and F. Nuez (eds.), *Vegetables I (Handbook of Plant Breeding)*, 283–315. Dordrecht: Springer.

———. 2013. "Phenotypic Diversity in Wild and Cultivated Melons (*Cucumis melo*)." *Plant Biotechnology* 30, 273–78.

Pitrat, M., P. Hanelt, and K. Hammer. 2000. "Some Comments on Infraspecific Classification of Cultivars of Melon." *Acta Horticulturae* 510 (Proceedings of Cucurbitaceae 2000), 29–36.

Pittman, H. 2013. "New Evidence for Interaction between the Iranian Plateau and the Indus Valley: Seals and Sealings from Konar Sandal South." In S. Abraham et al. (eds.), *Connections and Complexity: New Approaches to the Archaeology of South Asia*, 63–90. Walnut Creek, CA: Left Coast Press.

Plu, A. 1985. "Bois et graines." In L. Balout and C. Roubet (eds.), *La momie de Ramsès II: Contribution scientifique à l'égyptologie*, 166–74. Paris: Editions Recherche sur les Civilisations.

Pokharia, A. K. 2011. "Palaeoethnobotany at Lahuradewa: A Contribution to the 2nd Millennium BC Agriculture of the Ganga Plain, India." *Current Science* 101.12, 1569–78.

Pokharia, A. K., J. N. Pal, and A. Srivastava. 2009. "Plant Macro-Remains from Neolithic Jhusi in Ganga Plain: Evidence for Grain-Based Agriculture." *Current Science* 97(4), 564–72.

Pollard, E. A. 2009. "Pliny's Natural History and the Flavian Templum Pacis: Botanical Imperialism in First-Century CE. Rome." *Journal of World History* 20.3, 309–38.

Pollmann, B. 2003. *Archäobotanische Makrorestanalysen und molekulararchäologische Untersuchungen an botanischen Funden aus dem römischen vicus Tasgetium (Eschenz/Kanton Thurgau/CH).* Doctoral dissertation, University of Basel.

Pollock, S. 2005. "Archaeology Goes to War at the Newsstand." In S. Pollock and R. Bernbeck (eds.), *Archaeologies of the Middle East: Critical Perspectives*, 78–96. Oxford: Blackwell.

Pommerening, T., E. Marinova, and S. Hendrickx. 2010. "The Early Dynastic Origin of the Water-Lily Motif." *Chronique d'Egypte* 85, 14–40.

Ponchia, S. 2014. "Management of Food Resources in the Assyrian Empire: Data and Problems." In L. Milano (ed.), *Paleonutrition and Food Practices in the Ancient Near East*, 385–412. Padua: S.A.R.G.O.N.

Portères, R. 1960. "La sombre Aroidée cultivée: *Colocasia antiquorum* Schott ou taro de Polynésie. Essai d'étymologie sémantique." *Journal d'agriculture tropicale et de botanique appliquée* 7, 169–92.

Possehl, G. L. 1994. "Of Men." In J. M. Kenoyer (ed.), *From Sumer to Meluhha: Contributions to the Archaeology of South and West Asia in Memory of George F. Dales, Jr*, 179–86. Department of Anthropology, University of Wisconsin.

———. 1996. "Meluhha." In J. Reade (ed.), *The Indian Ocean in Antiquity*, 133–208. London: Kegan Paul International.

———. 1998. "The Introduction of African Millets to the Indian Subcontinent." In H. D. V. Prendergast et al. (eds.), *Plants for Food and Medicine: Proceedings of the Joint Conference of the Society for Economic Botany and the International Society for Ethnopharmacology, London, 1–6 July 1996*, 107–21. London: Royal Botanic Gardens, Kew.

———. 2002. *The Indus Civilization: A Contemporary Perspective.* Walnut Creek, CA: Rowman Altamira.

Postgate, J. N. 1984. "Introduction." *Bulletin on Sumerian Agriculture* 1, 1–7.

———. 1987. "Notes on Fruit in the Cuneiform Sources." *Bulletin on Sumerian Agriculture* 3, 115–44.

———. 1992. "Trees and Timber in the Assyrian Texts." *Bulletin on Sumerian Agriculture* 6, 177–92.

Potts, D. T. 1990. *The Arabian Gulf in Antiquity,* 2 vols. Oxford: Clarendon Press.

———. 1997. *Mesopotamian Civilization: The Material Foundations.* London: Athlone Press.

———. 2002. "Total Prestation in Marhashi-Ur Relations." *Iranica Antiqua* 37, 343–57.

———. 2004a. "Camel Hybridization and the Role of *Camelus bactrianus* in the Ancient Near East." *Journal of the Economic and Social History of the Orient* 47.2, 143–65.

———. 2004b. "Exit Aratta: Southeastern Iran and the Land of Marhashi." *Name-ye Iran-e Bastan* 4.1, 1–11.

———. 2005. "In the Beginning: Marhashi and the Origins of Magan's Ceramic Industry in the Third Millennium BC." *Arabian Archaeology and Epigraphy* 16.1, 67–78.

———. 2006. "Elamites and Kassites in the Persian Gulf." *Journal of Near Eastern Studies* 65.2, 111–19.

———. 2007a. "Differing Modes of Contact between India and the West: Some Achaemenid and Seleucid Examples." In D. T. Potts and H. P. Ray (eds.), *Memory as History: The Legacy of Alexander in Asia*, 122–30. New Delhi: Aryan Books International.

———. 2007b. "Revisiting the Snake Burials of the Late Dilmun Building Complex on Bahrain." *Arabian Archaeology and Epigraphy* 18.1, 55–74.

———. 2008. "The Persepolis Fortification Texts and the Royal Road." In P. Briant, W. Henkelman, and M. Stolper (eds.), *L'archive des fortifications de Persépolis: état des questions et perspectives de recherches*, 275–301. Paris: De Boccard.

———. 2019. "Wild Water Buffalo (*Bubalus arnee* [Kerr, 1792]) in the Ancient Near East." In S. Valentini and G. Guarducci (eds.), *Between Syria and the Highlands: Studies in Honor of Giorgio Buccellati and Marilyn Kelly-Buccellati*, 341–51. Rome: Arbor Sapientiae Editore.

Potts, T. F. 1993. "Patterns of Trade in Third-Millennium BC Mesopotamia and Iran." *World Archaeology* 24.3, 379–402.

Powell, M. A. 1987. "The Tree Section of Ur_5(= HAR)-ra = *hubullu*." *Bulletin on Sumerian Agriculture* 3, 145–51.

Prellwitz, W. 1905. *Etymologisches Wörterbuch der griechischen Sprache.* Göttingen: Vandenhoeck und Ruprecht.

Prescott-Allen, R., and C. Prescott-Allen. 1990. "How Many Plants Feed the World?" *Conservation Biology* 4.4, 365–74.

Psomiadis, D., E. Dotsika, K. Albanakis, B. Ghaleb, and C. Hillaire-Marcel. 2018. "Speleothem Record of Climatic Changes in the Northern Aegean Region (Greece) from the Bronze Age to the Collapse of the Roman Empire." *Palaeogeography, Palaeoclimatology, Palaeoecology* 489, 272–83.

Purcell, N. 2012. "Pliny the Elder." In S. Hornblower, A. Spawforth, and E. Eidinow (eds.), *The Oxford Classical Dictionary.* Oxford University Press.

Quillien, L. 2014. "Flax and Linen in First Millennium BC Babylonia." In M. Harlow et al. (eds.), *Prehistoric, Ancient Near Eastern and Aegean Textiles and Dress: An Interdisciplinary Anthology*, 271–95. Oxford: Oxbow.

———. 2019. "Dissemination and Price of Cotton in Mesopotamia during the 1st Millennium BCE." *Revue d'ethnoécologie* 15, 1–20.

Rabin C. 1966. "Rice in the Bible." *Journal of Semitic Studies* 11, 2–9.

Rabinowitz, L. I. 2007. "Four Species." In M. Berenbaum and F. Skolnik (ed.), *Encyclopaedia Judaica,* Vol. 7, 139–40. Detroit: Macmillan Reference USA.

Rachmati, G. R. 1932. "Zur Heilkunde der Uiguren II." *Sitzungsberichte der Preußischen Akademie der Wissenschaften* 26, 401–48.

Radivojević, M., T. Rehren, J. Kuzmanović-Cvetković, M. Jovanović, and J. P. Northover. 2013. "Tainted Ores and the Rise of Tin Bronzes in Eurasia, c. 6500 years Ago." *Antiquity* 87, 1030–45.

Radner, K. 2003. "An Assyrian View on the Medes." In G. B. Lanfranchi, M. Roaf, and R. Rollinger (eds.), *Continuity of Empire(?): Assyria, Media, Persia*, 37–64. Padua: S.A.R.G.O.N.

———. 2004. "Fressen und gefressen werden: Heuschrecken als Katastrophe und Delikatesse im Alten Orient." *Die Welt des Orients* 34, 7–22.

———. 2013. "Assyria and the Medes." In D. Potts (ed.), *The Oxford Handbook of Ancient Iran*, 442–56. Oxford University Press.

Rahmani, L. Y. 1967. "Jason's Tomb." *Israel Exploration Journal* 17.2, 61–100.

Rajan, K. 2011. "Emergence of Early Historic Trade in Peninsular India." In P-Y. Manguin, A. Mani, and G. Wade (eds.), *Early Interactions between South and Southeast Asia: Reflections on Cross-Cultural Exchange*, 177–96. Singapore: Institute of Southeast Asian Studies.

Ramadugu, C., et al. 2013. "A Six Nuclear Gene Phylogeny of *Citrus* (Rutaceae) Taking into Account Hybridization and Lineage Sorting." *PLoS One* 8.7, 1–15.

Ramón-Laca, L. 2003. "The Introduction of Cultivated Citrus to Europe via Northern Africa and the Iberian Peninsula." *Economic Botany* 57.4, 502–14.

Rangan, H., E. A. Alpers, T. Denham, C. A. Kull, and J. Carney. 2015. "Food Traditions and Landscape Histories of the Indian Ocean World: Theoretical and Methodological Reflections." *Environment and History* 21.1, 135–57.

Rangan, H., J. Carney, and T. Denham. 2012. "Environmental History of Botanical Exchanges in the Indian Ocean World." *Environment and History* 18.3, 311–42.

Raschke, M. G. 1975. "Papyrological evidence for Ptolemaic and Roman trade with India." *Proceedings of the XIV International Congress of Papyrologists, Oxford, 24–31 July 1974*, 241–46.

Raschke, M. G. 1978. "New Studies in Roman Commerce with the East." *Aufstieg und Niedergang der römischen Welt* 9.2, 604–1378.

Ratnagar, S. 2004. *Trading Encounters: From the Euphrates to the Indus in the Bronze Age.* Oxford University Press.

Ray, A., D. Chakraborty, and S. Ghosh. 2020. "A Critical Evaluation Revealed the Proto-Indica Model Rests on a Weaker Foundation and Has a Minimal Bearing on Rice Domestication." *Ancient Asia* 11:8, 1–8.

Ray, H. P. 2003. *The Archaeology of Seafaring in Ancient South Asia.* Cambridge University Press.

Reade, J. 1997. "Sumerian origins." In I. Finkel and M. J. Geller (eds.), *Sumerian Gods and Their Representations*, 221–29. Groningen: Styx.

———. 1978. "Studies in Assyrian Geography. Part I: Sennacherib and the Waters of Nineveh." *Revue d'assyriologie et d'archéologie orientale* 72.1, 47–72.

———. 2001. "Assyrian King-Lists, the Royal Tombs of Ur, and Indus Origins." *Journal of Near Eastern Studies* 60.1, 1–29.

———. 2004. "The Assyrians as Collectors: From Accumulation to Synthesis." In G. Frame (ed.), *From the Upper Sea to the Lower Sea: Studies on the History of Assyria and Babylonia in Honour of A.K. Grayson*, 255–68. Leiden: Nederlands Instituut voor het Nabije Oosten.

———. 2008. "The Indus-Mesopotamia Relationship Reconsidered." In R. Olijdam and R. H. Spoor (eds.), *Intercultural Relations between South and Southwest Asia: Studies in Commemoration of E.C.L. During Caspers, 1934–1996*, 12–18. Oxford: Archaeopress.

Reed, K., and T. Leleković. 2019. "First Evidence of Rice (*Oryza* cf. *sativa* L.) and Black Pepper (*Piper nigrum*) in Roman Mursa, Croatia." *Archaeological and Anthropological Sciences* 11.1, 271–78.

Reeves, T.G., G. Thomas, and G. Ramsay. 2016. *Save and Grow in Practice: Maize, Rice, Wheat: A Guide to Sustainable Cereal Production.* Rome: Food and Agriculture Organization of the United Nations.

Renfrew, J.M. 1985. "Preliminary Report on the Botanical Remains." In B.J. Kemp, *Amarna Reports* II, 175–90. London: Egypt Exploration Society.

———. 1987. "A Note on Vegetables from Ancient Iraq." *Bulletin on Sumerian Agriculture* 3, 162.

Renner, S.S., H. Schaefer, and A. Kocyan. 2007. "Phylogenetics of *Cucumis* (Cucurbitaceae): Cucumber (*C. sativus*) Belongs in an Asian/Australian Clade far from Melon (*C. melo*)." *BMC Evolutionary Biology* 7, 58–69.

Reuther, O. 1926. *Die Innenstadt von Babylon (Merkes).* Leipzig: J.C. Hinrichs.

Richardson, L., Jr. 1992. *A New Topographical Dictionary of Ancient Rome.* Johns Hopkins University Press.

Richardson, N., and M. Dorr. 2003. *The Craft Heritage of Oman.* Dubai: Motivate.

Rickman, G. 1971. *Roman Granaries and Store Buildings.* Cambridge University Press.

Riello, G., and T. Roy, eds. 2009. *How India Clothed the World: The World of South Asian Textiles 1500–1850.* Leiden: Brill.

Rivnay, D. 1962. *Field Crop Pests in the Near East.* The Hague: Uitgeverij Dr. W. Junk.

Robbins, G. 1931. *The Botany of Crop Plants*, 3rd ed. Philadelphia: Blakiston and Son.

Rocher, L. 1986. *The Purāṇas.* Wiesbaden: Harrassowitz.

Rokach, I., and G. Shaked. 2007. "Citrus." In M. Berenbaum and F. Skolnik (eds.), *Encyclopaedia Judaica,* Vol. 4, 737–38. Detroit: Macmillan Reference USA.

Roller, D.W. 2008. "Juba of Mauretania (275)." *Brill's New Jacoby,* online edition.

———. 2016. "Megasthenes: His Life and Work." In J. Wiesehöfer, H. Brinkhaus, and R. Bichler (eds.), *Megasthenes und seine Zeit,* 119–27. Wiesbaden: Harrassowitz.

Rosen, S.A., and B.A. Saidel. 2010. "The Camel and the Tent: an Exploration of Technological Change among Early Pastoralists." *Journal of Near Eastern Studies* 69.1, 63–77.

Roth, M.T. 1989/90. "The Material Composition of the Neo-Babylonian Dowry." *Archiv für Orientforschung* 36/37, 1–55.

Rowley-Conwy, P. 1989. "Nubia A.D. 0–550 and the 'Islamic' Agricultural Revolution: Preliminary Botanical Evidence from Qasr Ibrim, Egyptian Nubia." *Archeologie du Nil Moyen* 3, 131–38.

Ruas, M. P. 1996. "Éléments pour une histoire de la fructiculture en France: données archéobotaniques de l'Antiquité au XVIIe siècle." *Actes des congrès de la Société d'archéologie médiévale* 5.1, 92–105.

Ruas, M-P., et al. 2015. "Regard pluriel sur les plantes de l'héritage arabo-islamique en France médiévale." In C. Richarté, R-P. Gayraud, and J-M. Poisson (eds.), *Héritages arabo-islamiques dans l'Europe méditerranéenne*, 347–76. Paris: La Découverte-INRAP.

Rubinson, K. S. 1990. "Carpets vi. Pre-Islamic Carpets." *Encyclopaedia Iranica*, 4.7, 858–61 (http://www.iranicaonline.org/articles/carpets-vi).

Rumor, M. 2018. "At the Dawn of Plant Taxonomy: Shared Structural Design of Herbal Descriptions in *Šammu šikinšu* and Theophrastus' *Historia plantarum* IX." In S. V. Panayotov and L. Vacín (eds.), *Mesopotamian Medicine and Magic: Studies in Honor of Markham J. Geller*, 446–61. Leiden: Brill.

Ryan, S. E., et al. 2021. "Strontium Isotope Evidence for a Trade Network between Southeastern Arabia and India during Antiquity." *Scientific Reports* 11:303, 1–10.

Sabato, D., et al. 2019. "Molecular and Morphological Characterisation of the Oldest *Cucumis melo* L. Seeds Found in the Western Mediterranean Basin." *Archaeological and Anthropological Sciences* 11, 789–810.

Sadori, L., et al. 2009. "The Introduction and Diffusion of Peach in Ancient Italy." In J-P. Morel and A. M. Mercuri (eds.), *Plants and Culture: Seeds of the Cultural Heritage of Europe*, 45–61. Bari: Edipuglia.

Safrai, Z. 1994. *Economy of Roman Palestine*. Abingdon: Routledge.

Salaman, R. N., and W. G. Burton. 1985. *The History and Social Influence of the Potato*. Cambridge University Press.

Sallares, R. 1991. *The Ecology of the Ancient Greek World*. London: Duckworth.

Salles, J-F. 1996. "Achaemenid and Hellenistic Trade in the Indian Ocean." In J. Reade (ed.), *The Indian Ocean in Antiquity*, 251–67. London: Kegan Paul.

Salomon, R. 1998. *Indian Epigraphy: A Guide to the Study of Inscriptions in Sanskrit, Prakrit and the Other Indo-Aryan Languages*. Oxford University Press.

Samuel, D. 2001. "Archaeobotanical Evidence and Analysis." In. S. Berthier et al. (eds.), *Peuplement rural et aménagements hydroagricoles dans la moyenne vallée de l'Euphrate, fin VIIe-XIXe siècle: Région de Deir ez Zōr-Abu Kemāl, Syrie*, 347–481. Damascus: Institut français d'études arabes de Damas.

Sanderson, H. 2005. "Roots and Tubers." In G. Prance and M. Nesbitt (eds.), *The Cultural History of Plants*, 61–76. Abingdon: Routledge.

Sanlaville, P. 2002. "The Deltaic Complex of the Lower Mesopotamian Plain and Its Evolution through Millennia." In E. Nicholson and P. Clark (eds.), *The Iraqi Marshlands*, 94–109. London: Politico's.

Santhanam, V., and J. B. Hutchinson. 1974. "Cotton." In J. Hutchinson (ed.), *Evolutionary Studies in World Crops: Diversity and Change in the Indian Subcontinent*, 89–100. Cambridge University Press.

Sapir-Hen, L., and E. Ben-Yosef. 2013. "The Introduction of Domestic Camels to the Southern Levant: Evidence from the Aravah Valley." *Tel Aviv* 40:2, 277–85.

Saraswat, K. S. 1997. "Plant Economy of Barans at Ancient Sanghol (ca. 1900–1400 BC), Punjab." *Pragdhara* 7, 97–114.

———. 2014. "Botanical Studies of Sanghol Excavations." In C. Margabandhu, *Excavations at Sanghol 1985–6 and 1989–90*. New Delhi: Archaeological Survey of India.

Saraswat, K. S., and A. K. Pokharia. 2002–03. "Palaeoethnobotanical Investigations at Ancient Kunal, Haryana." *Pragdhara* 13, 105–39.

Sarianidi, V. 1971. "The Lapis Lazuli Route in the Ancient East." *Archaeology* 24, 12–15.

Sarkar, A., et al. 2016. "Oxygen Isotope in Archaeological Bioapatites from India: Implications to Climate Change and Decline of Bronze Age Harappan Civilization." *Scientific Reports* 6, 1–8.

Sarkar, H. 1981. "A Geographical Introduction to South-East Asia." *Bijdragen tot de taal-, land- en volkenkunde* 137, 297–301.

Sarna, N. M., et al. 2007. "Bible." In M. Berenbaum and F. Skolnik (eds.), *Encyclopaedia Judaica,* Vol. 3, 572–679. Detroit: Macmillan Reference USA.

Sataev, R. M. 2021. "Animal Burials at Gonur Depe." In B. Lyonnet and N. A. Dubova (eds.), *The World of the Oxus Civilization*, 386–404. Abingdon: Routledge.

Sato, Y. I. 2005. "Rice and Indus Civilization." In T. Osada (ed.), *Linguistics, Archaeology and Human Past*, 213–14. Kyoto: Research Institute for Humanity and Nature.

———. 2015. *Sugar in the Social Life of Medieval Islam.* Leiden: Brill.

Sauvage, C. 2014. "Spindles and Distaffs: Late Bronze and Early Iron Age Eastern Mediterranean Use of Solid and Tapered Ivory/Bone Shafts." In M. Harlow et al. (eds.), *Prehistoric, Ancient Near Eastern and Aegean Textiles and Dress: An Interdisciplinary Anthology*, 184–226. Oxford: Oxbow.

Sax, M. 1991. "The Composition of the Materials of First Millennium BC Cylinder Seals from Western Asia." In P. Budd et al. (eds.), *Archaeological Sciences 1989*, 104–14. Oxford: Oxbow.

Scarborough, J. 1985. "Galen's Dissection of the Elephant." *Korot* 8, 123–34.

Schaefer, H., C. Heibl, and S. S. Renner. 2009. "Gourds Afloat: A Dated Phylogeny Reveals an Asian Origin of the Gourd Family (Cucurbitaceae) and Numerous Oversea Dispersal Events." *Proceedings of the Royal Society of London B: Biological Sciences* 276, 843–51.

Schafer, E. H. 1963. *The Golden Peaches of Samarkand: A Study of T'ang Exotics.* University of California Press.

———. 1967. *The Vermilion Bird: T'ang Images of the South.* University of California Press.

Schilman, B., M. Bar-Matthews, A. Almogi-Labin, and B. Luz. 2001. "Global Climate Instability Reflected by Eastern Mediterranean Marine Records during the late Holocene." *Palaeogeography, Palaeoclimatology, Palaeoecology* 176.1–4, 157–76.

Schmidt, R. 1913. "Beiträge zur Flora Sanscritica: Der Lotus in der Sanskrit-Literatur." *Zeitschrift der Deutschen Morgenländischen Gesellschaft* 67.3, 462–70.

Schmidt-Colinet, A., and A. Stauffer. 2000. *Die Textilien aus Palmyra.* Mainz: Philipp von Zabern.

Schmitt, R. 2002a. "Greece xi–xii. Persian Loanwords and Names in Greek." *Encyclopedia Iranica* (http://www.iranicaonline.org/articles/greece-xi-xii).

———. 2002b. "Zoroaster: The Name." *Encyclopaedia Iranica* (http://www.iranicaonline.org/articles/zoroaster-i-the-name).

Schneider, P. 2012. "Les rois hellénistiques et la transplantation des plantes à aromates: héritage proche-oriental et spécificité hellène." *Phoenix* 66, 272–97.

Schoch, W. H. 1993. "Die Bedeutung von Holz- und Holzkohlenanalysen für Erkenntnisse zur Holzwirtschaft im antiken Zypern." In A. J. Kalis and J. Meurers-Balke (eds.), *7000*

Jahre bäuerliche Landschaft: Entstehung, Erforschung, Erhaltung, 89–104. Cologne: Rheinland.

Schrader, S. 2019. *Activity, Diet and Social Practice: Addressing Everyday Life in Human Skeletal Remains*. Cham: Springer.

Schwartz, M. 1985. "The Old Eastern Iranian World View according to the Avesta." In I. Gershevitch (ed.), *The Cambridge History of Iran: Median and Achaemenian Periods*, 640–63. Cambridge University Press.

Schweinfurth, G. 1884. "Further Discoveries in the Flora of Ancient Egypt." *Nature* 29, 312–15.

Scora, R. W. 1975. "On the History and Origin of Citrus." *Bulletin of the Torrey Botanical Club* 102, 369–75.

Scott, A., et al. 2020. "Exotic Foods Reveal Contact between South Asia and the near East during the Second Millennium BCE." *Proceedings of the National Academy of Sciences* 118(2), e2014956117.

Scott, J. C. 2017. *Against the Grain: A Deep History of the Earliest States*. Yale University Press.

Scullard, H. H. 1974. *The Elephant in the Greek and Roman World*. London: Thames and Hudson.

Sebastian, P., H. Schaefer, I. R. Telford, and S. S. Renner. 2010. "Cucumber (*Cucumis sativus*) and Melon (*C. melo*) Have Numerous Wild Relatives in Asia and Australia, and the Sister Species of Melon Is from Australia." *Proceedings of the National Academy of Sciences* 107, 14269–73.

Secord, J. 2016. "Introduced Species, Hybrid Plants and Animals, and Transformed Lands in the Hellenistic and Roman Worlds." In R. F. Kennedy and M. Jones-Lewis (eds.), *The Routledge Handbook of Identity and the Environment in the Classical and Medieval Worlds*, 210–29. Abingdon: Routledge.

Segal, S. J. 2003. *Under the Banyan Tree: A Population Scientist's Odyssey*. Oxford University Press.

Seland, E. H. 2010. *Ports and Political Power in the Periplus: Complex Societies and Maritime Trade on the Indian Ocean in the First Century AD*. Oxford: Archaeopress.

———. 2014. "The Organisation of the Palmyrene Caravan Trade." *Ancient West and East* 13, 197–211.

Seldeslachts, E. 1998. "Translated Loans and Loan Translations as Evidence of Graeco-Indian Bilingualism in Antiquity." *L'Antiquité classique* 67, 273–99.

Serrano, R. A. 1997. "Al-Buhturī's poetics of Persian Abodes." *Journal of Arabic Literature* 28.1, 68–87.

Serres-Giardi, L., and C. Dogimont. 2012. "How Microsatellite Diversity Helps to Understand the Domestication History of Melon." In N. Sari, I. Solmaz, and V. Aras (eds.), *Cucurbitaceae 2012: Proceedings of the Xth EUCARPIA Meeting on Genetics and Breeding of Cucurbitaceae*, 254–63. Adana: Çukurova University.

Shaffer, L. 1994. "Southernization." *Journal of World History* 5, 1–21.

Shah, U. P. 1987. *Jaina-Rupa Mandana: Jaina Iconography*, Vol. 1. New Delhi: Abhinav.

Shaked, S. 1986. "From Iran to Islam: On Some Symbols of Royalty." *Jerusalem Studies in Arabic and Islam* 7, 75–91.

Shamir, O. 2001. "Cotton." In A. Negev and S. Gibson (eds.), *Archaeological Encyclopaedia of the Holy Land*, 125–26. Jerusalem: Jerusalem.

Shan, Y. 2016. *Canned Citrus Processing: Techniques, Equipment, and Food Safety.* London: Academic Press.

Sharma, P. V. 1979. *Fruits and Vegetables in Ancient India.* Varanasi: Chaukhambha Orientalia.

Sharma, S., S. K. Manjul, A. Manjul, P. C. Pande, and A. K. Pokharia. 2020. "Dating Adoption and Intensification of Food-Crops: Insights from 4MSR (Binjor), an Indus (Harappan) Site in Northwestern India." *Radiocarbon* 62(5), 1349–69.

Sharples, R. W. 1989. "Review of The Budé of Theophrastus HP." *Classical Review* (New Series) 39.2, 197–98.

Sharples, R. W., and D. W. Minter. 1983. "Theophrastus on Fungi: Inaccurate Citations in Athenaeus." *Journal of Hellenic Studies* 103, 154–56.

Shastri, H. G. 2000. *Gujarat under the Maitrakas of Valabhi.* Vadodara: Oriental Institute.

Sheffer, A., and H. Granger-Taylor. 1994. "Textiles from Masada." In J. Aviram, G. Foerster, and E. Netzer (eds.), *Masada 4: The Yigael Yadin Excavations 1963–1965, Final Reports,* 153–250. Jerusalem: Israel Exploration Society and the Hebrew University of Jerusalem.

Sheffer, A., and A. Tidhar. 1991. "The Textiles from the 'En-Boqeq Excavation in Israel." *Textile History* 22, 3–46.

Sherratt, A. 1983. "The Secondary Exploitation of Animals in the Old World." *World Archaeology* 15.1, 90–104.

———. 1999. "Cash-Crops before Cash: Organic Consumables and Trade." In C. Gosden and J. G. Hather (eds.), *The Prehistory of Food.* Abingdon: Routledge.

———. 2006. "The Trans-Eurasian Exchange: The Prehistory of Chinese Relations with the West." In V. H. Mair (ed.), *Contact and Exchange in the Ancient World.* University of Hawaii Press.

Sheriff, A. 2010. *Dhow Culture of the Indian Ocean: Cosmopolitanism, Commerce and Islam.* Columbia University Press.

Sherwin, S. 2003. "*Hamsukkān* in Isaiah 40:20: Some Reflections." *Tyndale Bulletin* 56.1, 145–49.

Shev, E. T. 2016. "The Introduction of the Domesticated Horse in Southwest Asia." *Archaeology, Ethnology and Anthropology of Eurasia* 44, 123–36.

Shishegar, A. 2015. *Āramgāh-e do bānuy-e ʿīlāmi az ḫāndān-e šutur nahunte, pesar-e indada (dore-ye ʿīlam-e no. marhale-ye 3b, ḥodud-e 585 tā 539 p.m)* [Tomb of the two Elamite princesses of the house of King Shutur-Nahunte son of Indada. Neo-Elamite period, phase IIIB (ca. 585–539 B.C.)]. Tehran: Pažuhešgāh-e Sāzmān-e Mirāṣe Farhangi.

Shishlina, N. I., O. V. Orfinskaya, and V. P. Golikov. 2003. "Bronze Age Textiles from the North Caucasus: New Evidence of Fourth Millennium BC Fibres and Fabrics." *Oxford Journal of Archaeology* 22.4, 331–44.

Sidebotham, S. E. 1986. *Roman Economic Policy in the Erythra Thalassa.* Leiden: Brill.

———. 2011. *Berenike and the Ancient Maritime Spice Route.* University of California Press.

Silk, J. A. 2008. *Managing Monks: Administrators and Administrative Roles in Indian Buddhist Monasticism.* Oxford University Press.

Silva, F., A. Weisskopf, C. Castillo, C. Murphy, E. Kingwell-Banham, L. Qin, and D. Q. Fuller. 2018. "A Tale of Two Rice Varieties: Modelling the Prehistoric Dispersals of *japonica* and proto-*indica* rices." *Holocene* 28.11, 1745–58.

Simoons, F. J. 1991. *Food in China: A Cultural and Historical Inquiry.* Boca Raton: CRC Press.

Simpson, St. J. 2015. "The Land behind Ctesiphon: The Archaeology of Babylonia during the Period of the Babylonian Talmud." In M. J. Geller (ed.), *The Archaeology and Material Culture of the Babylonian Talmud*, 6–38. Leiden: Brill.

Singh, A. K. 2017. *Wild Relatives of Cultivated Plants in India*. Singapore: Springer.

Singh, T. B. 1999. *Glossary of Vegetable Drugs in Bṛhattrayī*. Varanasi: Chaukhamba Amarabharati Prakashan.

Skjaervo, P. O. 2011. *The Spirit of Zoroastrianism*. Yale University Press.

Small, E., and P. M. Catling. 2005. "Blossoming Treasures of Biodiversity. 15: Asian Lotus (*Nelumbo nucifera*) and American Lotus (*N. lutea*), Spiritual Symbols, Valuable Foods, and Successful Invaders." *Biodiversity* 6.1, 25–30.

Smereka, J. 1936. *Studia Euripidea*, Vol. I. Lwow: Sumptibus Societatis Litterarum.

Smith, J. E. 1814. *An Introduction to Physiological and Systematic Botany*. London: Longman.

Smith, M. L. 2006. "The Archaeology of Food Preference." *American Anthropologist* 108.3, 480–93.

———. 2013. "The Substance and Symbolism of Long-Distance Exchange: Textiles as Desired Trade-Goods in the Bronze Age Middle Asian Interaction Sphere." In S. Abraham et al. (eds.), *Connections and Complexity: New Approaches to the Archaeology of South Asia*, 143–60. Walnut Creek, CA: Left Coast Press.

Sołtysiak, A. 2006. "Physical Anthropology and the 'Sumerian Problem.'" *Studies in Historical Anthropology* 4, 145–58.

Sołtysiak, A. 2016. "Drought and the Fall of Assyria: Quite Another Story." *Climatic Change*, 136(3–4), 389–394.

Sonnini, C. S. 1799. *Voyage dans la haute et basse Égypte*, Vol. I. Paris: Chez F. Buisson.

Sood, R. 2020. "Asafoetida (*Ferula asafoetida*): A High-Value Crop Suitable for the Cold Desert of Himachal Pradesh, India." *Journal of Applied and Natural Science* 12(4), 607–17.

Sorrel, P., and M. Mathis. 2016. "Mid-to Late-Holocene Coastal Vegetation Patterns in Northern Levant (Tell Sukas, Syria): Olive Tree Cultivation History and Climatic Change." *Holocene* 26(6), 858–73.

Šoštarić, R., and H. Küster. 2001. "Roman Plant Remains from Veli Brijun (Island of Brioni), Croatia." *Vegetation History and Archaeobotany* 10.4, 227–33.

Southworth, F. 2005. *The Linguistic Archaeology of South Asia*. Abingdon: Routledge.

Spengler, R. N. 2019. *Fruit from the Sands: The Silk Road Origins of the Foods We Eat*. University of California Press.

Sperl, S., and C. Shackle, eds. 1996. *Qasida Poetry in Islamic Asia and Africa: Eulogy's Bounty, Meaning's Abundance: An Anthology*. Vol. 2. Leiden: Brill.

Squatriti, P. 2014. "Of Seeds, Seasons, and Seas: Andrew Watson's Medieval Agrarian Revolution Forty Years Later." *Journal of Economic History* 74.4, 1205–20.

Stauffer, A. 2000. "The Textiles from Palmyra: Technical Analyses and Their Evidence for Archaeological Research." In D. Cardon and M. Feugère (eds.), *Archéologie des textiles des origines au Ve siècle*, 247–51. Montagnac: Editions Monique Mergoil.

Stein, P. 2000. *Die mittel- und neubabylonischen Königsinschriften bis zum Ende der Assyrerherrschaft: Grammatische Untersuchungen*. Wiesbaden: Harrassowitz.

Steinkeller, P. 1982. "The Question of Marḫaši: A Contribution to the Historical Geography of Iran in the Third Millennium BC." *Zeitschrift für Assyriologie und vorderasiatische Archäologie* 72.2, 237–65.

———. 1987–99. "Maništūšu." In D.O. Edzard et al. (eds.), *Reallexikon der Assyriologie und Vorderasiatischen Archäologie,* Vol. VII, 334–35. Berlin: de Gruyter.

———. 2006. "New Light on Marhaši and Its Contacts with Makkan and Babylonia." *Journal of Magan Studies* 1, 1–17.

———. 2009. "Camels in Ur III Babylonia?" In J.D. Schloen (ed.), *Exploring the Longue Durée: Essays in Honor of Lawrence E. Stager*, 415–19. Winona Lake, IN: Eisenbrauns.

———. 2013. "Trade Routes and Commercial Networks in the Persian Gulf during the Third Millennium BC." In C. Faizee (ed.), *Collection of Papers Presented at the Third International Biennial Conference of the Persian Gulf (History, Culture and Civilisation)*, 413–31. Tehran University Press.

———. 2014. "Marhaši and Beyond: The Jiroft Civilization in a Historical Perspective." In C.C. Lamberg-Karlovsky and B. Genito (eds.), *"My Life is Like the Summer Rose": Maurizio Tosi e l'Archeologia come modo di vivere: Papers in Honour of Maurizio Tosi for his 70th Birthday*, 691–707. Oxford: Archaeopress.

Stevens, C.J., C. Murphy, R. Roberts, L. Lucas, F. Silva, and D.Q. Fuller. 2016. "Between China and South Asia: A Middle Asian Corridor of Crop Dispersal and Agricultural Innovation in the Bronze Age." *Holocene* 26, 1–15.

Stevens, P.T. 1976. *Colloquial Expressions in Euripides.* Wiesbaden: Steiner.

Stol, M. 1985. "Remarks on the Cultivation of Sesame and the Extraction of Its Oil." *Bulletin of Sumerian Agriculture* 2, 119–26.

———. 1987. "The Cucurbitaceae in the Cuneiform Texts." *Bulletin on Sumerian Agriculture* 3, 81–92.

Stolarczyk, J., and J. Janick. 2011. "Carrot: History and Iconography." *Chronica Horticulturae* 51.2, 13–18.

Stöllner, T., et al. 2011. "Tin from Kazakhstan: Steppe Tin for the West." In Ü. Yalçın (ed.), *Anatolian Metal V*, 231–52. Deutschen Bergbau-Museum Bochum.

Stonehouse, J., S.M. Sadeed, A. Harvey, and G.S. Haiderzada. 2008. "*Myiopardalis pardalina* in Afghanistan." In R. Sugayama et al. (eds.), *Proceedings of the 7th International Symposium on Fruit Flies of Economic Importance: From Basic to Applied Knowledge*, 1–12. Salvador: SBPC.

Strand, E.A., K.M. Frei, M. Gleba, U. Mannering, M-L. Nosch, and I. Skals. 2010. "Old Textiles—New Possibilities." *European Journal of Archaeology* 13.2, 149–73.

Strauch, I., ed. 2012. *Foreign Sailors on Socotra: The Inscriptions and Drawings from the Cave Hoq.* Bremen: Ute Hempen.

Subbarayalu, Y. 2014. "Early Tamil Polity." In N. Karashima (ed.), *A Concise History of South India: Issues and Interpretations*, 47–55. Oxford University Press.

Subramanian, T.S. 2012. "Potsherd with Tamil-Brahmi Script Found in Oman." *The Hindu*, October 28 (http://www.thehindu.com/news/national/potsherd-with-tamilbrahmi-script-found-in-oman/article4038866.ece).

Swamy, K.R.M. 2017. "Origin, Distribution and Systematics of Culinary Cucumber (*Cucumis melo* subsp. *agrestis* var. *conomon*). *Journal of Horticultural Sciences* 12, 1–22.

Sykes, N. 2012. "A Social Perspective on the Introduction of Exotic Animals: The Case of the Chicken." *World Archaeology* 44.1, 158–69.

Tackhölm, V., and M. Drar. 1950. *Flora of Egypt*, Vol. II. Cairo: Fouad I University Press.

Täckholm, V., and G. Täckholm. 1941. *Flora of Egypt*, Vol. I. Cairo: Fouad I University Press.

Tafażżolī, A. 1993. "Dādestān ī Mēnōg ī xrad." *Encyclopedia Iranica,* 6, 554–55 (http://www.iranicaonline.org/articles/dadestan-i-menog).

Talebi, K. S., T. Sajedi, and M. Pourhashemi. 2014. *Forests of Iran: A Treasure from the Past, a Hope for the Future.* Dordecht: Springer.

Tavernier, J. 2007. *Iranica in the Achaemenid Period (ca. 550–330 BC): Lexicon of Old Iranian Proper Names and Loanwords Attested in Non-Iranian Texts.* Leuven: Peeters.

———. 2008. "KADP 36: Inventory, Plant List, or Lexical Exercise?" In R. D. Briggs, J. Myers, and M. T. Roth (eds.), *Proceedings of the 51st Rencontre Assyriologique Internationale,* 191–202. Chicago: Oriental Institute.

Tengberg, M. 1999. "Crop Husbandry at Miri Qalat Makran, SW Pakistan (4000–2000 BC)." *Vegetation History and Archaeobotany* 8.1–2, 3–12.

———. 2002. "The Importation of Wood to the Arabian Gulf in Antiquity: The Evidence from Charcoal Analysis." *Proceedings of the Seminar for Arabian Studies* 32, 75–81.

Tengberg, M. and C. Moulhérat. 2008. "Les 'arbres à laine': Origine et histoire du coton dans l'Ancien Monde." *Nouvelles de l'Archéologie* 114, 42–46.

Tengberg, M., and D. T. Potts. 1999. "ĝišmes-má-gan-na (*Dalbergia sissoo* Roxb.) at Tell Araq." *Arabian Archaeology and Epigraphy* 10, 129–33.

Tengberg, M., D. T. Potts, and H. P. Francfort. 2008. "The Golden Leaves of Ur." *Antiquity* 82, 925–36.

Tengberg, M., and S. Thiébault. 2003. "Vegetation History and Wood Exploitation in Pakistani Baluchistan from the Neolithic to the Harappan Period: The Evidence from Charcoal Analysis." In S. A. Weber, *Indus Ethnobiology: New Perspectives from the Field,* 21–64. Lanham: Lexington Books.

Tewari, R., et al. 2008. "Early Farming at Lahuradewa." *Pragdhara* 18, 347–73.

Thanheiser, U. 1999. "Plant Remains from Kellis: First Results." In C. A. Hope and A. J. Mills (eds.), *Dakhleh Oasis Project: Preliminary Reports on the 1992–1993 and 1993–1994 Field Seasons,* 89–103. Oxford: Oxbow Books.

———. 2002. "Roman Agriculture and Gardening in Egypt as Seen from Kellis." In C. A. Hope and G. E. Bowen, *Dakhleh Oasis Project: Preliminary Reports on the 1994–1995 to 1998–1999 Field Seasons,* 299–310. Oxford: Oxbow.

Thapar, R. 1978. *Ancient Indian Social History: Some Interpretations.* New Delhi: Orient Longman.

———. 1997. "Early Mediterranean Contacts with India." In F. D. Romanis and A. Tchernia (eds.), *Crossings: Early Mediterranean Contacts with India,* 11–40. Delhi: Manohar.

———. 2002. *Early India: From the Origins to AD 1300.* University of California Press.

———. 2012. *Aśoka and the Decline of the Mauryas,* 3rd ed. Oxford University Press.

Thavapalan, S. 2020. *The Meaning of Color in Ancient Mesopotamia.* Leiden: Brill.

Thiselton-Dyer. 1918. "On Some Ancient Plant Names." *Journal of Philology* 34, 290–312.

Thomas, A. P. 1987. "Pillow Stuffings from Amarna?" *Journal of Egyptian Archaeology* 73, 211–13.

Thomas, R. I. 2012. "Port Communities and the Erythraean Sea Trade." *British Museum Studies in Ancient Egypt and Sudan* 18, 169–99.

Thomason, A. K. 2005. *Luxury and Legitimation: Royal Collecting in Ancient Mesopotamia.* Farnham: Ashgate.

Thompson, R. C. 1939. "*ᵁKurangu* and *ᵁlal(l)angu* as Possibly 'Rice' and 'Indigo' in Cuneiform." *Iraq* 6.2, 180–83.

———. 1949. *A Dictionary of Assyrian Botany.* London: British Academy.

Thornton, C. P. 2013. "Mesopotamia, Meluhha, and Those in Between." In H. Crawford (ed.), *The Sumerian World*, 600–19. Abingdon: Routledge.

Tomber, R. 2008. *Indo-Roman Trade: From Pots to Pepper.* London: Duckworth.

Toray Industries. 1996. "Report on the Analyses of Textiles Uncovered at the Nimrud Tomb-Chamber." *Al-Rāfidān* 17, 199–206.

Tosi, M. 1975. "Hasanlu Project 1974: Palaeobotanical Survey." *Iran* 13, 185–86.

Totelin, L., ed. 2009. *Hippocratic Recipes: Oral and Written Transmission of Pharmacological Knowledge in Fifth- and Fourth-Century Greece.* Leiden: Brill.

———. 2012. "Botanizing Rulers and Their Herbal Subjects: Plants and Political Power in Greek and Roman Literature." *Phoenix* 66, 122–44.

———. 2016. "Technologies of Knowledge: Pharmacology, Botany, and Medical Recipes." In *Oxford Handbook Topics in Classical Studies* (https://academic.oup.com/edited-volume/43505/chapter/364128871). Oxford University Press.

———. 2018. "Tastes in Ancient Botany, Medicine and Science." In K. C. Rudolph (ed.), *Taste and the Ancient Senses*, 60–71. Abingdon: Routledge.

Townsend, C. C. 1980. "Rutaceae." In C. C. Townsend, E. Guest, and S. A. Omar, *Flora of Iraq, Volume IV: Cornaceae to Resedaceae*, 454–73. Baghdad: Ministry of Agriculture.

Toynbee, A. 1884. *Lectures on the Industrial Revolution of the Eighteenth Century in England.* London: Rivingtons.

Trentacoste, A. 2020. "Fodder for Change: Animals, Urbanisation, and Socio-Economic Transformation in Protohistoric Italy." *Theoretical Roman Archaeology Journal* 3.1.

Tubbs, J. N. 1988. "Tell es-Sa'idiyeh: Preliminary Report on the First Three Seasons of Renewed Excavations." *Levant* 20, 23–88.

Tucker, E. 2007. "Greek and Iranian." In A-F. Christidis (ed.), *A History of Ancient Greek*, 773–85. Cambridge University Press.

Turnheim, Y. 2002. "Nilotic Motifs and the Exotic in Roman and Early Byzantine Eretz Israel." *Assaph: Studies in Art History* 7, 17–40.

Twiss, K. C. 2019. *The Archaeology of Food: Identity, Politics, and Ideology in the Prehistoric and Historic Past.* Cambridge University Press.

Tzitzikas, E. N., A. J. Monforte, A. Fatihi, Z. Kypriotakis, A. T. Iacovides, I. M. Ioannides, and P. Kalaitzis. 2009. "Genetic Diversity and Population Structure of Traditional Greek and Cypriot Melon Cultigens (*Cucumis melo* L.) Based on Simple Sequence Repeat Variability." *Hortscience* 44.7, 1820–24.

Ubaydli, A. 1993. "The Agrarian Economy of Oman (132–280/749–893) in Arabic Sources." *Journal of Islamic Studies* 4.1, 33–51.

Uesugi, A. 2019. "A Note on the Interregional Interactions between the Indus Civilization and the Arabian Peninsula during the Third Millennium BCE." In S. Nakamura, T. Adachi, and M. Abe (eds.), *Decades in Deserts: Essays on Near Eastern Archaeology in Honour of Sumio Fujii*, 337–55. Tokyo: Rokuichi Syobou.

Ulian, T., et al. 2020. "Unlocking Plant Resources to Support Food Security and Promote Sustainable Agriculture." *Plants, People, Planet* 2(5), 421–45.

Upadhye, A. N. 1983. "Bṛhat-Kathākośa." In A. N. Upadhye, *Papers*, 1–107. University of Mysore.

Ur, J. 2005. "Sennacherib's Northern Assyrian Canals: New Insights from Satellite Imagery and Aerial Photography." *Iraq* 67.1, 317–45.

van Driem, G. 2017. "The Domestications and the Domesticators of Asian Rice." In M. Robbeets and A. Savelyev (eds.), *Language Dispersal beyond Farming*, 183–214. Amsterdam: John Benjamins.

van Ess, M., and F. Pedde. 1992. *Uruk: Kleinfunde II (Ausgrabungen in Uruk-Warka; Bd. 7)*. Mainz am Rhein: Philipp von Zabern.

van de Mieroop, M. 1992. "Wood in the Old Babylonian Texts from Southern Babylonia." *Bulletin of Sumerian Agriculture* 6, 155–62.

———. 1999. *The Ancient Mesopotamian City*. Oxford University Press.

———. 2011. *A History of Ancient Egypt*. London: Wiley.

van Oppen, B. F. 2019. "Monsters of Military Might: Elephants in Hellenistic History and Art." *Arts* 8(4), 1–37.

van Ruymbeke, C. 2007. *Science and Poetry in Medieval Persia: The Botany of Nizami's Khamsa*. Cambridge University Press.

van der Veen, M. 1998. "Gardens in the Desert." In O. Kaper (ed.), *Life on the Fringe: Living in Southern Egyptian Deserts during the Roman and early Byzantine Periods*, 221–42. Leiden: CNWS.

———. 2001. "The Botanical Evidence." In V. Maxfield and D. Peacock, *Mons Claudianus 1987–1993, Vol. II, Excavations: Part 1*, 173–247. Cairo: Institut français d'archéologie orientale.

———. 2010. "Agricultural Innovation: Invention and Adoption or Change and Adaptation?" *World Archaeology* 42.1, 1–12.

———. 2011. *Consumption, Trade and Innovation: Exploring the Botanical Remains from the Roman and Islamic ports at Quseir al-Qadim, Egypt*. Frankfurt: Africa Magna.

van der Veen, M., C. Bouchaud, R. Cappers, and C. Newton. 2018. "Vie romaine dans le désert Oriental d'Égypte: Alimentation, puissance impériale et géopolitique." In J-P. Brun et al. (eds.), *Le désert oriental d'Égypte durant la période gréco-romaine: bilans archéologiques*. Paris: Collège de France (https://books.openedition.org/cdf/5170).

van der Veen, M., and H. Tabinor. 2007. "Food, Fodder and Fuel at Mons Porphyrites: The Botanical Evidence." In D. Peacock and V. Maxfield, *The Roman Imperial Quarries: Survey and Excavation at Mons Porphyrites 1994–1998. Vol. 2: The Excavations*, 83–142. London: Egypt Exploration Society.

van Wyck, B-E. 2005. *Food Plants of the World: An Illustrated Guide*. Portland: Timber Press.

van Zeist, W. 2008. "Comments on Plant Cultivation at Two Sites on the Khabur, North-Eastern Syria." In H. Kühne (ed.), *Umwelt und Subsistenz der Assyrischen stadt Dūr-Katlimmu am unteren Ḫābūr*, 133–48. Wiesbaden: Harrassowitz.

van Zeist, W., S. Bottema, and M. van der Veen. 2001. *Diet and Vegetation at Ancient Carthage: The Archaeobotanical Evidence*. Groningen Institute of Archaeology.

van Zeist, W., and G. J. de Roller. 1993. "Plant Remains from Maadi, a Predynastic Site in Lower Egypt." *Vegetation History and Archaeobotany* 2.1, 1–14.

van Zeist, W., W. Waterbolk-van Rooijen, R. M. Palfenier-Vegter, and G. Jan de Roller. 2003. "Plant Cultivation at Tell Hammam et-Turkman." In W. van Zeist (ed.), *Reports on Archaeobotanical Studies in the Old World*, 61–114. Groningen University Press.

Vanderhoeven, A., et al. 1993. "Het oudheidkundig bodemonderzoek aan de Veemarkt te Tongeren." *Archeologie in Vlaanderen* 3, 127–205.

Vandorpe, P. 2010. *Plant Macro Remains from the 1st and 2nd Cent. A.D. in Roman Oedenburg/Biesheim-Kunheim: Methodological Aspects and Insights into Local Nutrition,*

Agricultural Practices, Import and the Natural Environment. Doctoral dissertation, University of Basel.

Vandorpe, P., Ö. Akeret, and S. Deschler-Erb. 2017. "Crop Production and Livestock Breeding from the Late Iron Age to the Late Roman Period in North Western Switzerland." In S. Lepetz and V. Zech-Matterne (eds.), *Productions agro-pastorales, pratiques culturales et élevage dans le nord de la Gaule du deuxième siècle avant J.-C. à la fin de la période romaine*, 135–52. Quint-Fonsegrives: Éditions Mergoil.

Vanstiphout, H. L. J. 2003. *Epics of Sumerian Kings: The Matter of Aratta.* Atlanta: Society of Biblical Literature.

Varadarajan, L. 1993. "Indian Boat Building Traditions: The Ethnological Evidence." *Topoi* 3.2, 547–68.

Varisco, D. M. 1994. *Medieval Agriculture and Islamic Science: The Almanac of a Yemeni Sultan.* University of Washington Press.

Varisco, D. M. 2002. "Agriculture in Rasulid Zabīd." In J. F. Healey and V. Porter (eds.), *Studies on Arabia in Honour of Professor G. Rex Smith.* Oxford University Press.

Vartavan, C. D., and V. A. Amorós. 1997. *Codex of Ancient Egyptian Plant Remains.* London: Triade Exploration.

Vats, M. S. 1940. *Excavations at Harappa.* New Delhi: Government of India.

Vedeler, H. T. 2002. *The ḪAR.GUD Commentary and Its Relationship to the ḪAR-Ra = Ḫubullu Lexical List.* MA thesis, Yale University.

Veenhof, K. R., and J. Eidem. 2008. *Mesopotamia: The Old Assyrian Period.* Fribourg: Academic Press.

Vermaak, P. S. 2012. "The Foreign Triangle in South-Eastern Mesopotamia." *Journal for Semitics* 21.1, 91–105.

Versluys, M. J. 2002. *Aegyptiaca Romana: Nilotic Scenes and the Roman Views of Egypt.* Leiden: Brill.

Vidale, M. 2004. "Growing in a Foreign World: For a History of 'Meluhha Villages' in Mesopotamia in the 3rd Millennium BC." In A. Panaino and A. Piras (eds.), *Melammu Symposia IV: Schools of Oriental Studies and the Development of Modern Historiography*, 261–80. Milan: ISIAO.

Vidale, M., and D. Frenez. 2014. "Translated Symbols: Indus Reminiscences in a Carved Chlorite Artefact of the Halil Rud Civilization." *Rivista di Archeologia* 38, 7–18.

Vink, M. P. 2007. "Indian Ocean Studies and the 'New Thalassology'." *Journal of Global History* 2.1, 41–62.

Virgilio, M., et al. 2010. "Macrogeographic Population Structuring in the Cosmopolitan Agricultural Pest *Bactrocera cucurbitae* (Diptera: Tephritidae)." *Molecular Ecology* 19, 2713–24.

Vogelsang, W. 1990. "The Achaemenids and India." In H. Sancisi-Weerdenburg and A. Kuhrt (eds.), *Achaemenid History IV: Centre and Periphery*, 93–110. Leiden: Nederlands Instituut voor het nabije oosten.

Vogelsang-Eastwood, G. 2006. "A Preliminary Survey of Iranian Archaeological Sites with Textiles." In S. Schrenk (ed.), *Textiles in Situ: Their Find Spots in Egypt and Neighbouring Countries in the First Millennium CE*, 221–40. Riggisberg: Abegg-Stiftung.

Völling, E. 2008. *Textiltechnik im alten Orient: Rohstoffe und Herstellung.* Würzburg: Ergon.

Vosmer, T. 2003. "The Naval Architecture of Early Bronze Age Reed-Built Boats of the Arabian Sea." In D. T. Potts, H. Al Naboodah, and P. Hellyer (eds.), *Archaeology of the United Arab Emirates: Proceedings of the First International Conference of the U.A.E.*, 152–57. London: Trident Press.

Waetzoldt, H. 1980–83. "Leinen." In D. O. Edzard et al. (eds.), *Reallexikon der Assyriologie und Vorderasiatischen Archäologie 6*, 583–94. Berlin: de Gruyter.

Wagler, P. 1897. "Baumwolle." In *Paulys Realencyclopädie der classischen Altertumswissenschaft Band III, 1*, 167–73. Stuttgart: Metzler.

Wagner, G. 1987. *Les oasis d'Égypte à l'époque grecque, romaine et byzantine d'après les documents grecs: recherches de papyrologie et d'épigraphie grecques.* Cairo: Institut français d'archéologie orientale du Caire.

Walter, H., and S. W. Breckle. 1986. *Ecological Systems of the Geobiosphere. Vol. 2: Tropical and Subtropical Zonobiomes.* Dordrecht: Springer Science.

Waltham, T. 1999. "The Ruby Mines of Mogok." *Geology Today* 15.4, 143–49.

Warrier, P. K., V. P. K. Nambiar, and C. Ramankutty, eds. 1994. *Indian Medicinal Plants: A Compendium of 500 Species*, Vol. II. Chennai: Orient Longman.

Warinner, C., K. K. Richter, and M. J. Collins. 2022. "Paleoproteomics." *Chemical Reviews* 122(16), 13401–446.

Watanabe, K. 1987. *Die adê-Vereidigung anlässlich der Thronfolgeregelung Asarhaddons.* Berlin: Gebr. Mann.

Watson, A. M. 1974. "The Arab Agricultural Revolution and Its Diffusion, 700–1100." *Journal of Economic History* 34, 8–35.

———. 1981. "A Medieval Green Revolution: New Crops and Farming Techniques in the Early Islamic World." In A. Udovitch (ed.), *The Islamic Middle East, 700–1900: Studies in Economic and Social History*, 29–58. Princeton, NJ: Darwin Press.

———. 1983. *Agricultural Innovation in the Early Islamic World.* Cambridge University Press.

Watson, W. G. 2004. "A Botanical Snapshot of Ugarit: Trees, Fruit, Plants and Herbs in the Cuneiform Text." *Aula orientalis: revista de estudios del Próximo Oriente Antiguo* 22.1, 107–55.

Weidner, S. 1985. *Lotos im alten Ägypten.* Pfaffenweiler: Centaurus.

Weisskopf, A., and D. Q. Fuller. 2013a. "Apricot: Origins and Development." In C. Smith (ed.), *Encyclopedia of Global Archaeology*, 294–96. New York: Springer.

———. 2013b. "Citrus Fruits: Origins and Development." In C. Smith (ed.), *Encyclopedia of Global Archaeology*, 1479–83. New York: Springer.

Welch, P. B. 1997. *Chinese New Year.* Oxford University Press.

Wellmann, M. 1907. *Pedanii Dioscuridis Anazarbei De materia medica libri quinque,* Vol. I. Berlin: Weidmann.

Wendel, J. F. 1995. "Cotton." In J. Smartt and N. W. Simmonds (eds.), *Evolution of Crop Plants*, 2nd ed., 358–66. London: Longman.

Wendrich, W. Z., et al. 2003. "Berenike Crossroads: The Integration of Information." *Journal of the Economic and Social History of the Orient* 46.1, 46–87.

Wheatley, P. 1961. *The Golden Khersonese: Studies in the Historical Geography of the Malay Peninsula before A.D. 1500.* University of Malaya Press.

White, D. G. 1991. *Myths of the Dog-Man.* University of Chicago Press.

Wicks, Y. 2012. *Bronze "Bathtub" Coffins*. Honours dissertation, Department of Archaeology, University of Sydney.

———. 2015. *Bronze "Bathtub" Coffins: In the Context of 8th–6th century BC Babylonian, Assyrian and Elamite Funerary Practices*. Oxford: Archaeopress.

Wiethold, J. 2003. "How to Trace 'Romanisation' of Central Gaule by Archaeobotanical Analysis? Some Considerationes on New Archaeobotanical Results from France Centre-Est." In F. Favory and A. Vignot (eds.), *Actualité de la recherche en histoire et archéologie agraires: Actes du colloque international AGER V, septembre 2000. Annales Littéraires 764; Environnement, sociétés et archéologie* 5, 269–82. Besançon: ALUFC.

Wiggermann, F. A. M. 1992. *Mesopotamian Protective Spirits: The Ritual Texts*. Groningen: STYX and PP.

Willcox, G. H. 1977. "Exotic Plants from Roman Waterlogged Sites in London." *Journal of Archaeological Science* 4.3, 269–82.

———. 1992. "Timber and Trees: Ancient Exploitation in the Middle East." *Bulletin of Sumerian Agriculture* 6, 1–32.

Wild, F. C., and A. J. Clapham. 2007. "Irrigation and the Spread of Cotton Growing in Roman Times." *Archaeological Textiles Newsletter* 44, 16–18.

Wild, F. C., and J. P. Wild. 2001. "Sails from the Roman Port at Berenike, Egypt." *International Journal of Nautical Archaeology* 30.2, 211–20.

Wild, J. P. 1997. "Cotton in Roman Egypt: Some Problems of Origin." *Al-Rāfidān* 18, 287–98.

Wild, J. P., and F. C. Wild. 2006. "Qasr Ibrim: Study Season 2005." *Archaeological Textiles Newsletter* 43, 16–19.

———. 2008. "Early Indian Cotton Textiles from Berenike." In E. M. Raven (ed.), *South Asian Archaeology 1999*, 229–33. Groningen: Egbert Forsten.

———. 2014a. "Through Roman Eyes: Cotton Textiles from Early Historic India." In S. Bergerbrant and S. H. Fossøy (eds.), *A Stitch in Time: Essays in Honour of Lise Bender Jørgensen*, 209–235. Department of Historical Studies, Gothenburg University.

———. 2014b. "Berenike and Textile Trade on the Indian Ocean." In K. Droß-Krüpe (ed.), *Textile Trade and Distribution in Antiquity*, 91–109. Wiesbaden: Harrassowitz.

Wild, J. P., F. C. Wild, and A. J. Clapham. 2008. "Roman Cotton Revisited." In C. Alfaro and L. Karali (ed.), *Purpureae Vestes II. Vestidos, textiles y tintes: Estudios sobre la producción de bienes de consumo en la Antigüedad*, 143–48. University of Valencia.

Wilkin, S., et al. 2021. "Dairying Enabled Early Bronze Age Yamnaya Steppe Expansions." *Nature* 598, 629–33.

Wilkinson, T. J., and L. Rayne. 2010. "Hydraulic Landscapes and Imperial Power in the Near East." *Water History* 2.2, 115–44.

Wilson, J. K. 2005. "The Assyrian Pharmaceutical Series URU.AN.NA: MAŠTAKAL." *Journal of Near Eastern Studies* 64.1, 45–52.

Winter, I. 1994. "Radiance as an Aesthetic Value in the Art of Mesopotamia (with Some Indian Parallels)." In B. N. Saraswati et al. (eds.), *Art: the Integral Vision. A Volume of Essays in Felicitation of Kapila Vatsyayan*, 123–32. New Delhi: N. K. Bose Memorial Foundation.

———. 2007. "Review of Ornamental Wall Painting in the Art of the Assyrian Empire by Pauline Albenda." *Journal of the American Oriental Society* 127.3, 378–80.

———. 2010. "The Aesthetic Value of Lapis Lazuli in Mesopotamia." In I. Winter, *On Art in the Ancient Near East, Vol. II: From the Third Millennium BCE*, 291–306. Leiden: Brill.

Winter, J. G., and H. C. Youtie. 1944. "Cotton in Graeco-Roman Egypt." *American Journal of Philology* 65.3, 249–58.
Winternitz, M. 1983. *A History of Indian Literature II*, trans. V. S. Sarma. Delhi: Motilal Banarsidass.
Wiseman, D. J. 1983. "Mesopotamian Gardens." *Anatolian Studies* 33, 137–44.
Witas, H. W., J. Tomczyk, K. Jędrychowska-Dańska, G. Chaubey, and T. Płoszaj. 2013. "mtDNA from the Early Bronze Age to the Roman Period Suggests a Genetic Link between the Indian Subcontinent and Mesopotamian Cradle of Civilization." *PloS One* 8(9), e73682.
Witzel, M. 1999a. "Early Sources for South Asian Substrate Languages." *Mother Tongue*, special issue (October), 1–76.
———. 1999b. "Substrate Languages in Old Indo-Aryan." *Electronic Journal of Vedic Studies* 5.1, 1–67.
———. 2001. "Autochthonous Aryans? The Evidence from Old Indian and Iranian Texts." *Electronic Journal of Vedic Studies* 7.3, 1–115.
Woelk, D. 1965. *Agatharchides von Knidos: Über das Rote Meer, Überzetsung und Kommentar.* Doctoral dissertation, University of Freiburg.
Woenig, F. 1897. *Die Pflanzen im alten Ägypten.* Leipzig: A. Heitz.
Woods, C. E. 2004. "The Sun-God Tablet of Nabû-apla-iddina Revisited." *Journal of Cuneiform Studies* 56, 23–103.
Woolley, L., and M. Mallowan. 1976. *Ur Excavations VII: The Old Babylonian Period.* London: British Museum.
Wright, R. P. 2010. *The Ancient Indus: Urbanism, Economy, and Society.* Cambridge University Press.
Wu, G. A., et al. 2018. "Genomics of the Origin and Evolution of Citrus." *Nature* 554, 311–16.
Yang, X., H. Li, M. Liang, Q. Xu, L. Chai, and X. Deng. 2015. "Genetic Diversity and Phylogenetic Relationships of Citron (*Citrus medica* L.) and Its Relatives in Southwest China." *Tree Genetics and Genomes* 11.6, 1–13.
Yvanez, E., and M. M. Wozniak. 2019. "Cotton in Ancient Sudan and Nubia: Archaeological Sources and Historical Implications." *Revue d'ethnoécologie* 15, 1–53.
Zach, B. 2002. "Vegetable Offerings on the Roman Sacrificial Site in Mainz, Germany: Short Report on the First Results." *Vegetation History and Archaeobotany* 11, 101–06.
Zadok, R. 1977. "Iranians and Individuals Bearing Iranian Names in Achaemenian Babylonia." *Israel Oriental Studies* 7, 89–138.
Zarins, J. 2008. "Magan Shipbuilders at the Ur III Lagash State Dockyards 2062–2025 BC." In E. Olijdam and R. H. Spoor (eds.), *Intercultural Relations between South and Southwest Asia: Studies in commemoration of E.C.L. During Caspers, 1934–1996*, 209–29. Oxford: Archaeopress.
Zawadzki, S. 2005. "Šamaš Visit to Babylon." *Nouvelles Assyriologiques Brèves et Utilitaires* 9.
———. 2006. *Garments of the Gods: Studies on the Textile Industry and the Pantheon of Sippar according to the Texts from the Ebabbar Archive.* Academic Press Fribourg.
Zeiss, A., and F. Bachmayer. 1986. "Zum Alter der Ernstbrunner Kalke (Tithon; Niederösterreich) Vorläufige Mitteilung." *Annalen des Naturhistorischen Museums in Wien: Serie A für Mineralogie und Petrographie, Geologie und Paläontologie, Anthropologie und Prähistorie* 90, 103–09.

Zhang, T., S. Hu, G. Zhang, L. Pan, X. Zhang, I. S. Al-Mssallem, and J. Yu. 2012. "The Organelle Genomes of Hassawi Rice (*Oryza sativa* L.) and Its Hybrid in Saudi Arabia: Genome Variation, Rearrangement, and Origins." *PloS One* 7(7), e42041.

Zhao, G., et al. 2019. "A Comprehensive Genome Variation Map of Melon Identifies Multiple Domestication Events and Loci Influencing Agronomic Traits." *Nature Genetics* 51.11, 1607–15.

Ziffer, I. 2019. "Pinecone or Date Palm Male Inflorescence: Metaphorical Pollination in Assyrian Art." *Israel Journal of Plant Sciences* 66.1–2, 19–33.

Zisis, V. G. 1955. "Cotton, Linen and Hemp Textiles from the Fifth Century BC." *Praktiká tēs Akadēmías Athēnṓn* 29, 587–93.

Zohary, D. 1998. "The Diffusion of South and East Asian and of African Crops into the Belt of Mediterranean Agriculture." In H. D. V. Prendergast et al. (eds.), *Plants for Food and Medicine: Proceedings of the Joint Conference of the Society for Economic Botany and the International Society for Ethnopharmacology, London 1–6 July 1996*, 123–34. London: Royal Botanic Gardens, Kew.

Zohary, D., M. Hopf, and E. Weiss. 2012. *Domestication of Plants in the Old World: The Origin and Spread of Domesticated Plants in Southwest Asia, Europe, and the Mediterranean Basin.* Oxford University Press.

Zohary, M. 1982. *Plants of the Bible.* Cambridge University Press.

Zöldföldi, J., and Z. Kasztovsky. 2009. "Provenance Study of Lapis Lazuli by Non-Destructive Prompt Gamma Activation Analysis." In Y. Maniatis (ed.), *Proceedings of the 7th International Conference of the Association for the Study of Marble and Other Stones in Antiquity*, 677–91. Athens: École française d'Athènes.

Zubaida, S. 1994. "Rice in the Culinary Cultures of the Middle East." In S. Zubaida and R. Tapper (eds.), *Culinary Cultures of the Middle East*, 93–104. London: I. B. Tauris.

INDEX

Printed in the USA
CPSIA information can be obtained
at www.ICGtesting.com
LVHW040324070823
754491LV00003B/269